T0233682

CISM COURSES AND LECTURES

Series Editors:

The Rectors of CISM
Sandor Kaliszky - Budapest
Horst Lippmann - Munich
Mahir Sayir - Zurich

The Secretary General of CISM
Giovanni Bianchi - Milan

Executive Editor
Carlo Tasso - Udine

The series presents lecture notes, monographs, edited works and proceedings in the field of Mechanics, Engineering, Computer Science and Applied Mathematics.
Purpose of the series in to make known in the international scientific and technical community results obtained in some of the activities organized by CISM, the International Centre for Mechanical Sciences.

INTERNATIONAL CENTRE FOR MECHANICAL SCIENCES

COURSES AND LECTURES - No. 332

EVALUATION OF GLOBAL BEARING CAPACITIES OF STRUCTURES

EDITED BY

G. SACCHI LANDRIANI
POLYTECHNIC OF MILAN, MILAN

and

J. SALENÇON
ECOLE POLYTECHNIQUE, PALAISEAU

SPRINGER-VERLAG WIEN GMBH

Le spese di stampa di questo volume sono in parte coperte da
contributi del Consiglio Nazionale delle Ricerche.

This volume contains 143 illustrations.

This work is subject to copyright.
All rights are reserved,
whether the whole or part of the material is concerned
specifically those of translation, reprinting, re-use of illustrations,
broadcasting, reproduction by photocopying machine
or similar means, and storage in data banks.
© 1993 by **Springer-Verlag Wien**
Originally published by Springer Verlag Wien-New York in 1993

In order to make this volume available as economically and as rapidly as possible the authors' typescripts have been reproduced in their original forms. This method unfortunately has its typographical limitations but it is hoped that they in no way distract the reader.

ISBN 978-3-211-82493-1 **ISBN 978-3-7091-2752-0 (eBook)**
DOI 10.1007/978-3-7091-2752-0

PREFACE

This book collects the lectures given at CISM during June 1992, devoted to the evaluation of global bearing capacities of structures. The course aimed at presenting a unified frame for yield design, limit analysis and optimal design theories.

After a brief introduction on the basic results and the general comments on the validity of cach theory from the practical point of view, typical and illustrative applications in various domains are given, in order to point out their versatility.

Based upon simple arguments of convex analysis, the theory of yield design makes possible to determine the global bearing capacities of a structure from the sole knowledge of the local strength of the constitutive materials.

Soil mechanics problems are considered, such as stability analysis of reinforced structures, where classical limit analysis cannot be considered as adequate. Efficient methods have been derived from the theory of yield design, some of them making use also of the theory of homogenization.

On a different scale, composite materials with long fibers as reinforcement are considered with the purpose of determining their strength capacities. Interesting results have been obtained through this approach, in good agreement with experiments.

In the second part of the book the "classical" limit analysis theory is presented and it is underlined how its use makes it possible to determine the global bearing capacities of a class of structure, on the basis of the local strength and the plastic associated flow rul e of the constituent material. Structural mechanics problems are considered, such as the limit state of steel structures or reinforced concrete plates and shells, where classical limit analysis can be fully applied. Efficient methods are derived from the combined use of limit analysis and convex analysis theories.

In the third part a general formulation of optimal design theory is proposed. Particular emphasis is given to this method as a tool making it possible to determine the best distribution of the employed material in a structure, with a given layout, to achieve a minimum total weight at prescribed limit plastic load or buckling load. In such a way, optimal design problems can be conceived as inverse problems in comparison to the yield design and limit analysis formulation.

Finally some optimal design problems for structures with given layout and prescribed elastic compliance are presented. The aim of these contributions is to point out some analogies existing between some plastic and elastic problems. Also, opportunity is taken for proposing some numerical applications.

The unifying frame in which yield design, limit analysis and optimal design theories have been presented is intended to be a clarifying reasoning tool leading to various new relevant developments, the necessity of experiments being of course the ultimate and unavoidable validation test.

Giannantonio Sacchi Landriani

Jean Salençon

CONTENTS

YIELD DESIGN: A SURVEY OF THE THEORY

J. Salençon
Ecole Polytechnique, Palaiseau, France

ABSTRACT

The theory of Yield Design is based upon the obvious necessary condition for the stability of a structure that the equilibrium of that structure and the resistance of its constituents should be compatible. The static approach of the yield design theory proceeds directly from this condition, leading to lower estimates of the extreme loads. The kinematic approach is derived by dualizing the static approach through the principle virtual work, thus ensuring full mechanical consistency. The treatment of a classical example illustrates these arguments. Present and possibly future domains of practical applications of the theory are reviewed, including the full adequacy between the Yield Design Theory and the Ultimate Limit State Design concept of safety.

1.- SOME HISTORICAL LANDMARKS

The concept of yield (or limit) design is most popular among civil engineers who have made an extensive use of it for centuries when designing structures or earthworks. As a matter of fact, one usually refers to Galileo's study [1] of the cantilever beam as the first written explicitation of the concept : the maximum load the beam may withstand is derived from the only knowledge of the strength of its constituents, namely the fibers (Fig.1).

Figure 1 - Galileo's analysis of the cantilever beam.

Coulomb's celebrated memoir [2] appears then as the reference where such problems as the compression of a column, the stability of a retaining wall or of a masonry arch, etc., are considered from the same point of view combining the equilibrium of the structure and the resistance of its constituents. Papers by Heyman [3-7] and by Delbecq [8, 9] may help getting a comprehensive view of Coulomb's analysis and of the subsequent works (e.g. Méry [10], Durand-Claye [11, 12], ...) concerning masonry works.

As regards soil mechanics, the original Coulomb analysis has been followed by others in the same "spirit" : the general equilibrium, in terms of resulting moments and resulting forces, of a bulk of soil, defined in the considered earthwork by one or a few parameters, is checked under the condition that the soil cannot withstand stresses outside its strength criterion usually defined as a Coulomb criterion. Among many, one must quote the Culmann method [13], and Fellenius' famous analysis for slope stability where the bulks of soil whose equilibrium is checked are limited by circular lines. When considering such circular lines in the case of a frictional soil, the reasoning can only be carried out by introducing complementary assumptions as in the slices method for instance [14-17]. Rendulic [18] introduced the use of logspirals instead of circular lines in that case, which removes the difficulty and preserves the full significance of the analysis as it will be shown later on. Analogous analyses have been performed in order to investigate the stability of surface foundations.

For what concerns the stability of surface foundations and the determination of their bearing capacity, the limit equilibrium methods (as classified by Chen [19] for instance) have also been used, especially in the case of plane problems : the equations were first established by Massau [20] and Kötter [21, 22], and important developments may be found in the famous books by Sokolovski [23-25] and Berezancew [26]. Since the limit equilibrium equations for plane problems are homologous to the equations appearing in the study of plane strain problems for perfectly plastic materials obeying Tresca's yield criterion with associated flow rule [27], they are often associated with the concept of plasticity : in other words they are improperly said to rely on the assumption of a rigid perfectly plastic soil obeying Coulomb's criterion with associated flow rule. As a matter of fact, it should be clear to everybody that limit equilibrium methods are but another technique to perform the reasoning of the yield design theory, and that they do not require any more assumption about the constitutive law of the considered soil than the data of its strength criterion.

Another classical domain of application for the yield design theory concerns the analysis of metallic plates and reinforced concrete slabs, with the use of Johansen's criterion and yield line theory [28, 29] and other methods, as explained for instance in Save and Massonnet [30].

Recently the combination of the homogeneization theory with the yield design theory has opened the access to new types of investigations and has already led to interesting results such as the determination of the "macroscopic" homogenized strength criterion of a composite based upon the data of the strength criteria of its constituent materials, including the interfaces between them, which is to be introduced in the homogenized yield design analysis of a structure made of that composite material. From the practical point of view the domain of relevance ranges from mechanical engineering with, for instance, the "classical" long fiber/epoxy matrix composites [31, 32] to civil engineering where soils reinforced by inclusions such as nails, geotextiles and geogrids, or metallic strips, etc., are now being extensively used [33-35].

The purpose of the presentation of the yield design theory to be given hereafter is to cover those different types of applications, and others which have not been listed to make short. Starting from the simple example of a structure, in order to introduce and discuss the first fundamental concepts, it will then be developed within the framework of classical continuum mechanics with the help of a few basic notions in convex analysis.

Through its generality it is hoped to show the unity between existing methods in the various domains of applications of the yield design theory, together with their efficiency, and to arouse the interest for devising new developments which may take advantage of the everyday increasing possibilities offered by numerical techniques.

2.- AN INTRODUCTION TO THE YIELD DESIGN APPROACH.

2.1.- Definition of the problem under consideration

The square truss ABCD shown in Fig. 2 is made of six bars with pivot hinges in A, B, C and D, and no connection between bars AC and BD. This structure is submitted to an external force acting on joint C, in the direction of AC with intensity Q, while joint A is fixed. No other force is applied to the truss and Q appears naturally as the load parameter for the problem.

The internal efforts in the bars obviously reduce to tensile or compressive forces, namely $N_1, N_2, \dots, N_6$ in bars AB, BC, CB, DA, AC and BD respectively.

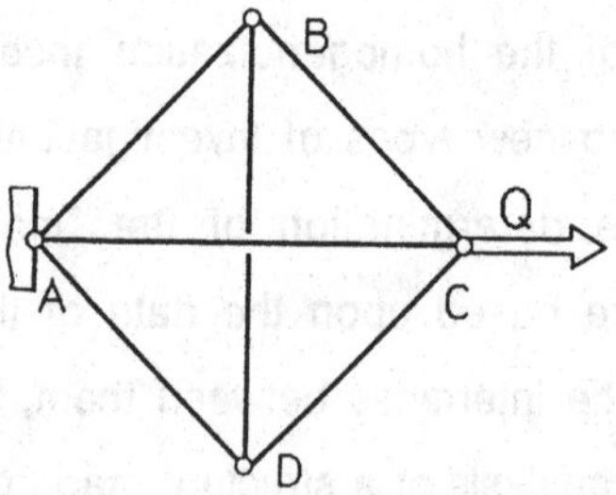

Figure 2 - One parameter loading of a simple structure.

It is assumed that the six bars of the truss exhibit the same resistance in tension and in compression, equal to L, so that condition (2.1) has to be satisfied :

$$|N_i| \leq L \quad , \qquad \forall\, i = 1, 2, ..., 6. \qquad (2.1)$$

The question to be answered is to determine the maximum and minimum values of parameter Q the structure can withstand under the strength conditions (2.1) imposed to the bars.

2.2.- Compatibility between equilibrium and resistance.

The equilibrium equations for the structure in Figure 2 are written

$$N_1 = N_2 = N_3 = N_4 \,,$$
$$N_5 + N_1\sqrt{2} = Q \,, \qquad (2.2)$$
$$N_6 + N_1\sqrt{2} = 0 \,.$$

A necessary condition for that structure to withstand load Q under the strength conditions (2.1) is therefore that Eq. (2.2) be mathematically compatible with Ineq. (2.1). It follows that the intensity Q of the applied force should not exceed the conditions :

$$|Q| \leq 2L \qquad (2.3)$$

This leads to the following statements :

- if the applied load Q is such that $|Q| > 2L$, it is sure that the structure cannot be in equilibrium together with the strength conditions being satisfied at the same time ;
- if the applied load Q is such that $|Q| \leq 2L$, it is possible that the structure be in equilibrium and the strength conditions be satisfied at the same time.

Briefly speaking, introducing the termes "stable" and "unstable" to characterize those circumstances :

$|Q| > 2L$: the structure cannot be stable under Q ,
it is *surely unstable* under Q ;
(2.4)
$|Q| \leq 2L$: the structure may be stable under Q,
it is *potentially stable* under Q.

The loads $Q^+ = 2L$ and $Q^- = -2L$ will be called the extreme loads of the structure in the loading mode defined in Fig. 2.

2.3.- Comments.

Statement (2.4) deserves many comments, most of which will remain relevant in the general case.

2.3.1.

The analysis has been performed on the given geometry of the structure : no geometry changes are taken into account.

2.3.2.

The only data required for the analysis given in Section (2.2) are the strength conditions imposed on the bars. It follows that results (2.4) hold whatever the initial internal forces in the structure for Q = O, whatever the loading path, whatever the constitutive equations for the bars provided they are consistent with the strength conditions (2.1) and with the preceding comment regarding the geometry where the equations are written.

2.3.3.

Statement (2.4) is but a partial answer to the original question asked in Section (2.1) since it only offers a guarantee of instability (!) when $|Q| > 2L$ and a presumption of stability when $|Q| \leq 2L$.

It should be understood that such a conclusion is the maximum one can logically derive starting from the only available data (2.1) on the resistance of the constituent elements of the structure imposed as a constraint. In order to be able to assert the stability of the structure under a given load Q it would be necessary that complementary information regarding

- the mechanical behaviour of the constituent elements, through their constitutive equations for instance,
- the initial state of self-equilibrated interior forces,
- the loading history followed to reach the actual load Q from the unloaded initial state

be available.

As a matter of fact it can be easily perceived that the condition for the structure to withstand an extreme load is that the strength capacities required to equilibrate that load can be actually mobilized simultaneously in the concerned elements, at the end of the concerned loading path, starting from the given initial state of self equilibrated interior forces, and with the given constitutive equations.

Fig. 4 recalls an example from [36] where the response of the structure introduced in Fig.2 is studied for a monotous increasing loading path, starting from the zero self equilibrated initial state of interior forces, under two different assumptions for the constitutive equations of the elements (Fig.3) :

a) the bars are elastic and perfectly plastic in tension and in compression,

b) the bars are elastic and perfectly plastic in compression, elastic and brittle in tension.

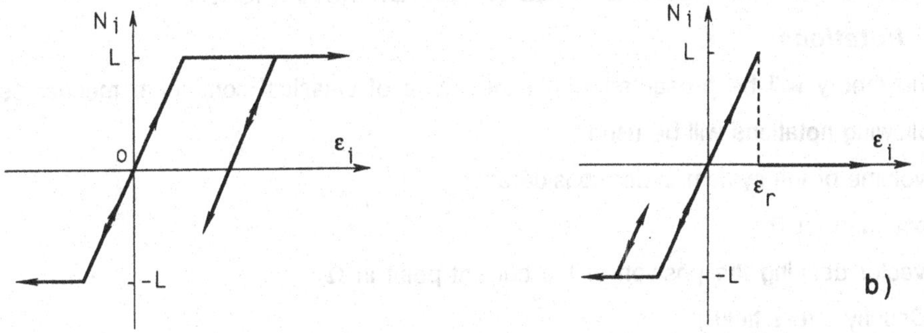

Figure 3 - Elastic perfectly plastic (a), and elastic brittle/perfectly plastic (b) behaviours for the bars in Fig. 2.

Although the latter assumption might seem somewhat schematic, the results shown in Fig. 4 point out that in the first case the structure will actually withstand any load between Q^- and Q^+, whereas in the second case the structure will break when Q reaches the level $Q = L\sqrt{2}$.

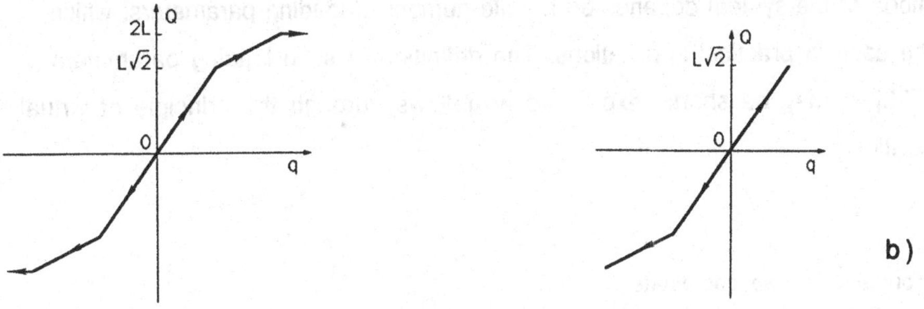

Figure 4 - Response of the truss in Fig. 2 with the constitutive equations shown in Fig. 3 for the bars.

The preeminent role of ductility in the mechanical behaviour of the bars is quite apparent here. Generally speaking such ideal behaviours as shown in Fig. 3 will scarcely be encountered when dealing with practical applications. Nevertheless it is clear that the idea expressed by Jewell [37] that the deformations necessary to fully mobilize the required resistance of the elements, in the structure, along the considered loading path, must be "compatible" with each other must be retained.

3.- PRINCIPLES OF THE THEORY OF YIELD DESIGN : STATIC APPROACH

3.1.- Notations

The theory will be presented within the frame of classical continuum mechanics. The following notations will be used :

Ω : volume of the system under consideration ;

$\partial\Omega$: boundary of Ω ;

$\mathbf{x}$: vector defining the position of the current point in Ω ;

$\underline{\sigma}$: Cauchy stress field ;

$\mathbf{U}$: velocity field ; $\mathbf{U}$: virtual velocity field ;

$\underline{\mathbf{d}}$: strain rate field derived from $\mathbf{U}$, ($\underline{\mathbf{d}}$ from $\mathbf{U}$) ;

[ ] : symbol for the jump accross surface Σ following unit normal $\mathbf{n}$;

. : symbol for the contracted product (e.g. $\underline{\sigma} . \mathbf{n} = \sigma_{ij} n_j$ and $\mathbf{T.U} = T_i U_i$ ([1])) ;

: : symbol for the twice contracted product (e.g. $\underline{\sigma} : \underline{\mathbf{d}} = \sigma_{ij} d_{ji}$).

3.2.- Principle of virtual work (powers). Loading parameters.

The theory of yield design is most conveniently formulated in the case when the loading mode of the system depends on a finite number of loading parameters, which is always the case in practical applications. The definition of such loading parameters is given in [36]. It may be shortly expressed as follows, through the principle of virtual work in statics.

([1]) Orthonormal cartesian coordinates.

For any stress field $\underline{\sigma}$, statically admissible for the problem in the loading mode, for any virtual velocity field $\hat{\mathbf{U}}$, kinematically admissible for the problem in the loading mode,
the following equation holds

$$\int_{\Omega} \underline{\sigma}(\mathbf{x}) : \underline{\hat{d}}(\mathbf{x})\, d\Omega + \int_{\Sigma} [\hat{\mathbf{U}}(\mathbf{x})] . \underline{\sigma}(\mathbf{x}) . \mathbf{n}(\mathbf{x})\, d\Sigma = Q_j(\underline{\sigma}) . \dot{q}_j(\hat{\mathbf{U}}) = \mathbf{Q}(\underline{\sigma}) . \dot{\mathbf{q}}(\hat{\mathbf{U}}) \tag{3.1}$$

where the applications

$$\underline{\sigma} \rightarrow \mathbf{Q} = \mathbf{Q}(\underline{\sigma}) \in \mathbf{R}^n \tag{3.2}$$

$$\hat{\mathbf{U}} \rightarrow \hat{\dot{\mathbf{q}}} = \dot{\mathbf{q}}(\hat{\mathbf{U}}) \in \mathbf{R}^n \tag{3.3}$$

are linear.

The Q_j, components of the n-dimension vector **Q**, are the *loading* parameters of the problem, **Q** being called the *load*. The q_j, components of the n-dimension vector $\dot{\mathbf{q}}$, are the associated kinematic parameters.

It is recalled that the statically admissible stress fields $\underline{\sigma}$ in Eq. (3.1) are piecewise continuous with continuous derivatives, and satisfy the equilibrium equations on Ω and the boundary conditions on the stresses on $\partial\Omega$. The kinematically admissible virtual velocity field are piecewise continuous with continuous derivative on Ω and satisfy the boundary conditions on the velocity on $\partial\Omega$; it must be emphasized that no restriction are imposed on the velocity discontinuities $[\hat{\mathbf{U}}]$.

Classical examples of loading parameters are given in [36] ; de Buhan [38], dealing with systems made of composite materials through the homogenization theory, gives interesting developments on the concept of loading parameters regarding both the structure itself and the unit representative cell.

3.3.- Statement of the problem.

The general formulation of the yield design problem is similar to that given in Section (2.1) in the particular case of the truss in Fig.2.

The system under consideration is geometrically defined through Ω. It is loaded according to a loading mode depending on n parameters Q_j. The strength condition, homologous to Ineq. (2.1) must then be defined. Obviously, such a condition will now refer to the value of the stress tensor $\underline{\sigma}(\mathbf{x})$ at each point $\mathbf{x}$ in Ω .

The strength domain $G(x)$ defining the admissible stress states at point $\mathbf{x}$ is given as the only data to characterize the constituent material mechanically. It depends explicitly on $\mathbf{x}$ since the material needs not be homogeneous. The strength condition will be written :

$$\forall \mathbf{x} \in \Omega \quad , \quad \underline{\underline{\sigma}}(\mathbf{x}) \in G(\mathbf{x}) \subset \mathbb{R}^6 \tag{3.4}$$

As a rule $G(\mathbf{x})$ exhibits the following properties

- $\underline{\underline{\sigma}}(\mathbf{x}) = \underline{\underline{0}} \in G(\mathbf{x})$,
- $G(\mathbf{x})$ is star-shaped with respect to $\underline{\underline{0}}$.

As a matter of fact $G(\mathbf{x})$ is usually convex (that implies the latter property), which will assumed from now on.

The question is then to decide whether the system can or cannot withstand a given value of load $\mathbf{Q}$, under the constraint of the strength condition (3.4).

3.4.- Potentially safe loads, extreme loads.

The answer follows from the same considerations as in chapter 2.

For the system to withstand load $\mathbf{Q}$ under the strength condition (3.4) it is necessary that :

$$\left.\begin{array}{l} \exists\, \underline{\underline{\sigma}} \ , \text{ statically admissible for the problem} \\ \quad \text{in the loading mode with load } \mathbf{Q} \text{ ,} \\ \quad \text{complying with the strength condition (3.4).} \end{array}\right\} \tag{3.5}$$

Under such a load $\mathbf{Q}$, the system will be said *potentially stable*, and the load itself will be called *potentially safe.*

The set generated by all potentially safe loads will be denoted K :

$$\mathbf{Q} \in K \subset \mathbb{R}^n \Leftrightarrow \exists\, \underline{\underline{\sigma}} \left| \begin{array}{l} \text{S.A. equilibrating } \mathbf{Q} \\ \underline{\underline{\sigma}}(\mathbf{x}) \in G(\mathbf{x}) \quad , \quad \forall \mathbf{x} \in \Omega \ ; \end{array}\right. \tag{3.6}$$

stress field $\underline{\underline{\sigma}}$ is associated with load $\mathbf{Q}$ through the linear application (3.2).

The properties of K are derived immediately from definition (3.6) and the linearity of application (3.2). It comes out that :

$$\begin{array}{l} \mathbf{Q} = \mathbf{0} \in K \ , \\ K \text{ is convex} \ , \end{array} \tag{3.7}$$

what proves that the properties of the G's at the level of the constituent material are transferred to K at the level of the system.

The loads on the convex boundary of K are called the *extreme loads* of the system, which recalls that any load out of K is *certainly unsafe* and will induce instability of the system.

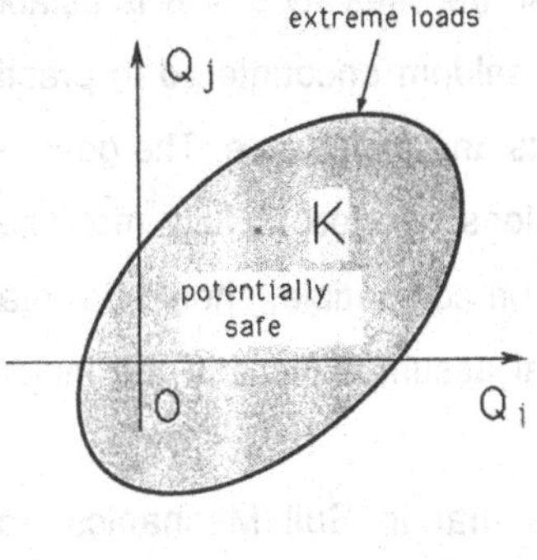

Figure 5 - Domain K : potentially safe loads, extreme loads.

3.5.- Comments.

Comments homologous to those expressed in section (2.3) should be made. The answer illustrated in Fig.5 is but a partial one to the original question : domain K offers no guarantee of stability for the system, which would require the data of the initial state, of the loading history, and of the mechanical behaviour of the constituent material.

It is hoped that the above given presentation made it clear that the definition of domain K is only derived from the

$$\textit{Compatibility} \text{ between } \left\{ \begin{array}{l} \textit{equilibrium} \text{ (of the system)} \\ \textit{resistance} \text{ (of the material)} \end{array} \right. \qquad (3.8)$$

no other data being available. Therefore, what may appear a shortcoming in the result can also be considered positive : domain K is independent of the initial state, of the loading history of the system and of the mechanical behaviour of the constituent material apart from the strength condition (3.4). Since this information may be either totally or partially missing in some cases, or improperly assessed in others, it is important that such a general result as K be available.

Anyhow the question of the relevance of K for practical applications cannot be disregarded.

From the theoretical point of view it has been proved [39] (cf. [40]) that a system whose constituent material is elastic and perfectly plastic and obeys the maximum plastic work principle will withstand any load **Q** inside K (which generalizes the result in Fig. 4a). In that case the relevance of K is established.

This theoretical situation is seldom encountered in practice and the relevance of K must be assessed by experiments and experience. The general idea already expressed in para (2.3.2.), that the deformations needed to fully mobilize the required strengths in the element of the system must be compatible, should be retained. Attention should also be paid to the initial geometrical assumptions and the incidence of geometry changes should be considered.

As an example it appears that in Soil Mechanics, for usual stability analysis purposes, the reasoning based upon the compatibility stated in (3.8) which is the only and very argument of the Yield Design Theory, has proved to be relevant in most cases, but is not valid when dealing for instance with the stability of deep tunnels or deep cavities.

Such validations of the theory support the determination of K and, more precisely, that of the exteme loads.

3.6.- Static approach from inside.

The construction of domain K comes directly from definition (3.6) : any stress-field $\underline{\underline{\sigma}}$, statically admissible in the loading mode and complying with the strength condition in Ω produces a potentially safe load $\mathbf{Q} = \mathbf{Q}(\underline{\underline{\sigma}})$ through application (3.2). In other words :

$$\left.\begin{array}{l}\underline{\underline{\sigma}} \text{ S.A. in the mode} \\ \underline{\underline{\sigma}}(\mathbf{x}) \in G(\mathbf{x}) \;, \; \forall \, \mathbf{x} \in \Omega\end{array}\right| \overset{(3.2)}{\longrightarrow} \mathbf{Q}(\underline{\underline{\sigma}}) \in K \,. \qquad (3.9)$$

Drawing the convex envelope of such **Q** gives an interior approach of K and "lower bounds" for the extreme loads (Fig. 6).

The exact determination of K requires exploring the whole set of $\underline{\underline{\sigma}}$ satisfying (3.9). This is most often impossible from the practical point of view, except for the case of simple structures (as in Section 2). It follows that one will usually be satisfied with the interior approach derived through the implementation of construction (3.9) with a reduced number of stress-fields chosen for an efficient balance between simplicity in the procedure and quality of the obtained results.

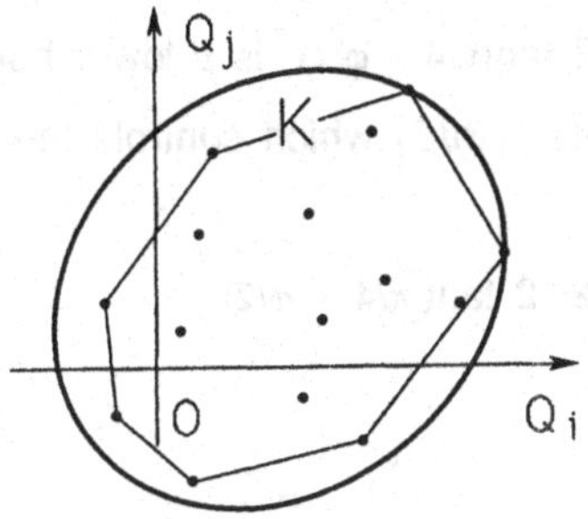

Figure 6 - Static approach from inside.

4.- AN EXAMPLE OF THE YIELD DESIGN STATIC APPROACHES.

4.1.- Stability analysis of a vertical cut.

The stability of the homogeneous vertical cut represented in Fig. 7 is investigated. The load is defined by the gravity forces acting in the bulk of soil, the corresponding loading parameter being the voluminal weight usually denoted γ in Soil Mechanics.

The strength condition of the soil is characterized through Coulomb's isotropic strength criterion with cohesion C and friction angle φ, written in the form

$$f\,(\underline{\sigma}(\mathbf{x})) = \underset{i,\,j = 1,\,2,\,3}{\text{Sup}} \left\{\sigma_i\,(1 + \sin\varphi) - \sigma_j\,(1 - \sin\varphi) - 2\,C\cos\varphi\right\} \le 0 \qquad (4.1)$$

where σ_i denote the principal values of $\underline{\sigma}(\mathbf{x})$.

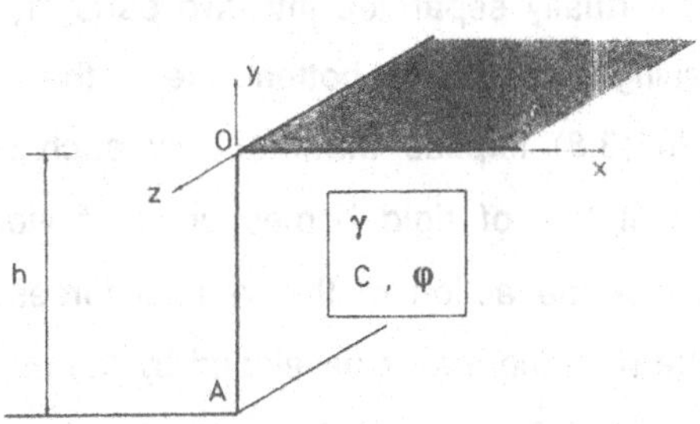

Figure 7 - Stability of a homogeneous vertical cut under its own weight.

4.2.- Static approach from inside.

A stress-field for the static approach from inside was given by Drucker and Prager [41] and proved to comply with Ineq. (4.1) as long as

$$(\gamma h/C) \leq 2 \tan(\pi/4 + \varphi/2) \; ; \qquad (4.2)$$

it is presented in Fig.8.

It follows that the value $2 \tan(\pi/4 + \varphi/2)$ is a lower bound for the extreme value of the non-dimensional parameter $(\gamma h/C)$ which controls the stability of the cut :

$$(\gamma h/C)^+ \geq 2 \tan(\pi/4 + \varphi/2) \; . \qquad (4.3)$$

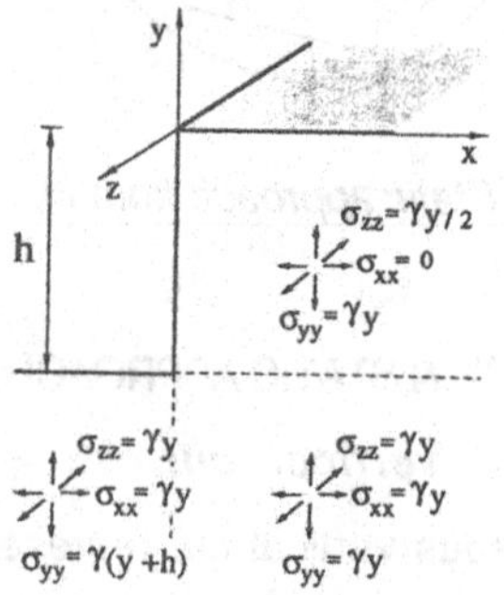

Figure 8 - Static approach for the problem of the vertical cut [41].

4.3.- Static approach from outside.

Still starting from definition (3.6) of K it will now be shown that the static reasoning may be performed in a different way in order to provide an upper bound for $(\gamma h/C)^+$.

Let the bulk of soil be virtually separated into two parts (1) and (2) by means of an arbitrary plane (P) passing ghrough the bottom line of the cut as drawn in Fig.9. The compatibility stated in (3.8) implies that, for any such plane (P), the global equilibrium –i.e. from the statics of rigid bodies point of view– of volume (1) or volume (2) be possible under the action of the vertical forces due to gravity (bulk weight of the soil) and of the resisting forces developed by the material.

The cut being considered of infinite length along Oz, the resisting forces developed on the sections parallel to OAB at both ends at infinity are negligible compared with those along the section by plane (P) : the analysis can therefore be restricted to checking the stability of a slice with thickness D along Oz as sketched in Fig.9 assuming no resisting forces on OAB and O'A'B'.

Considering the equilibrium of volume (1) submitted to gravity forces acting throughout the volume and to normal and tangential stresses acting on the surface ABB'A', the following equations shall be satisfied :

$$\frac{\gamma h^2 D}{2} \tan\alpha \sin\alpha = - \int_{ABB'A'} \sigma \, dS \qquad (4.4)$$

$$\frac{\gamma h^2 D}{2} \sin\alpha \, \mathbf{t} = - \int_{ABB'A'} \boldsymbol{\tau} \, dS \qquad (4.5)$$

where σ and τ (a vector) are the normal and tangential components of the stress vector acting at each point of ABB'A', tensile stresses are counted positive, and **t** is the unit vector along **BA**.

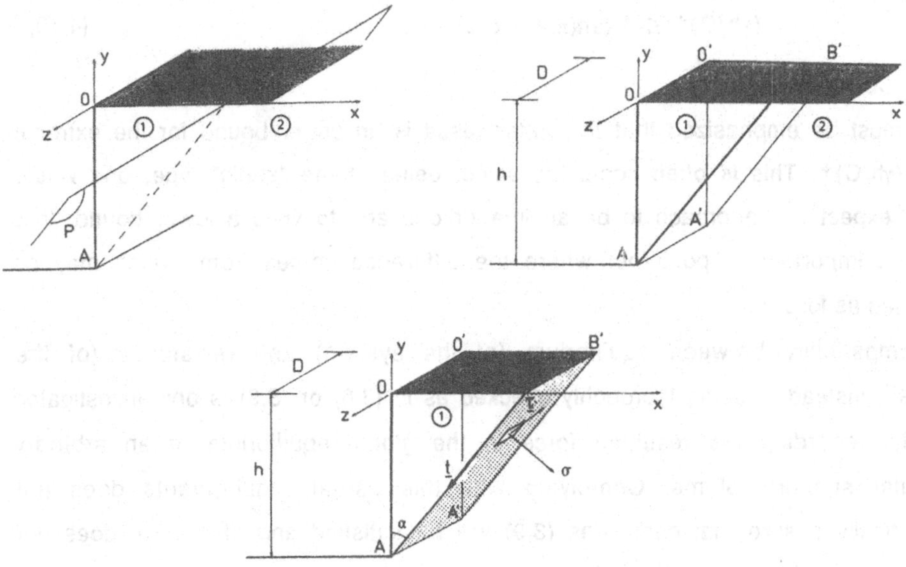

Figure 9 - Global equilibrium analysis of volume (1).

Coulomb's strength condition (4.1) implies that whatever the point in ABB'A', σ and τ must comply with the inequality

$$|\tau| \le C - \sigma \tan\varphi \ . \qquad (4.6)$$

It follows from Eq. (4.5) and (4.6) that

$$\frac{\gamma h^2 D}{2} \sin\alpha \leq \int_{ABB'A'} |\tau| \, dS \leq \int_{ABB'A'} (C - \sigma \tan\varphi) dS \quad . \qquad (4.7)$$

then through Eq. (4.4)

$$\gamma h/C \leq 2 \cos\varphi / \sin\alpha \cos(\alpha + \varphi) \quad , \qquad (4.8)$$

establishing a necessary condition for the stability of the cut, to be satisfied whatever the plane (P).

The minimum of the second hand of Ineq. (4.8) with respect to α which provides the strongest necessary condition on $\gamma h/C$, is obtained for $\alpha = \pi/4 - \varphi/2$ which yields the inequality

$$\gamma h/C \leq 4 \tan(\pi/4 + \varphi/2) \qquad (4.9)$$

with the conclusion that *any cut for which* $(\gamma h/C)$ *is greater than* $4 \tan(\pi/4 + \varphi/2)$ *will be unstable* ; thence :

$$(\gamma h/C)^+ \leq 4 \tan(\pi/4 + \varphi/2) \ . \qquad (4.10)$$

4.4.- Comments.

It must be emphasized that this latter result is an upper bound for the extreme value $(\gamma h/C)^+$. This is often confusing since, being of the "static" type, one would readily expect this approach to be an interior one and to yield a lower bound. It is therefore important to point out where the difference comes from, which may be expressed as follows.

Compatibility between equilibrium (of the system) and resistance (of the material), instead of being thoroughly checked as in (3.6) or (3.9) is only investigated partially, regarding the resulting force in the global equilibrium of an arbitrary triangular shaped volume. Complying with this partial requirements does not automatically ensure that conditions (3.9) will be satisfied and, therefore, does not provide an interior approach of K, i.e. a lower bound for $(\gamma h/C)^+$. Conversely, not complying with this requirement implies that conditions (3.9) cannot be satisfied simultaneously, which yields an exterior approach of K, i.e. an upper bound for $(\gamma h/C)^+$.

The importance of getting such an exterior approach of K is evident from what has been said about the practical application of the static approach form inside in Section

(3.6). It is also apparent on the above given example that such a static approach from outside

- requires great care in its implementation,
- remains rather crude : it would certainly be worth carrying on thoroughly with the same idea, within the framework of continuum mechanics, that means proving incompatibility between equilibrium and resistance without restricting the analysis to "global" considerations.

The following chapter will present a method wich meets this latter goal together with being systematic as regards its implementation.

5.- PRINCIPLES OF THE THEORY OF YIELD DESIGN : KINEMATIC APPROACH.

5.1.- Fundamental statement of the kinematic approach.

Starting from definition (3.6) of domain K, Eq. (3.1) expressing the principle of virtual work may be written for any potentially safe load **Q**, considering one stress field $\underline{\sigma}$ referred to in (3.6) and any k.a. virtual velocity field $\hat{\mathbf{U}}$:

$$\int_\Omega \underline{\sigma}(\mathbf{x}) : \hat{\underline{\mathbf{d}}}(\mathbf{x})\, d\Omega + \int_\Sigma [\hat{\mathbf{U}}(\mathbf{x})] . \underline{\sigma}(\mathbf{x}) . \mathbf{n}(\mathbf{x})\, d\Sigma = \mathbf{Q}(\underline{\sigma}) . \dot{\mathbf{q}}(\hat{\mathbf{U}}) = P_e(\mathbf{Q}, \hat{\mathbf{U}}) \qquad (5.1)$$

The following "π functions" will be introduced :

$$\pi(\mathbf{x}, \hat{\underline{\mathbf{d}}}(\mathbf{x})) = \operatorname{Sup}\{\underline{\sigma}'(\mathbf{x}) : \hat{\underline{\mathbf{d}}}(\mathbf{x}) \mid \underline{\sigma}'(\mathbf{x}) \in G(\mathbf{x})\} \qquad (5.2)$$

$$\pi(\mathbf{x}, \mathbf{n}(\mathbf{x}), [\hat{\mathbf{U}}(\mathbf{x})]) = \operatorname{Sup}\{[\hat{\mathbf{U}}(\mathbf{x})] . \underline{\sigma}'(\mathbf{x}) . \mathbf{n}(\mathbf{x}) \mid \underline{\sigma}'(\mathbf{x}) \in G(\mathbf{x})\} \quad . \qquad (5.3)$$

Since stress field $\underline{\sigma}$ in Eq. (5.1) complies with strength condition (3.4) in Ω, it follows from Eq. (5.1 to 5.3) that :

$$\forall \mathbf{Q} \in K \,,\ \forall \hat{\mathbf{U}} \text{ k.a} \,, \qquad (5.4)$$
$$P_e(\mathbf{Q}, \hat{\mathbf{U}}) = \mathbf{Q} . \dot{\mathbf{q}}(\hat{\mathbf{U}}) \le \int_\Omega \pi(\mathbf{x}, \hat{\underline{\mathbf{d}}}(\mathbf{x}))\, d\Omega + \int_\Sigma \pi(\mathbf{x}, \mathbf{n}(\mathbf{x}), [\hat{\mathbf{U}}(\mathbf{x})])\, d\Sigma \quad .$$

The second hand of Ineq. (5.4) will be denoted $P_{mr}(\hat{\mathbf{U}})$:

$$P_{mr}(\hat{\mathbf{U}}) = \int_\Omega \pi(\mathbf{x}, \hat{\underline{\mathbf{d}}}(\mathbf{x}))\, d\Omega + \int_\Sigma \pi(\mathbf{x}, \mathbf{n}(\mathbf{x}), [\hat{\mathbf{U}}(\mathbf{x})])\, d\Sigma \quad . \qquad (5.5)$$

and called the *maximum resisting power* in virtual velocity field $\hat{\mathbf{U}}$.

The obtained result will thus be written in the form :

$$\forall \hat{\mathbf{U}} \text{ k.a } , \quad K \subset \left\{ \mathbf{Q} \cdot \dot{\mathbf{q}}(\hat{\mathbf{U}}) - P_{mr}(\hat{\mathbf{U}}) \leq 0 \right\} \quad . \tag{5.6}$$

5.2.- "π functions".

5.2.1.- Mechanical significance of the π functions.

Statement (5.6) clearly relies on the introduction of the π functions as defined by Eq. (5.2). The significance of these functions must now be investigated.

Eq. (5.2) shows that, given a symmetric tensor $\hat{\underline{\mathbf{d}}}(\mathbf{x})$ which may be interpreted as a virtual strain rate at point $\mathbf{x}$, $\pi(\mathbf{x}, \hat{\underline{\mathbf{d}}}(\mathbf{x}))$ represents the maximum value of the work $\underline{\sigma}'(\mathbf{x}) : \hat{\underline{\mathbf{d}}}(\mathbf{x})$ which may be developed by any stress tensor $\underline{\sigma}'(\mathbf{x})$ complying with strength condition (3.4) at point $\mathbf{x}$. Function $\pi(\mathbf{x}, \hat{\underline{\mathbf{d}}}(\mathbf{x}))$ therefore appears as the *maximum resisting power density* in strain rate $\hat{\underline{\mathbf{d}}}(\mathbf{x})$ under strength condition (3.4) defined by $G(\mathbf{x})$.

From the mathematical point of view $\pi(\mathbf{x}, \bullet)$ is the support function of convex $G(\mathbf{x})$ [42] and it may be proved that the data of $\pi(\mathbf{x}, \bullet)$ is equivalent to that of $G(\mathbf{x})$ (through a strength criterion for instance), which means that $\pi(\mathbf{x}, \bullet)$ carries, in a dual formulation, all information contained in $G(\mathbf{x})$.

The origin of the word "resisting" in the terminology is evident from Eq. (3.1) written as a balance between the power of the external forces and the resisting power (opposite to the power of the internal forces).

The same interpretation as given above holds for $\pi(\mathbf{x}, \mathbf{n}(\mathbf{x}), [\hat{\mathbf{U}}(\mathbf{x})])$ since it may be easily verified that

$$\pi\,(\mathbf{x}, \mathbf{n}(\mathbf{x}),\; [\hat{\mathbf{U}}(\mathbf{x})]) = \frac{1}{2}\,\pi\,(\mathbf{x}, \mathbf{n}(\mathbf{x}) \otimes [\hat{\mathbf{U}}(\mathbf{x})] + [\hat{\mathbf{U}}(\mathbf{x})] \otimes \mathbf{n}(\mathbf{x})) \quad . \tag{5.7}$$

It is worth pointing out that the fundamental idea of the kinematic approach, that is defining the strength domain by duality through the π function, is apparent in a paper by Prager [43].

5.2.2.- Calculation of the π functions.

The values of the π functions are obtained from Eq. (5.2) and (5.3). Considering Eq. (5.7), only the case of function $\pi(\mathbf{x}, \hat{\mathbf{d}}(\mathbf{x}))$ will be investigated.

Fig.10, schematically presents the two typical circumstances which are encountered when calculating $\pi(\mathbf{x}, \hat{\underline{\mathbf{d}}}(\mathbf{x}))$.

1°- If convex G(x) is bounded in any direction of $\mathbf{R}^6$, then :

$$\pi\,(\mathbf{x}, \hat{\underline{\mathbf{d}}}(\mathbf{x})) = \underline{\sigma}^*(\mathbf{x}) : \hat{\underline{\mathbf{d}}}(\mathbf{x}) \tag{5.8}$$

where $\underline{\sigma}^*(\mathbf{x})$ is any stress tensor on the boundary of G(x) where the outward normal is colinear to $\hat{\underline{\mathbf{d}}}(\mathbf{x})$. It follows that, in that case, $\pi(\mathbf{x}, \hat{\underline{\mathbf{d}}}(\mathbf{x}))$ is finite whatever $\hat{\underline{\mathbf{d}}}(\mathbf{x})$.

2°- If convex G(x) is not bounded in any direction, let (I) be the convex cone of the directions in which it extends to infinity in $\mathbf{R}^6$. Then :

• if $\hat{\underline{\mathbf{d}}}(\mathbf{x})$ belongs to the convex cone orthogonal to (I), $\pi(\mathbf{x}, \hat{\underline{\mathbf{d}}}(\mathbf{x}))$ has a finite value

$$\pi\,(\mathbf{x}, \hat{\underline{\mathbf{d}}}(\mathbf{x})) = \underline{\sigma}^*(\mathbf{x}) : \hat{\underline{\mathbf{d}}}(\mathbf{x}) \tag{5.9}$$

where $\underline{\sigma}^*(\mathbf{x})$ is any stress tensor on the boundary of G(x) where the outward normal is colinear to $\hat{\underline{\mathbf{d}}}(\mathbf{x})$, as in the preceding case ;

• if $\hat{\underline{\mathbf{d}}}(\mathbf{x})$ does not belong to that cone, $\pi(\mathbf{x}, \hat{\underline{\mathbf{d}}}(\mathbf{x}))$ is infinite

$$\pi(\mathbf{x}, \hat{\underline{\mathbf{d}}}(\mathbf{x}) = +\infty \tag{5.10}$$

Figure 10 - Calculation of $\pi(\mathbf{x}, \hat{\mathbf{d}}(\mathbf{x}))$.

Within the framework of classical continuum mechanics, strength domains under general three dimensional stress condition are usually of the 2nd type presented in Fig. 10 (e.g. Tresca's, von Mises', Coulomb's, ... strength criteria), and the corresponding π functions take finite or infinite values according to the given $\hat{\underline{\mathbf{d}}}(\mathbf{x})$. It is obvious that only in the case when G(x) is unbounded in any direction, that is the ideal rigid material, will $\pi(\mathbf{x}, \hat{\underline{\mathbf{d}}}(\mathbf{x}))$ be infinite whatever $\hat{\underline{\mathbf{d}}}(\mathbf{x})$.

Strength domains for plane stress loading of the material are bounded in any direction, as are the strength domains referring to bending moments in the yield design

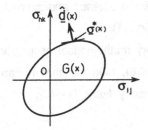

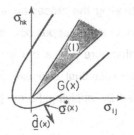

of plates and the strength domains for bending moments, normal and shear forces, etc. in the yield design of beams.

A comprehensive, but of course non exhaustive, list of strength criteria and corresponding π functions is given in [36]. A shorter list appears in Section 5.4.

5.2.3.- Properties of the π functions.

The following properties are evident from Eq. (5.2) and the assumptions on $G(\mathbf{x})$:

- $\pi(\mathbf{x}, \bullet)$ is non negative,
- $\pi(\mathbf{x}, \bullet)$ is positively homogeneous of degree 1,
- $\pi(\mathbf{x}, \bullet)$ is convex.

5.3.- The kinematic approach from outside.

5.3.1.- Exterior approach.

Statement (5.6) provides a powerful exterior approach of K.

For any virtual k.a. velocity field $\hat{\mathbf{U}}$ the maximum resisting power $P_{mr}(\hat{\mathbf{U}})$ is computed referring to the relevant expressions of the π functions, while the power of the external forces $P_e(\mathbf{Q}, \hat{\mathbf{U}}) = \mathbf{Q} \cdot \dot{\mathbf{q}}(\hat{\mathbf{U}})$ is a linear form in $\mathbf{Q}$ where $\dot{\mathbf{q}}(\mathbf{U})$ is known. It follows that domain K is included in the halfspace of $\mathbf{R}^n$ defined by Ineq. (5.6).

Repeating the procedure with different $\dot{\mathbf{q}}$ rapidly gives an *exterior approach* of the boundary of K and *upper bounds* for the extreme loads (Fig. 11).

Furthermore it was proved, with some complementary mathematical assumptions, that the exact dual definition of K is given by Form. (5.6), meaning that K can be generated by applying Form. (5.6) to all k.a. virtual velocity fields [44-46].

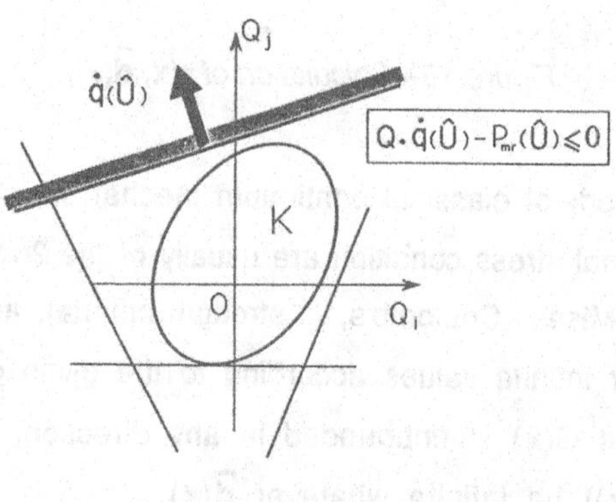

Figure 11 - Kinematical approach from outside.

5.3.2.- Relevant velocity fields.

For Ineq. (5.6) to give a non trivial result the k.a. virtual velocity field must be chosen in such a way that :

- the position of the external forces be non zero
- the maximum resisting power be finite.

The first condition does not need any comment since it meets common sense. The latter means that $\hat{\mathbf{U}}$ should be selected in order to obtain finite values for the corresponding π functions (cf. para. 5.2.2.).

Such virtual velocity fields will be said *relevant*. It must be emphasized that this concept comes straight from the dual formulation of the strength condition and has nothing to do with any flow rule or other constitutive assumption regarding the way the material deforms or collapses when the strength condition is saturated.

5.4.- Some Classical strength criteria and corresponding π functions.

5.4.1.- Definition of G(x) through a strength criterion.

As a rule, the strength domain G(**x**) is characterized, from the practical point of view, by the data of a strength criterion through a scalar function f of $\underline{\sigma}(\mathbf{x})$, which depends explicitly on **x** as does G(**x**), such that :

$$\begin{array}{lll} f(\mathbf{x}, \underline{\sigma}(\mathbf{x})) < 0 & \Leftrightarrow \ \underline{\sigma}(\mathbf{x}) & \text{interior to } G(\mathbf{x}) \ , \\ f(\mathbf{x}, \underline{\sigma}(\mathbf{x})) = 0 & \Leftrightarrow \ \underline{\sigma}(\mathbf{x}) & \text{on the boundary of } G(\mathbf{x}) \ , \\ f(\mathbf{x}, \underline{\sigma}(\mathbf{x})) > 0 & \Leftrightarrow \ \underline{\sigma}(\mathbf{x}) & \text{exterior to } G(\mathbf{x}) \ , \end{array} \qquad (5.11)$$

For practical applications, function f(**x**, •) is usually chosen a convex, continuous, with piecewise continuous derivatives, function of $\underline{\sigma}(\mathbf{x})$. it is clear anyhow that such a function f(**x**, •) for a given G(**x**) is not unique.

With this definition of G(**x**), Eq. (5.2) may now be written, with simplified notations, in the form :

$$\pi(\mathbf{x}, \underline{d}) = \underset{\underline{\sigma}}{\mathrm{Sup}}\left\{\underline{\sigma} : \underline{d} \ \middle| \ f(\mathbf{x}, \underline{\sigma}) \leq 0\right\} \ . \qquad (5.12)$$

Conversely, the data of function π(**x**, •) as the dual definition of G(**x**) provides an expression for function f(**x**, •) :

$$f(\mathbf{x}, \underline{\sigma}) = \underset{\underline{d}}{\mathrm{Sup}}\left\{\underline{\sigma} : \underline{d} - \pi(\underline{d}) \ \middle| \ \mathrm{tr}\, \underline{d}^2 = 1\right\} \qquad (5.13)$$

this expression may not be the most popular one for the considered G(**x**) as regards practical applications.

Staying within the framework of classical continuum mechanics some usual strength criteria and corresponding π functions for isotropic materials will now be presented, together with the strength criteria and π functions of the most commonly encountered isotropic interfaces. In order to simplify the expressions, the dependence on **x** will not be explicited and the notation **V** will be used for $[\hat{\mathbf{U}}]$.

5.4.2.- Von Mises's strength criterion and π functions.

$$f(\underline{\underline{\sigma}}) = \sqrt{\mathrm{tr}\,\underline{\underline{s}}^2/2} - k \tag{5.14}$$

where $\underline{\underline{s}}$ is the deviatoric part of tensor $\underline{\underline{\sigma}}$.

$$\begin{aligned} \pi(\underline{\underline{d}}) &= +\infty && \text{if } \mathrm{tr}\,\underline{\underline{d}} \neq 0\ ,\\ \pi(\underline{\underline{d}}) &= k\sqrt{2\,\mathrm{tr}\,\underline{\underline{d}}^2} && \text{if } \mathrm{tr}\,\underline{\underline{d}} = 0\ ; \end{aligned} \tag{5.15a}$$

thence :

$$\begin{aligned} \pi(\mathbf{n}, \mathbf{V}) &= +\infty && \text{if } \mathbf{V}\,.\,\mathbf{n} \neq 0\ ,\\ \pi(\mathbf{n}, \mathbf{V}) &= k\,|\mathbf{V}| && \text{if } \mathbf{V}\,.\,\mathbf{n} = 0\ . \end{aligned} \tag{5.15b}$$

5.4.3.- Tresca's strength criterion and π functions.

$$f(\underline{\underline{\sigma}}) = \sup_{i,j=1,2,3,} \{\sigma_i - \sigma_j - \sigma_0\} \tag{5.16}$$

where σ_i, σ_j denote the principal stresses.

$$\begin{aligned} \pi(\underline{\underline{d}}) &= +\infty && \text{if } \mathrm{tr}\,\underline{\underline{d}} \neq 0\ ,\\ \pi(\underline{\underline{d}}) &= \frac{\sigma_0}{2}\,(|d_1| + |d_2| + |d_3|) && \text{if } \mathrm{tr}\,\underline{\underline{d}} = 0\ ; \end{aligned} \tag{5.17a}$$

where the d_i's are the principal components of $\underline{\underline{d}}$; thence :

$$\begin{aligned} \pi(\mathbf{n}, \mathbf{V}) &= +\infty && \text{if } \mathbf{V}\,.\,\mathbf{n} \neq 0\ ,\\ \pi(\mathbf{n}, \mathbf{V}) &= \frac{\sigma_0}{2}\,|\mathbf{V}| && \text{if } \mathbf{V}\,.\,\mathbf{n} = 0\ . \end{aligned} \tag{5.17b}$$

5.4.4.- Coulomb's strength criterion and π functions.

$$f(\underline{\underline{\sigma}}) = \sup_{i,j=1,2,3,} \{\sigma_i(1+\sin\varphi) - \sigma_j(1-\sin\varphi) - 2\,C\cos\varphi\} \tag{5.18}$$

$$\begin{aligned} \pi(\underline{\underline{d}}) &= +\infty && \text{if } \mathrm{tr}\,\underline{\underline{d}} < (|d_1| + |d_2| + |d_3|)\sin\varphi\ ,\\ \pi(\underline{\underline{d}}) &= C\cot\varphi\,\mathrm{tr}\,\underline{\underline{d}} && \text{if } \mathrm{tr}\,\underline{\underline{d}} \geq (|d_1| + |d_2| + |d_3|)\sin\varphi\ ; \end{aligned} \tag{5.19a}$$

thence :

$$\begin{aligned} \pi(\mathbf{n}, \mathbf{V}) &= +\infty && \text{if } \mathbf{V}\,.\,\mathbf{n} < |\mathbf{V}|\sin\varphi\ ,\\ \pi(\mathbf{n}, \mathbf{V}) &= C\cot\varphi\,\mathbf{V}\,.\,\mathbf{n} && \text{if } \mathbf{V}\,.\,\mathbf{n} \geq |\mathbf{V}|\sin\varphi\ . \end{aligned} \tag{5.19b}$$

It will be observed that these expressions are different from the homologous formulae given by Chen [19] which are only valid in the case when the inequalities in (5.19) are saturated.

5.4.5. - Drucker-Prager's criterion and π functions.

$$f(\underline{\sigma}) = \sqrt{\operatorname{tr} \underline{s}^2/2} - \frac{3 \sin\varphi}{\sqrt{3(3+\sin^2\varphi)}} (C \cot\varphi - \operatorname{tr} \underline{\sigma}/3) \quad . \tag{5.20}$$

$$\begin{aligned} \pi(\underline{d}) &= +\infty && \text{if } \operatorname{tr} \underline{d} < \sqrt{\frac{2\sin^2\varphi}{3+\sin^2\varphi}(3\operatorname{tr} \underline{d}^2 - (\operatorname{tr} \underline{d})^2)} \\ \pi(\underline{d}) &= C \cot\varphi \operatorname{tr} \underline{d} && \text{if } \operatorname{tr} \underline{d} \geq \sqrt{\frac{2\sin^2\varphi}{3+\sin^2\varphi}(3\operatorname{tr} \underline{d}^2 - (\operatorname{tr} \underline{d})^2)} \end{aligned} \tag{5.21a}$$

thence :

$$\begin{aligned} \pi(\mathbf{n}, \mathbf{V}) &= +\infty && \text{if } \mathbf{V} . \mathbf{n} < |\mathbf{V}| \sin\varphi \quad , \\ \pi(\mathbf{n}, \mathbf{V}) &= C \cot\varphi\, \mathbf{V} . \mathbf{n} && \text{if } \mathbf{V} . \mathbf{n} \geq |\mathbf{V}| \sin\varphi \quad . \end{aligned} \tag{5.21b}$$

5.4.6.- Tresca's strength criterion with T-tension cut off and π functions.

$$f(\underline{\sigma}) = \sup_{i, j = 1, 2, 3,} \{\sigma_i - \sigma_j - \sigma_0 \ , \ \sigma_i - T\} \quad . \tag{5.22}$$

$$\begin{aligned} \pi(\underline{d}) &= +\infty && \text{if } \operatorname{tr} \underline{d} < 0 \quad , \\ \pi(\underline{d}) &= \frac{\sigma_0}{2} (|d_1| + |d_2| + |d_3|) + (T - \frac{\sigma_0}{2}) \operatorname{tr} \underline{d} && \text{if } \operatorname{tr} \underline{d} \geq 0 \quad ; \end{aligned} \tag{5.23a}$$

thence :

$$\begin{aligned} \pi(\mathbf{n}, \mathbf{V}) &= +\infty && \text{if } \mathbf{V} . \mathbf{n} < 0 \quad , \\ \pi(\mathbf{n}, \mathbf{V}) &= \frac{\sigma_0}{2} |\mathbf{V}| + (T - \frac{\sigma_0}{2}) \mathbf{V} . \mathbf{n} && \text{if } \mathbf{V} . \mathbf{n} \geq 0 \quad . \end{aligned} \tag{5.23b}$$

When the material under consideration does not withstand any tensile stress the value of T is set equal to nought.

5.4.7.- Coulomb's strength criterion with T-tension cut off and π functions.

$$f(\underline{\sigma}) = \sup_{i, j = 1, 2, 3,} \{\sigma_i(1 + \sin\varphi) - \sigma_j(1 - \sin\varphi) - 2\, C \cos\varphi \, , \sigma_i - T\} \quad . \tag{5.24}$$

$$\pi(\underline{d}) = +\infty \qquad \text{if } \operatorname{tr} \underline{d} < (|d_1| + |d_2| + |d_3|)\sin\varphi \ ,$$

$$\pi(\underline{d}) = \frac{C\cos\varphi - T\sin\varphi}{1-\sin\varphi}(|d_1| + |d_2| + |d_3|) + \frac{T - C\cos\varphi}{1-\sin\varphi}\operatorname{tr}\underline{d} \tag{5.25a}$$

$$\text{if } \operatorname{tr}\underline{d} \geq (|d_1| + |d_2| + |d_3|)\sin\varphi \ ;$$

thence :

$$\pi(n, V) = +\infty \qquad \text{if } V.n < |V| \sin\varphi$$

$$\pi(n, V) = \frac{C\cos\varphi - T\sin\varphi}{1-\sin\varphi}|V| + \frac{T - C\cos\varphi}{1-\sin\varphi} V.n \tag{5.25b}$$

$$\text{if } V.n \geq |V| \sin\varphi \ ;$$

5.4.8.- Smooth or frictionless interface.

Vector **T** denotes the stress vector, with components (σ, τ), acting on the interface with outward normal **n**. Vector **V** stands for the velocity jump in the interface.

$$f(T) = \operatorname{Sup}(\sigma, \tau) \ . \tag{5.26}$$

$$\begin{aligned} \pi(V) &= +\infty \quad \text{if } V.n < 0 \ , \\ \pi(V) &= 0 \quad \text{if } V.n \geq 0 \ . \end{aligned} \tag{5.27}$$

5.4.9.- Isotropic interface with Tresca's friction and no resistance to traction.

$$f(T) = \operatorname{Sup}\{\sigma, \tau - C_i) \ . \tag{5.28}$$

where C_i denotes the shear strength of the interface.

$$\begin{aligned} \pi(V) &= +\infty \quad \text{if } V.n < 0 \ , \\ \pi(V) &= C_i V_t \quad \text{if } V.n \geq 0 \ . \end{aligned} \tag{5.29}$$

where V_t is the modulus of the component of **V** in the interface.

5.4.10.- Isotropic interface with Coulomb's dry friction.

$$f(T) = \tau - \sigma\tan\varphi_i \tag{5.30}$$

where φ_i denotes the friction angle of the interface.

$$\begin{aligned} \pi(V) &= +\infty \quad \text{if } V_n < V_t\tan\varphi_i \ , \\ \pi(V) &= 0 \quad \text{if } V_n \geq V_t\tan\varphi_i \ . \end{aligned} \tag{5.31}$$

5.5.- Implementation of the kinematic approach from outside.

5.5.1.- Classical applications.

Usually the practical implementation of the kinematic approach from outside makes use of *simple* relevant velocity fields depending on a reduced number of parameters. The optimization of the obtained results is gained through the minimization of $P_{mr}(\hat{U})$ with respect to these parameters while keeping $\dot{q}(\hat{U})$ constant. Examples of such classically used velocity fields may be given now.

Piecewise rigid body motion virtual velocity fields where one rigid block is moving and the rest of the structure is motionless, or more elaborated virtual "mechanisms" where several rigid blocks move with respect to one another, will be encountered (Fig. 12). In both cases the virtual velocity jumps will be governed by the condition that $\pi(\mathbf{x}, \mathbf{n}(\mathbf{x}), [\hat{\mathbf{U}}(\mathbf{x})])$ remains finite as explained in para. (5.3.2)

- Tresca's or von Mises' strength criterion.

For instance, when dealing with a material whose strength criterion is of the Tresca or von Mises types, it comes out from Eq. (5.15) and (5.17) that such k.a. relevant virtual velocity fields will only exhibit purely tangental velocity jumps ; examples are shown in Fig. 12:

$$[\hat{\mathbf{U}}(\mathbf{x})] \cdot \mathbf{n}(\mathbf{x}) = 0 \ . \qquad (5.32)$$

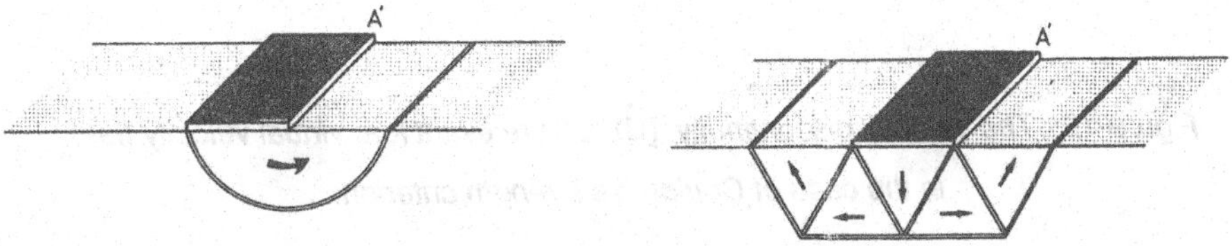

Figure 12 - Indentation of a half space by a rigid plate : relevant rigid body motion virtual velocity fields in the case of Tresca's or von Mises' strength criterion.

Relevant virtual velocity fields will also be constructed where the strain rate field is non zero : they will be governed by the condition that $\pi(\mathbf{x}, \hat{\underline{\mathbf{d}}}(\mathbf{x}))$ be finite.

Still taking the case of a Tresca or von Mises strength criterion as an example, the corresponding k.a. relevant virtual velocity fields will induce no volume change as can be seen from Eq. (5.15) and (5.17) :

$$\mathrm{tr}\,\hat{\underline{\mathbf{d}}}(\mathbf{x}) = 0 \ . \qquad (5.33)$$

Such fields as shear velocity fields will thus be relevant for those materials.

- Coulomb's strength condition -

The case of a material with a strength criterion of Coulomb's type, must now be considered for it will help emphasizing some essential points of the kinematic approach. Eq. (5.19) gives the conditions to be satisfied by relevant virtual velocity fields :

$$\mathrm{tr}\ \hat{\underline{\underline{d}}}(\mathbf{x}) \geq (|\hat{d}_1(\mathbf{x})| + |\hat{d}_2(\mathbf{x})| + |\hat{d}_3(\mathbf{x})|)\ \sin\varphi \quad , \qquad (5.34)$$

which establishes a minimum value for the virtual dilatancy ;

$$[\hat{\mathbf{U}}(\mathbf{x})]\ .\ \mathbf{n}(\mathbf{x}) \geq |[\hat{\mathbf{U}}(\mathbf{x})]|\ \sin\varphi \quad , \qquad (5.35)$$

which shows that the velocity jump, if non zero, cannot be purely tangential : $[\hat{\mathbf{U}}(\mathbf{x})]$ must be inclined over the jump surface Σ by an angle β at least equal to φ and be directed outwards, corresponding to a virtual separation between both sides of the jump surface (Fig.13).

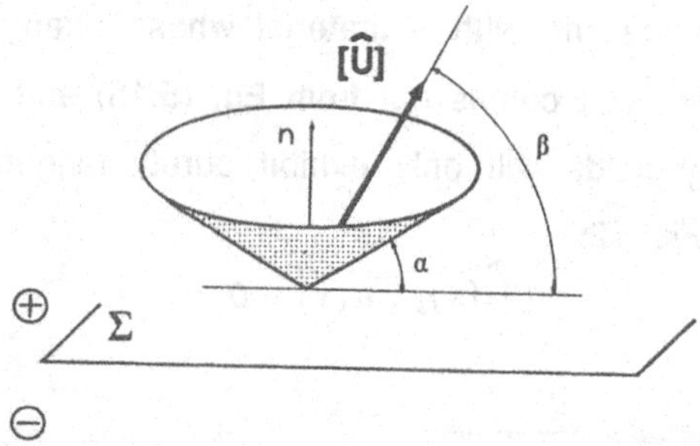

Figure 13 - The velocity discontinuity $[\hat{\mathbf{U}}]$ *of a relevant k.a. virtual velocity field in the case of Coulomb's strength criterion.*

It must be underlined once again that conditions (5.34, 5.35) are imposed on *virtual* velocity fields and do not claim any significance from the point of view of any flow rule,... ; this is often misunderstood. Moreover it must be recalled that *any* velocity jump is allowable for a virtual velocity field : velocity jump surfaces need not be slip surfaces where the velocity jump is tangential. Therefore conditions (5.35) and (5.32) are exactly on the same status, proceeding only from the strength condition of the considered material. The latter point has often proved a difficulty in the understanding of the kinematic approach from outside, in soil mechanics for instance, through it is the key of the efficiency of the method.

Obviously from Eq. (5.21, 5.23, 5.25), what has just been said for Coulomb's strength criterion will apply to the strength criteria considered in para. (5.4.5. - 5.4.7.).

- Plane strain relevant k.a. virtual velocity fields -

When the problem under consideration accepts plane strain velocity fields as kinematically admissible the simplification due to passing from dimension 3 to dimension 2 makes it possible to construct more elaborate relevant k.a. virtual velocity fields.

Considering first the case of Tresca's or von Mises' strength criterion (or any criterion independent of tr$\underline{\sigma}$), Eq. (5.33) applied to virtual k.a. velocity fields in plane strain parallel to Oxy results in generating the relevant virtual velocity fields in the following way.

Considering two families of mutually orthogonal curves (α- and β-lines), the components $\hat{U}_\alpha$ and $\hat{U}_\beta$ of $\hat{U}$ at the current point verify the differential equations [47]:

$$\begin{aligned} &d\hat{U}_\alpha - \hat{U}_\beta d\theta = 0 \quad \text{along the } \alpha\text{-lines} \\ &d\hat{U}_\beta + \hat{U}_\alpha d\theta = 0 \quad \text{along the } \beta\text{-lines} \end{aligned} \qquad (5.36)$$

where θ is defined as $\theta = (0x, e_\alpha)$, e_α and e_β being the unit vectors tangent to the α- and β-lines at the current point, $(e_\alpha, e_\beta) = +\pi/2$ (Fig. 14).

For Coulomb's strength criterion, equations homologous to (5.36) are established for the plane strain velocity fields which saturate Ineq. (5.34) (cf. [48]).

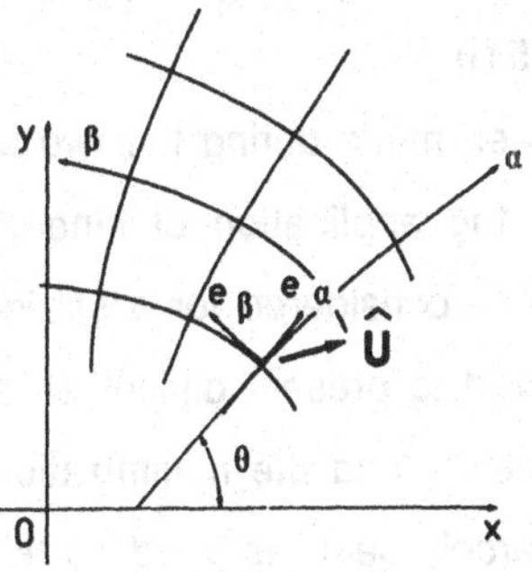

Figure 14 - Plane strain relevant virtual k.a. velocity fields for Tresca's strength criterion : Geiringer's equations [47].

A classical example of such a plane strain relevant k.a. virtual velocity field is recalled in Fig. 15 where volumes (1), (3), and (5) are given rigid body motions while in volumes (2) and (4) the velocity is constant and orthoradial ; the plane and cylindrical surfaces ABB'A', BCC'B', and CDD'C' and the symmetric ones are velocity discontinuity surfaces.

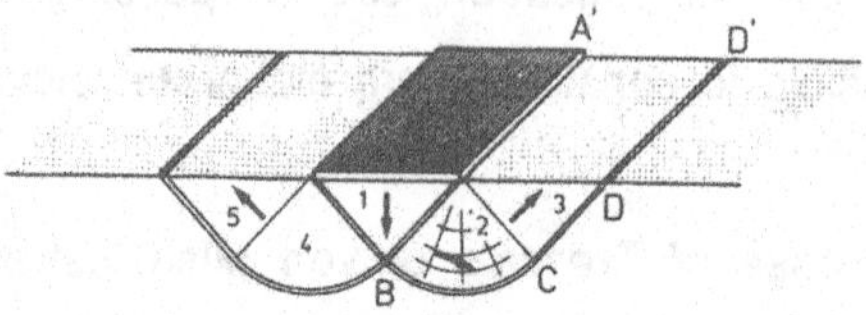

Figure 15 - Indentation of a half space by a rigid plate : Prandtl's velocity field in the case of Tresca's strength criterion.

It may be added that, due to the corresponding simplifications, many yield design problems are often treated within the plane strain assumption, at least for a first study. A detailed presentation as to the significance of this treatment within the framework of yield design may be found in [36] and many interesting solutions are available in numerous textbooks.

5.5.2.- Numerical implementation.

The efficiency of the methods which have just been sketched is highly improved when analytical methods for the construction of k.a. relevant virtual velocity fields are combined with numerical minimization procedures. This has been the basis of computer codes which have been developed for instance for application to soil mechanics stability analysis problems (e.g. [49 - 51]).

Many attempts have also been made during the two past decades aiming at using numerical methods directly for the application of kinematic approach from outside. Finite element methods have been considered for a straight forward application of the theory [52 - 56], which has proved to present difficulties as regards the generation of strictly relevant virtual velocity fields, and the minimization procedure. In many cases the results so obtained have scarcely been as good as those gained through classical procedures where discontinuous relevant virtual velocity fields can usually be generated and explored more easily.

Recently, in the case of Tresca's strength criterion, a numerical method based upon the generation of stream functions has been developed and seems quite efficient [57, 58].

6.- AN EXAMPLE OF THE YIELD DESIGN KINEMATIC APPROACH [59].

6.1.- Stability of a vertical cut for a purely cohesive soil.

The stability of the vertical cut introduced in Fig.7 is now considered, assuming the homogeneous constituent soil to be purely cohesive, i.e. with a Tresca strength criterion.

The relevant k.a. virtual velocity fields for such a material have already been discussed in para. (5.5.1).

Applying the same arguments as in Section (4.3) it appears that the kinematic approach for the problem can be restricted to a slice of arbitrary thickness D parallel to plane Oxy with no resisting forces nor resisting power developed in the parallel sections at both extremities. (A detailed presentation of this type of yield design problems –plane strain yield design problems– appear in [36] where it is shown how they can be studied through a two-dimensional yield design analysis).

For the slice of thickness D the following k.a. virtual velocity field can be retained in accordance with Eq. (5.32, 5.33) : the prismatic volume OABB'A'O' is given a virtual downwards translation motion with velocity V parallel to AB, while the rest of the soil mass remains motionless (Fig.16).

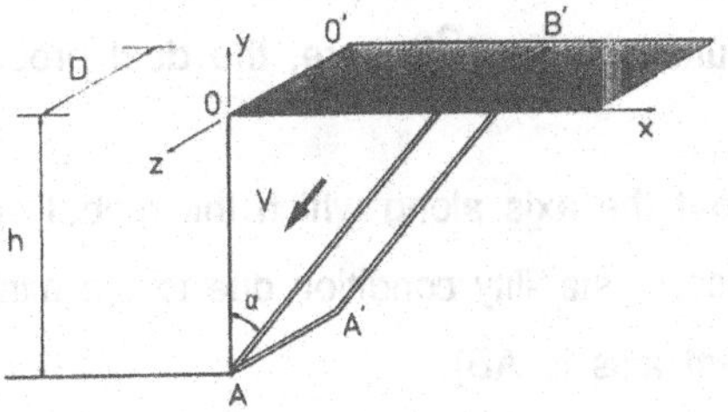

Figure 16 - Stability of a vertical cut in a purely cohesive soil : kinematic approach.

The power of the external forces performed in this virtual velocity field by the gravity forces writes :

$$Q\dot{q}(\hat{U}) = \frac{\gamma h^2 D}{2} V \sin\alpha \qquad (6.1)$$

while the maximum resisting power being developed along plane ABB'A' in the tangential velocity discontinuity reduces to :

$$P_{mr}(\hat{\underline{U}}) = CV\,h\,D/\cos\alpha \quad . \qquad (6.2)$$

Thence through Ineq. (5.4)

$$(\gamma h/C)^{+} \leq 4/\sin 2\alpha \quad . \qquad (6.3)$$

Looking for the minimum of the second hand of this upper bound with respect to α gives $\alpha = \pi/4$ and

$$(\gamma h/C)^{+} \leq 4 \qquad (6.4)$$

as the best upper estimate of $(\gamma h/C)^{*}$ for this class of one parameter k.a. virtual velocity fields.

The following comments can be made.

(a) Comparing the upper bound (6.4) with the value obtained in Section (4.3) (Ineq. 4.10) for this particular case where $\varphi = 0$ shows that both results are identical :

$$(\gamma h/C) \leq 4$$

is proved a necessary condition for the stability of the cut.

(b) This is not a mere coincidence!

As a matter of fact the kinematic reasoning performed here with a uniform translation motion given to volume OABB'A'O' results in checking the global equilibrium of that volume, as regards the resultant component in the direction of the motion, of the gravity forces in Ω and resisting forces developed on ABB'A'.

Through the use of function $\pi(\underline{n}, [\hat{\underline{U}}])$ here, the dual procedure in the kinematic approach

- automatically finds out the axis along which the global equilibrium equation will produce a non trivial necessary stability condition due to the expression of the strength condition (here this relevant axis is AB),

- selects the distribution(s) of the resisting forces which most favour(s) the global equilibrium of the volume so that external forces being not balanced under these conditions will necessarily mean the instability of the cut.

(c) One may question about the admissibility of the considered velocity field since the tip AA' of volume OABB'A'O' being given the velocity $\underline{V}$ parallel to AB would penetrate the horizontal surface of the motionless mass of the rest of the soil.

Many arguments have been put forward in order to solve this apparent paradox which occurs quite frequently in the applications of the kinematic approach (cf. Fig.12 and 15). For those which are relevant they appear adequate to the problem under consideration with no versatility to be applied to similar cases, and do actually miss the fundamental point.

The answer lies in the very significance of virtual k.a. velocity fields, already recalled in para. (5.5.1), that are the *test functions*, the "mathematical tools" in the dualization procedure. This means that they are piecewise continuous with piecewise continuous derivatives, but no condition is imposed on the possible velocity discontinuity. In the application of the kinematic approach, the "strength" of the material, through the corresponding $\pi(\mathbf{n}, [\hat{\mathbf{U}}])$ function, imposed the choice of tangential velocity discontinuity on the plane AA'B'B for the value of $\pi(\mathbf{n}, [\hat{\mathbf{U}}])$ to remain finite. *No such condition is to be found to impose any constraint on* $\hat{\mathbf{U}}$ *along* AA' and this can be understood from the following : the three-dimensional continuum model does not take lineal densities of forces into account (nor does the two-dimensional model with punctual ones) and therefore the static analysis performed in Section (4.3) did not consider any lineal resisting force at the vertex of volume (1) along AA', which is consistent with no term appearing on AA' in the dual formulation.

(d) Another result of the kinematic approach is to answer the question of whether the consideration of the moment equation of the global equilibrium of volume (1) in the static analysis performed at Section (4.3) would have led to an additional limitation on ($\gamma h/C$) for the stability necessary condition.

The answer is negative and is apparent from the dual formulation. Considering the moment equation corresponds, in the dual formulation, to a rigid body rotational motion of volume OABB'A'O' inducing a velocity discontinuity across plane ABB'A' which would not be tangential everywhere, unless the axis of rotation parallel to Oz be rejected at infinity in the direction normal to AB which is the case already examined.

This illustrates the fact that, but for that particular case which leads to the final result already obtained, the strength capacities of the soil will have no limiting consequance on the balance of the moment equation of global equilibrium of volume (1).

Thus Ineq. (6.4) expresses the best result to be derived from the consideration of the global equilibrium of triangular prismatic volumes.

(e) The same reasoning, through the kinematic approach as the dual formulation of the static analysis, shows that circular cylindrical volumes may also be considered for the "global equilibrium check" and produce significant results. In that case the relevant equation which provides a necessary stability condition on $(\gamma h/C)$ turns out to be the moment equation with respect to the geometrical axis of the cylinder (Fig. 17).

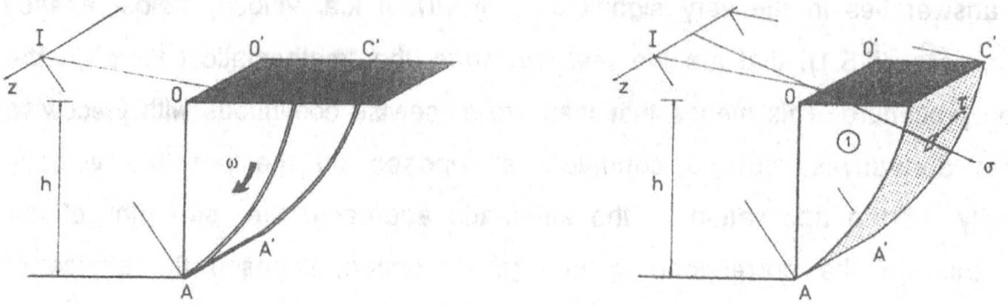

Figure 17 - Stability of a vertical cut in purely cohesive soil : kinematic approach with a rigid body rotational motion and "global equilibrium check".

The result obtained by looking for the strongest necessary condition over all cylindrical volumes is classical [16, 17] and is given by

$$(\gamma h/C)^{+} \leq 3.83 \ . \tag{6.5}$$

This upper estimate is better than condition (6.4) ; this is no surprise since the analyses with triangular prismatic volumes are but particular cases of the present ones.

(f) Finally it also appears from the dual formulation that no other cylindrical volumes can be considered significantly for a static analysis through the "global equilibrium check" : they would not result in any further constraint on $(\gamma h/C)^{+}$.

The upper bound resulting from Ineq. (6.5) for $(\gamma h/C)^{+}$ has long been the best available for this problem, despite numerous attempts where even sophisticated numerical methods have been used. De Buhan, Dormieux and Maghous [60] have quite recently produced a better result, obtained rigorously without heavy numerical procedure, through an ingenious relevant virtual velocity field :

$$(\gamma h/C)^{+} \leq 3.817 \ . \tag{6.6}$$

It is clear that the importance of this result will not be assessed from its consequences as regards practical applications : the analysis of the stability of a vertical cut is somewhat of a symbol where the "slip circle" has been baffling the endeavours of researchers for decades, although it could be proved that it was not to give the exact answer to the problem.

6.2.- Stability of a vertical cut for a soil exhibiting both cohesion and friction.

Assuming now the strength condition of the constituent soil to be of Coulomb's type, the following relevant k.a. virtual velocity fields will be used which satisfy Ineq. (5.34, 5.35) : the prismatic volume OABB'A'O' is given a virtual downward translation motion with the velocity V inclined at an angle β to AB with $\varphi \le \beta \le \pi - \varphi$ while the rest of the soil mass remains motionless (Fig. 18).

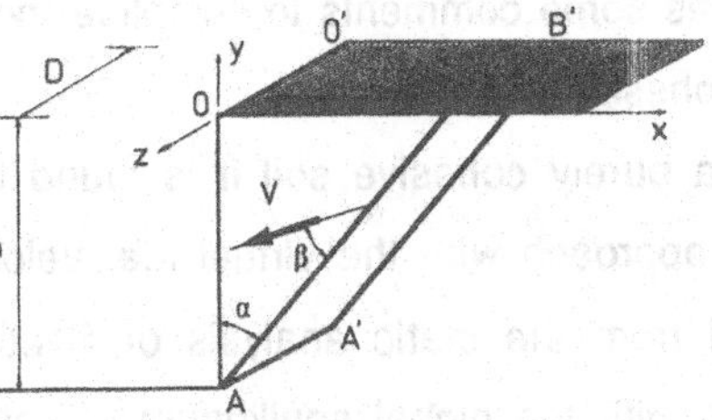

Figure 18 - Stability of a vertical cut in a frictional and cohesive soils : kinematic approach.

The power performed in this velocity field by the gravity forces becomes :

$$Q\dot{q}(\hat{U}) = \frac{\gamma h^2 D}{2} \tan\alpha \; V \cos(\alpha + \beta) \tag{6.7}$$

while the maximum resisting power being developed along plane ABB'A' in the velocity discontinuity is equal to

$$P_{mr}(\hat{U}) = \frac{C}{\tan\varphi} \frac{hD}{\cos\alpha} V \sin\beta \tag{6.8}$$

It follows through Ineq. (5.4)

$$\left(\frac{\gamma h}{C}\right)^{+} \le \frac{2\sin\beta}{\tan\varphi \; \sin\alpha \cos(\alpha+\beta)} \tag{6.9}$$

where α and β are two parameters with the constraints

$$\begin{gathered} 0 \le \alpha + \beta \le \pi/2 \\ \varphi \le \beta \le \pi - \varphi \,. \end{gathered} \tag{6.10}$$

Minimizing the second hand of Ineq. (6.9) with respect to β corresponds to

$$\beta = \varphi \tag{6.11}$$

for which Ineq. (6.9) becomes

$$(\gamma h/C)^{+} \le 2 \cos\varphi/\sin\alpha \, \cos(\alpha+\varphi) \tag{6.12}$$

(to be compared with Ineq. (4.8)).

Minimizing then with respect to α (as already done at Section 4.3) gives

$$(\gamma h/C)^{+} \le 4 \, \tan(\pi/4+\varphi/2) \tag{6.13}$$

as the best estimate of $(\gamma h/C)^{+}$ one may achieve for this class of two parameters k.a. virtual velocity fields.

This results also deserves some comments to complete those already made in the simpler case of the purely cohesive soil.

(a) As in the case of a purely cohesive soil it is found that the upper estimate derived from the kinematic approach with the virtual k.a. velocity fields of Fig. 18 is equal to the result derived from the static analysis of Section (4.3) for the same volume (1) in Fig. 9 dealing with the global equilibrium of volume (1) from the point of view of the resulting force.

(b) One may also notice that the optimality condition $\beta = \varphi$ in the kinematic approach means that, in the static analysis, the distributions of (σ, τ) along plane ABB'A' which most favour the global equilibrium of volume (1) correspond to the equality

$$|\tau| = C - \sigma \tan \varphi \qquad \text{with } \tau \text{ in direction AB}$$

(which is quite evident here).

It also shows that the significant global equilibrium equation which is to provide the stability constraint on $(\gamma h/C)$ is the resulting force equation in the direction at angle φ to AB (the direction of V).

(c) The optimality condition $\beta = \varphi$ found in this particular case deserves being looked at in a more general perspective since the constraints on [U] for the relevant virtual k.a. velocity fields derived from Ineq. (5.35) are weaker than for a purely

cohesive soil, namely

$$\varphi \leq \beta \leq \pi - \varphi \quad . \tag{6.14}$$

The proofs of the statements to appear hereafter can be found in the results given in [36] : they rely on the kinematic dual formulation of the "global equilibrium check" static analysis and can be generalized whatever the slope of the cut.

The idea would be to profit by the liberty left by condition (6.14) to perform the kinematic approach using more general piecewise rigid body motion velocity fields with any cross section for the cylindrical volume in motion. Such a rigid motion will be either a rotation around any axis Iz or a translation with any velocity **V** (Fig. 19).

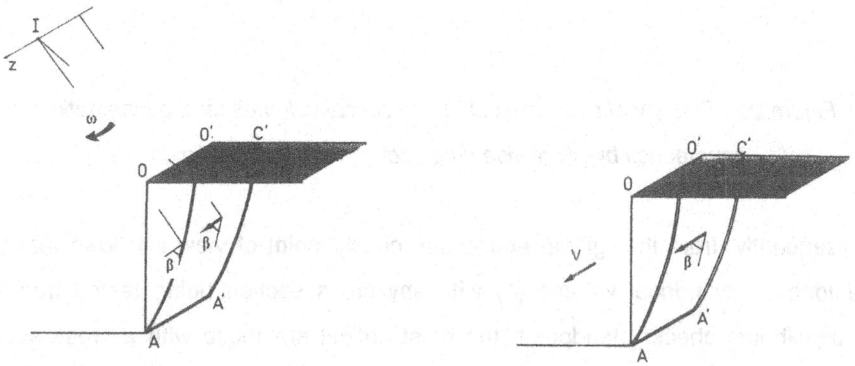

Figure 19 - Stability of a vertical cut : more general kinematic approach using piecewise rigid body motion velocity fields.

From the static analyis of view this would amount to the "global equilibrium check" of a cylindrical volume (1) = OACC'A'O', with any cross section, being considered thoroughly.

For the homogeneous vertical cut the results are as follows.

Starting from the kinematic approach it comes out that :

• for a given axis Iz defining a rotation with a velocity ω (Fig. 20) the ***most critical volume*** OACC'A'O', that is the volume leading to the strongest constraint on $(\gamma h/C)^+$, is obtained when AC is an arc of a "φ" logspiral with pole I, so that the induced velocity discontinuity be inclined at the angle $\beta = \varphi$ to the surface ACC'A' at each point ; such a "φ" logspiral is defined in polar coordinates (ρ, θ) attached to I by the expression $\rho = \rho_0 \exp(-\theta \tan \varphi)$;

• as a particular case, when a translation is considered parallel to velocity **V** the *most critical volume* OABB'A'O' is obtained when OAB is a triangle and **V** makes the angle φ with AB in the outwards direction (Fig. 20).

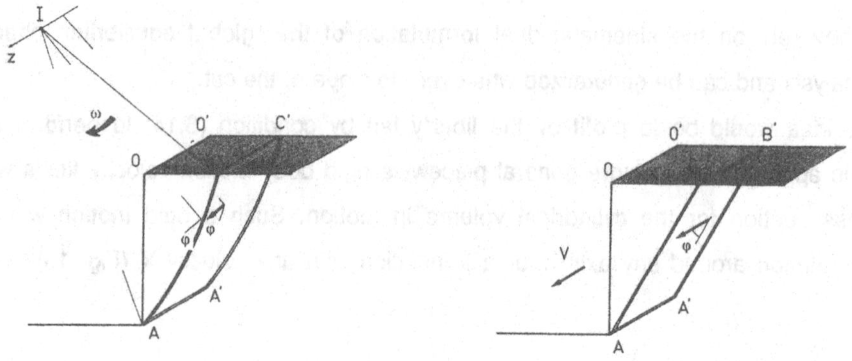

Figure 20 - Stability of a vertical cut : most critical volumes for the kinematic approaches by piecewise rigid body motion velocity fields.

Consequently, from the "global equilibrium check" point of view it follows that :

• among all cylindrical volume (1) with any cross section being tested from the "global equilibrium check" standpoint, the most critical are those with a cross section defined by "φ" logspiral arc AC or a straight line AB (Fig. 21) ;

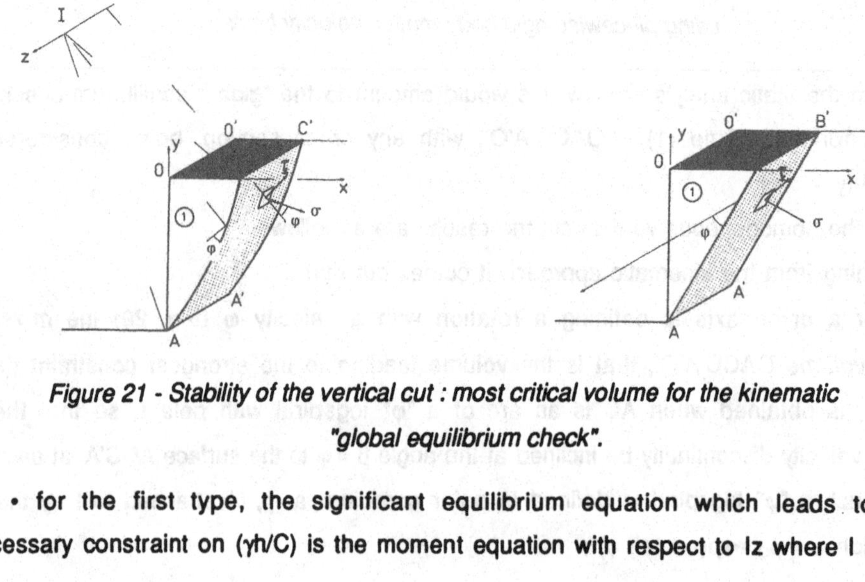

Figure 21 - Stability of the vertical cut : most critical volume for the kinematic "global equilibrium check".

• for the first type, the significant equilibrium equation which leads to the necessary constraint on ($\gamma h/C$) is the moment equation with respect to Iz where I is the pole of the considered logspiral ;

• for the second type, the significant equilibrium equation is the resulting force equation in the direction inclined at the outward angle φ to AB in the plane Oxy (Fig. 21).

The best result obtained through these equivalent analyses turns out approximately to fit the formula [19] :

$$(\gamma h/C)^{+} \leq 3.83 \ \tan(\pi/4+\varphi/2) \qquad (6.15)$$

(d) the difference between Eq. (5.32) and (5.33) on one side, and Ineq. (5.34) and(5.35) on the other, have led to conclusions concerning the general conditions to be fulfilled by relevant virtual k.a. velocity fields which might seem antinomic and make it difficult to imagine the possibility of passing continuously from one case to the other when $\varphi \to 0$.

This can be explained for instance for the relevant velocity discontinuities. For a frictional and cohesive soil ($\varphi \neq 0$, $C \neq 0$) it has been explained that the relevant $[\hat{U}]$ must belong to the cone drawn in Figure 22. Making φ tend to zero, the allowable cone for $[\hat{U}]$ becomes wider and flattens down into the upper half space (Fig. 15). At the same time it is seen from Eq. (5.19) where C is a constant that, keeping $|[\hat{U}]|$ constant, the value of $\pi(\mathbf{n}, [\hat{U}])$ tends to infinity when $\varphi \to 0$ unless $[\hat{U}]$ belongs to the very boundary of the cone. It follows that, φ tending to zero, the relevant $[\hat{U}]$ will tend to be confined in the velocity jump surface.

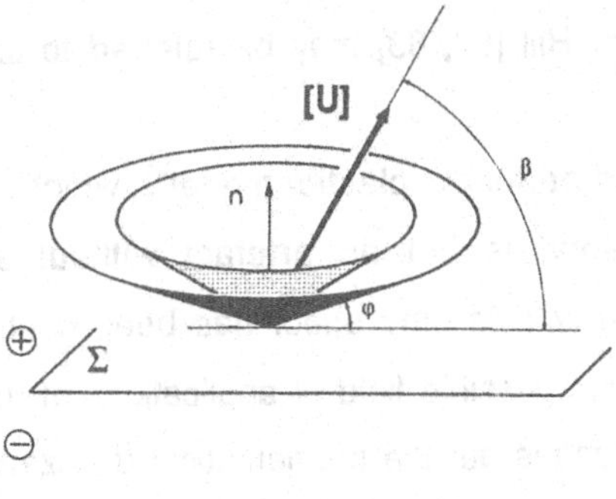

Figure 22 - The relevant velocity discontinuities as $\varphi \to 0$.

6.3.- Comments.

The stability analysis of the vertical cut has been chosen in this presentation with the purpose of giving an illustrative example of the kinematic approach being the dual formulation of the static one. This point of view is quite obvious in the comparison between Eq. (6.12) and Eq. (4.8) for instance.

Consequently, since only rigid motion virtual velocity fields have been used, this example is not the best suited to enhance the possibilities of the kinematic approach from outside, whose efficiency results from the variety of relevant k.a. virtual velocity fields which can be used : they make it possible and easy, through the dual formulation, to check the compatibility between equilibrium and resistance in a more elaborate way than the mere *global* equilibrium of one or a few volumes. As a matter of fact the upper bound (Ineq. 6.6) obtained in [60] is derived from such a kinematic approach where the virtual velocity field combines a rigid body motion velocity field and a shear velocity field. A good example to illustrate this point of view would be the indentation of a half space by a rigid plate (sketched in Fig. 12 and 15) with the classical results recalled in [36], and more recent developments, such as those presented in [61] in the case of an excentrated and inclined load on the plate acting on a purely cohesive soil with no resistance to tensile stresses.

7.- CONCLUDING REMARKS.

7.1.- The fundamental principles.

Concluding this survey of the Theory of Yield Design, the fundamental principles will first be recolled. The yield design theory relies on the necessary stability condition that "Equilibrium and Resistance should be Compatible". From this follows the *static approach.* Two papers by Hill [62, 63] may be referred to as early application of this concept.

The so called "rigid-perfectly plastic model", which has unfortunately been connected with this theory, is but an artefact without any physical reality nor theoretical or practical utility ; its only effect has been confusing ideas and giving a restrictive perspective of the possible field of applications of the yield design concept.

The *kinematic approach* is but the mathematical dualization of the static one. The involved velocity fields are *virtual* velocity fields. They do not claim any actuality as regards the true collapse modes of the considered system, even though experience seems to show some similarity between observed collapse modes and "good" relevant k.a. virtual velocity fields.

7.2.- Possible fields of applications in civil engineering.

The short historical review at the beginning of this paper already gave a list of various applications of the theory. A few words may now be added regarding recent or possibly future developments in soil mechanics, taking advantage of the versatility of the method :

stability analyses of seabed soils [64, 65], of tunnels [66, 67] ;

stability analyses assuming T or zero tension cut off strength criteria (para. 5.4.6 and 5.4.7) as first illustrated in the paper by Drucker [68], (e.g. [61]) ;

reinforced soil structures stability analyses through methods combining the yield design and the homogenization theories [33-35], or through mixed modelling, that is adopting a continuum mechanics model for the constitutive soil and a beam (or a strip) model for the reinforcing inclusions : the interaction of the two models is mechanically consistent and the strength criteria and π functions are available ; the sound mechanical bases make a clean sweep of all difficulties usually encountered in trying to adapt "tradtional" stability analysis methods to these new circumstances.

Moreover, as already announced in [59], the same sound mechanical bases of the theory make it the reliable way to build stability analysis methods and computer codes which thoroughly comply with both the spirit and the letter of the Ultimate Limit State Design (ULSD).

As a matter of fact the approach of safety which is conveyed in ULSD requires a clear distinction to be made between the given *loads* on the considered structure and the *resistances* which may be mobilized in the constituent materials (soil, reinforcement, interfaces,...). It then introduces partial safety factors to be applied : as multipliers on the loads, greater than unity when the concerned load is unfavourable to stability, lower than unity when it is favourable, they provide the design values of the loads ; as dividers on the strength parameters, greater than unity, they provide the design values of the strength parameters.

"According to the principles of Limit States Design, the design criterion is simply to design for equilibrium in the design limit state of failure. The design criterion could be expressed in the following way :

$$R_d \geq S_d \tag{7.1}$$

S_d is the design load effect... The design resistance effect R_d..." [69].

In order to take the adequacy of the design method into account, the design value of ratio R_d/S_d in Ineq. (6.16), denoted

$$\Gamma_s = R_d/S_d \tag{7.2}$$

and called "Method coefficient", will not be compared to unity but to a reference value, depending on the method used for the assessment of stability, which is usually greater than unity.

This short presentation of ULSD clearly shows how the Yield Design Theory is fully suited to this approach of safety which appeared as early as 1927 in Denmark and was fostered by Brinch Hansen [70].

Contrary to many traditional methods used for stability analyses (e.g. the method of slices,...) it does make the required clear distinction between loads and resistances, in the sense that the calculation of R_d on the "strength" side of Ineq. (7.1) does not, in any way, involve the loads which only appear in S_d on the "load" side of Ineq. (7.1) .

It will be easily recognized that the fundamental principle of ULSD, as quoted from [69] and expressed by Ineq. (7.1) , does not differ from the fundamental principle of the Yield Design Theory expressed in Formula (3.8), once the loads and the strength parameters have been modified through their respective partial safety factors.

Ineq. (5.4 or 5.6) and Ineq. (7.1) are not only similar but actually identical, which means that the kinematic approach clearly and rigorously results in the upper bound approach of Γ_s .

The reproach has sometimes been made to ULSD that, since all quantities are modified through partial safety factors, the design "mechanisms" might (or will) have nothing in common with those which may actually occur in case of collapse. Though questionable, for it may apply as well to the global safety factor approach, this remark may be partially answered to as follows. It turns out that calculation methods and computer codes based upon the Yield Design Theory are very efficient in the sense that they prove to be usually less time consuming than many traditional methods ; added to their rigorous mechanical basis, this makes it possible to perform parametric studies which are certainly the best way to investigate the behaviour of a structure.

REFERENCES

1. GALILEO G., *Discorzi e dimostrazioni matematiche intorno a due nueve scienze*, Dialogo secundo, Leyden 1638.
2. COULOMB C.A., *Essai sur une application des règles de Maximis et Minimis à quelques problèmes de statique relatifs à l'architecture*, Mémoire à l'Académie Royale des Sciences, Paris, 1773.
3. HEYMAN J., *The stone skeleton*, Int. Jl. Solids and Structures, vol.2, (1966), n°2, 249-279.
4. HEYMAN J., *The safety of masonry arches*, Int. Jl. Mech. Sc., vol.11, (1969), 363-385.
5. HEYMAN J., *Coulomb's memoir on statics : an essay in the history of civil engineering*, Cambridge, G.B., 1972.
6. HEYMAN J., *The estimation of the strength of masonry arches*, Proc. A.S.C.E., Part 2, n°69, (1980), 921-937.
7. HEYMAN J., *The masonry arch*, Ellis Horwood Ltd., John Wiley, 1982.
8. DELBECQ J.M., *Analyse de la stabilité des voûtes en maçonnerie de Charles Augustin Coulomb à nos jours*, Ann. Ponts et Chaussées, n°19, (1981), 36-43.
9. DELBECQ J.M., *Analyse de la stabilité des voûtes en maçonnerie par le calcul à la rupture*, J. Méc. Th. Appl., vol.1, (1982), n°1, 91-121.
10. MÉRY E., *Equilibre des voûtes en berceau*, Ann. Ponts et Chaussées, (1840), I, 50-70.
11 DURAND-CLAYE A., *Stabilité des voûtes en maçonnerie*, Ann. Ponts et Chaussées, (1867), I, 63-96.
12. DURAND-CLAYE A., *Stabilité des voûtes et des arcs*, Ann. Ponts et Chaussées, (1880), I, 416-440.
13. CULMANN K., *Die graphische Statik*, Zurich, 1866.
14. FELLENIUS W., *Calculation of the Stability of Earth Dams*, Trans. 2nd Cong. on Large Dams, vol.4, Washington D.C., 1936.
15. BISHOP A.W., *The use of the slip circle in the stability analysis of slopes*, Proc. Eur. Conf. Soil Mech. and Foundation Engineering, Stockholm, 1954.
16. TAYLOR D.W., *Stability of earth slopes*, Jl. Boston Soc. Civ. Eng., 24, (1937), 3, 337-386.
17. TAYLOR D.W., *Fundamentals of soil mechanics*, John Wiley, 1948.
18. RENDULIC L., *Ein Beitrag zur Bestimmung der Gleitsicherheit*, Der Bauingenieur n°19/20, 1935
19. CHEN W.F., *Limit analysis and soil plasticity*, Elsevier, 1975.
20. MASSAU J., *Mémoire sur l'intégration graphique des équations aux dérivées partielles ; chap. VI : Equilibre limite des terres sans cohésion*, Ann. Ass. Ing. Ecole de Gand (1899). Editions du Centenaire, Comité National de Mécanique, Bruxelles.
21. KÖTTER F., *Die Bestimmung des Druckes an gebrümmeten Blattflächen, eine Aufgabe aus des Lehre vom Erddruck*, Berl. Akad. Bericht, (1903), 229.
22. KÖTTER F., *Uber dem Druck von Sand*, Berl. Akad. Bericht, 1909.

23. SOKOLOVSKI V.V., *Theorie der Plastizizät*, VEB Verlag Technik, Berlin, 1955.
24. SOKOLOVSKI V.V., *Statics of soil media*, Butterworths, London, 1960, (after the 2nd edition, Moscow, 1954).
25. SOKOLOVSKI V.V., *Statics of granular media*, Pergamon Press, 1965.
26. BEREZANCEW B.G., *Problème de l'équilibre limite d'un milieu pulvérulent en symétrie axiale*, Moscow, 1952.
27. HILL R., *The mathematical theory of plasticity*, Clarendon Press, Oxford (G.B.), 1950.
28. JOHANSEN K.W., *Bruchmomente der kreuzweise bewehrten Platten*, Mémoires de l'A.I.P.C., vol.1, (1932), 277-296, Zurich.
29. JOHANSEN K.W., *Brudlinieteorier*, Copenhague, 1952.
30. SAVE M. et MASSONNET Ch., *Calcul plastique des constructions*, Vol.2, 2ème éd., Ed. CBLIA, Bruxelles, 1973.
31. de BUHAN P., SALENÇON J., TALIERCIO A., *Lower and upper bound estimates for the macroscopic strength criteria of fiber composite materials*, Proc. IUTAM Symp. on Inelastic Deformation of Composite Materials, Troy N.Y., 29 may-1st june 1990, 563-580, Springer-Verlag, 1991.
32. de BUHAN P., TALIERCIO A., *A homogenization approach to the yield strength of composite materials*. Eur. J. Mech., A/Solids, 10, (1991), 129-154.
33. de BUHAN P., SALENÇON J., SIAD L., *Critère de résistance pour le matériau "terre armée"*, C.R.Ac.Sc., Paris, t. 302, série II, (1986), 337-381.
34. SIAD L., *Analyse de stabilité des ouvrages en terre armée par une méthode d'homogénéisation*, Thèse Dr. École Nationale des Ponts et Chaussées, Paris, 1987.
35. de BUHAN P., MANGIAVACCHI R., NOVA R., PELLEGRINI G., SALENÇON J., *Yield design of reinforced earth walls by a homogenization method*, Géotechnique, vol.39, (1989), n°2, 189-201.
36. SALENÇON J., *Calcul à la rupture et analyse limite*, Presses de l'École Nationale des Ponts et Chaussées, Paris, 1983.
37. JEWELL R.A., *Compatibility, serviceability and design factors for reinforced soil walls*, Proc. Int. Geotech. Symp. on Theory and Practice of Earth Reinforcement, Fukuoka (Japan), Balkema publ., 1988, 611-616.
38. de BUHAN P., *Approche fondamentale du calcul à la rupture des ouvrages en sols renforcés*, Th. Dr. Sc., Paris, 1986.
39. BREZIS H., *Opérateurs maximaux monotones et semi-groupes de contractions dans les espaces de Hilbert*, North-Holland, Mathematic studies, 1973.
40. HALPHEN B., SALENÇON J., *Elasto-plasticité*, Presses de l'École Nationale des Ponts et Chaussées, Paris, 1987.
41. DRUCKER D.C., PRAGER W., *Soil mechanics and plastic analysis or limit design*, Quart. Appl. Math., 10, (1952), 157-165.
42. MOREAU J.J., *Fonctionnelles convexes*, Séminaire, 1966.
43. PRAGER W., Théorie générale des états limites d'équilibre, J. Math. Pures Appl., 34, (1955), 395-406.
44. NAYROLES B., *Essai de théorie fonctionnelle des structures rigides plastiques parfaites*, J. Méc., 9, (1970), 491-506.
45. FRÉMOND M., FRIAA A., *Analyse limite - Comparaison des méthodes statique et cinématique*, C.R.Ac.Sc., Paris, t. 286, (1978), A, 107-110.

46. FRIAA A., *La loi de Norton-Hoff généralisée en plasticité et viscoplasticité*, Thèse Dr. Sc., Univ. Pierre et Marie Curie, Paris, 1979.
47. GEIRINGER H., *Fondements mathématiques de la théorie des corps plastiques isotropes*. Mem. Sci. Math., 86, Gauthier-Villars, Paris, 1937.
48. SALENÇON J., *La théorie des charges limites dans la résolution des problèmes de plasticité en déformation plane*, Thèse Dr. Sc., Univ. Paris, 1969.
49. ANTHOINE A., *Mixed modelling of reinforced soils within the framework of the yield design theory*, Comp. Geotech., 7, (1989), 67-82.
50. ANTHOINE A., *Une méthode pour le dimensionnement à la rupture des ouvrages en sols renforcés*, Rev. Fr. Géotech., 50, (1990), 5-21.
51. ANTHOINE A., SALENÇON J., *Une optimisation d'ouvrages en sols renforcés*, Proc. XII Int. Conf. Soil Mech., Rio de Janeiro, Brasil, 1989, 1219-1220.
52. ANDERHEGGEN E. and KNOPFEL M., *Finite Element Limit Analysis using Linear Programming*, Int. J. Solids & Structures, 8, (1972), 1413-1431.
53. FRÉMOND M. and SALENÇON J., *Limit analysis by finite element method*, Proc. Symp. on the Role of Plasticity in Soil Mechanics, Cambridge, UK, 1973, 297-308.
54. DELBECQ J.M., FRÉMOND M., PECKER A., and SALENÇON J., *Eléments finis en plasticité et visco-plasticité*, J. Mécanique Appliquée, 1, (1977), 3, 267-304.
55. TURGEMAN S., *Contribution au calcul des charges limites en milieux isotropes et orthotropes de révolution par une approche cinématique numérique*, Thèse Dr. Sc., Grenoble, 1983.
56. PASTOR J., TURGEMAN S., *Approches numériques des charges limites pour un matériau orthotrope de révolution en déformation plane*, K. Méca. Théor. Appl., 2, (1983), 3, 393-416.
57. MAGHOUS S., *Détermination du critère de résistance macroscopique d'un matériau hétérogène à structure périodique - Approche numérique*, Th. Dr. École Natioanle des Ponts et Chaussées, Paris, 1991.
58. de BUHAN P. and MAGHOUS S., *Une méthode numérique pour la détermination du critère de résistance macroscopique de matériaux hétérogènes à structure périodique*, C.R.Ac.Sc., Paris, 313, (1991), II, 983-988.
59. SALENÇON J., *An introduction to the yield design theory and its applications to soil mechanics*, Eur. J. Mech., A/Solids, 9, (1990), n°5, 477-500.
60. de BUHAN P., DORMIEUX L., and MAGHOUS S., Private communication.
61. SALENÇON J. and PECKER A., *Capacité portante de fondations superficielles sous sollicitations inclinées et excentrées (étude théorique)*, Journée "Problèmes Scientifiques de l'Ingénieur", Hommage à D. Radenkovic, Palaiseau (France, 16 janvier 1992.
62. HILL R., *On the limits set by plastic yielding to the intensity of singularities of stress*, J. Mech. Phys. Solids, 2, (1954), 4, 278-285.
63. HILL R., *The extremal stress-field concept*, J. Mech. Phys. Solids, 14, (1966), 239-243.
64. DORMIEUX L., *Stability of a purely cohesive seabed soil under wave loading*, Géotechnique, 38, (1988), 121-123.
65. DORMIEUX L., *Influence de la houle sur la stabilité d'un massif sous-marin*, Thèse Dr. Ecole Nationale des Ponts et Chaussées, Paris, 1989.

66. LECA E., PANET M., *Application du calcul à la rupture à la stabilité du front de taille d'un tunnel*, Rev. Fr. Géotech., 43, (1988), 5-19.

67. CHAMBON P., CORTÉ J.F., *Stabilité du front de taille d'un tunnel en milieu frottant*, Rev. Fr. Géotech., 51, (1990), 51-69.

68. DRUCKER D.C., *Limit analysis of two and three dimensional soil mechanics problems*, J. Mech. Phys. Solids, 1, (1953), n°4, 217-226.

69. KREBS OVESEN N., *General Report, Session 30 : Codes and Standards*, Proc. XII Int. Conf. Soil Mech. Found. Eng., Rio de Janeiro, 1989, vol.4.

70. BRINCH HANSEN J., *Earth pressure calculation*, The Danish Technical Press, Copenhagen, 1953.

APPLICATION OF THE YIELD DESIGN THEORY TO THE MECHANICS OF REINFORCED SOILS

P. de Buhan
Ecole Polytechnique, Palaiseau, France

ABSTRACT

The considerable development of various reinforcement techniques in soil mechanics over the last few decades, has made it necessary to devise both reliable and efficient methods for analysing the stability of such composite soil structures. The Yield Design theory is perfectly well suited to undertaking such a task , since it provides a comprehensive mechanical framework. The purpose of this contribution is to present two different approaches to the problem.

The first part is devoted to the so-called *mixed modelling* approach , where the reinforcements are perceived as structural elements (rods, beams, plates etc.) embedded in the soil treated as a 3-D continuum. It is shown in particular how the kinematic method of yield design can be implemented, leading to practical developments such as a computer code for the design of reinforced soil retaining structures.

A complementary (and sometimes alternative) point of view is presented in the second part. It originates from the intuitive idea that, in many cases, the composite reinforced soil may be perceived as a homogeneous material on the scale of the structure. The Yield Design theory associated with a homogenization technique makes it then possible to derive explicit expressions for the macroscopic strength criterion of such reinforced soils, on the basis of which a design method is implemented. Possible extensions of this approach to some particular kinds of reinforcement are briefly outlined.

PART I : YIELD DESIGN OF REINFORCED SOIL STRUCTURES : A "MIXED MODELLING" APPROACH

1. FUNDAMENTALS OF THE "MIXED MODELLING" APPROACH

Most of the methods currently devised in the engineering practice for analysing the stability of reinforced soil structures, quite naturally refer to such a mixed modelling of the composite reinforced soil. The native soil is treated as a classical three dimensional continuum, while the reinforcing inclusions are perceived as either one dimensional (beams or rods) or two dimensional (plates) structural elements. Unfortunately, due to the lack of a mechanically consistent framework, they often come up against some difficulties when trying to set up reliable design methods aimed at predicting the actual performance of reinforced soil structures in a rational way. The aim of the present section is therefore to propose such a theoretical framework based upon the yield design reasoning.

1.1. STATEMENT OF THE PROBLEM

So as to help clarify the basic concepts to be introduced in the sequel, an illustrative example will be considered (Fig. 1). It concerns a vertical cut of height h, excavated in a homogeneous soil, whose stability is to be enhanced by putting half-way up a single array of infinitely long reinforcements, evenly spaced in a horizontal plane.

Referring to the yield design theory ([1], [2], [3]), the stability analysis of such a structure relies upon three types of data, namely :

- the geometry of the structure, i.e. the depth h of the excavation,
- the loading parameters, reduced here to the specific weight of the soil γ,
- and the strength properties of the constituent materials : soil and reinforcements.

While the soil will be modelled as a classical 3D continuum at every point of which the stress tensor will be denoted by $\underline{\underline{\sigma}}$ (tensile normal stresses are counted positive), the reinforcements will be treated as beams, i.e. 1D continuous media whose internal forces are defined as follows.

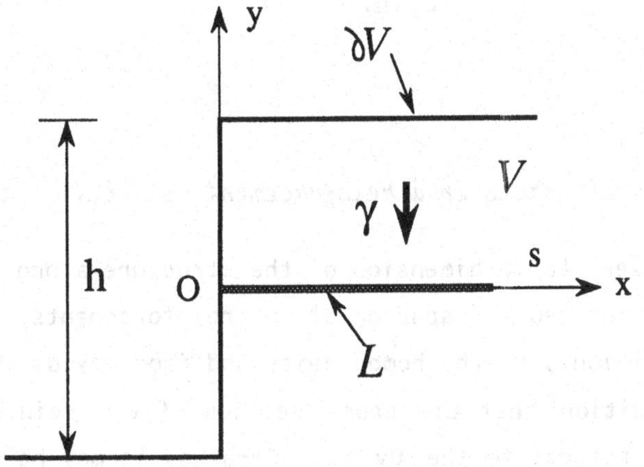

Fig. 1 : Yield design of a reinforced vertical cut.

A positive orientation is chosen along any reinforcing inclusion, (for instance from the left hand side to the right hand side of the figure), thereby defining the unit tangent vector $\underline{e}_x$ and the right handed frame $(\underline{e}_x, \underline{e}_y, \underline{e}_z)$ at every point P. Since it will be stated later on that the problem may be studied as a two dimensional one in the plane Oxy, the internal forces to be considered in the reinforcements reduce to the axial force N, the shear force V and the bending moment M, which turn out to be the components of the resultant force and the moment acting at point P on the "upstream" part of the beam due to the "downstream" part (Fig. 2).

The strength of the soil is defined by a cohesion C and an internal friction angle φ (Coulomb's strength criterion)

$$f^s(\underline{\underline{\sigma}}) = \sup_{i,j} \{\sigma_i (1 + \sin\varphi) - \sigma_j (1 - \sin\varphi) - 2C\cos\varphi\} \leq 0 \qquad (1)$$

where σ_i, σ_j are the principal stresses.

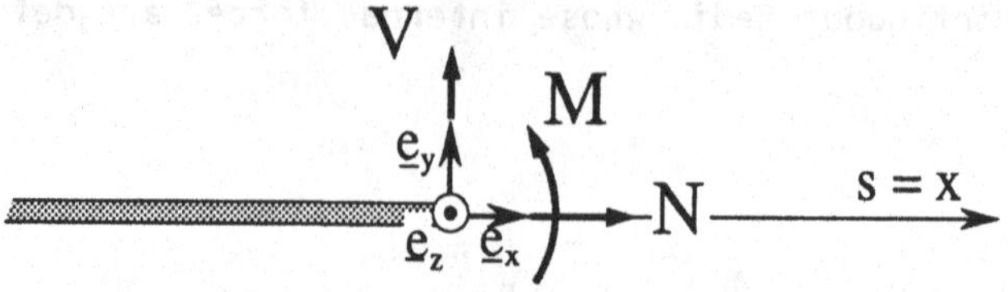

Fig. 2 : Internal forces in a reinforcement modelled as a beam.

Owing to the very large dimension of the structure along Oz with respect to h, to the regular spacing of the reinforcements, to the uniform loading conditions, to the homogeneity and isotropy of the native soil, and on condition that the cross section of the reinforcements be symmetric with respect to the Oy and Oz-axes, it may be proved that the 3D problem can be conveniently dealt with as a *two dimensional one* ("plane strain" yield design problem : cf. [1] for the rigorous definition of this intuitive concept). Consequently, the above defined N, V and M will be considered from now on as forces and moment per unit transverse length along Oz, and the following expression may be adopted for the strength condition of the reinforcements :

$$f(N, V, M) = (N/N_o)^2 + (V/V_o)^2 + |M/M_o| - 1 \leq 0 \qquad (2)$$

where N_o, V_o and M_o denote the ultimate values of N, V and M. Fig. 3 displays a geometrical representation of the corresponding strength domain in the region of space defined by $(N \geq 0, V \geq 0, M \geq 0)$.

Furthermore, as a first approach, *perfect bonding* will be assumed between the soil and the reinforcements.

According to the yield design reasoning, the above described reinforced structure will be termed (potentially) stable, if one can exhibit throughout the whole structure a system of *internal forces* (namely a stress field $\underline{\underline{\sigma}}$ within the soil, and a distribution of N, V, M along the reinforcement) in *equilibrium* with the loading (weight of the soil) while complying with the *strength criteria* of the respective constituents. It can be seen from single dimensional analysis arguments that the stability of the structure is governed by the following dimensionless parameters : $\gamma h/C$, N_o/Ch , V_o/N_o , $M_o/N_o h$ and φ , in such a way that the equivalence holds :

$$\text{Stability of the structure} \iff \gamma h/C \leq (\gamma h/C)^* = K^* \qquad (3)$$

where K^* depends merely on N_o/Ch , V_o/N_o , $M_o/N_o h$ and φ . From a practical viewpoint, it may be convenient to introduce the *factor of confidence* defined as the ratio between the highest possible value of the stability factor and its actual value :

$$\Gamma^* = (\gamma h/C)^* \;/\; (\gamma h/C) = (C/\gamma h)K^* \; . \qquad (4)$$

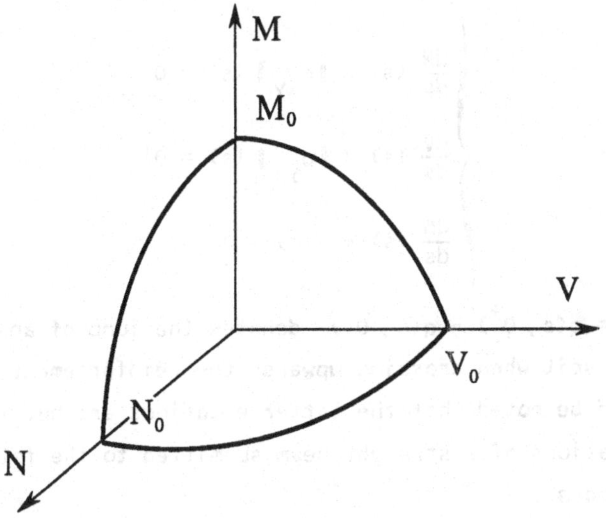

Fig. 3 : Sketch of the strength domain of a reinforcement.

Therefore the structure proves definitely unstable as soon as this factor drops below unity.

1.2. PRINCIPLE OF THE STATIC APPROACH

Implementing the static approach of the yield design theory requires the construction of statically admissible "mixed" stress fields which must satisfy the strength criteria everywhere. Any such stress field denoted by $\{\underline{\underline{\sigma}} ; (N, V, M)\}$ has therefore to comply with the equilibrium equations in the soil as well as along the reinforcement.

Thus :

- in the *soil* (V)

$$\text{div}\, \underline{\underline{\sigma}}(\underline{x}) - \gamma\, \underline{e}_y = 0 \qquad \text{whatever } \underline{x} \in V \tag{5}$$

or $[\![\underline{\underline{\sigma}}]\!] \cdot \underline{n} = 0$ on possible stress discontinuity surfaces (which reduce here to lines in the Oxy-plane), where $[\![\underline{\underline{\sigma}}]\!]$ represents the jump of $\underline{\underline{\sigma}}$ across such a surface when following its normal $\underline{n}$;

- at any point P of abscissa s along the *reinforcement* (L) the equilibrium equations write :

$$\left\{\begin{array}{ll} \dfrac{dN}{ds}(s) + [\![\sigma_{xy}]\!](s) = 0 & \quad (6\text{-}a) \\[2ex] \dfrac{dV}{ds}(s) + [\![\sigma_{yy}]\!](s) = 0 & \quad (6\text{-}b) \\[2ex] \dfrac{dM}{ds}(s) + V(s) = 0 & \quad (6\text{-}c) \end{array}\right.$$

where $[\![q]\!](s) = q(s, 0^+) - q(s, 0^-)$ denotes the jump of any quantity q relative to the soil when crossing upwards the reinforcement at point P(s) . It should be noted that the latter equations are but the classical equilibrium equations of a straight beam submitted to the following distribution of forces :

$$[\![\underline{\underline{\sigma}}]\!](s) \cdot \underline{e}_y = [\![\underline{T}(\underline{e}_y)]\!](s) = \underline{T}^+(s) - \underline{T}^-(s)$$

generated by the discontinuity of the stress vector $\underline{T}(\underline{e}_y)$ acting on both sides of the reinforcement.

These equations may easily be derived by expressing for instance the resultant and moment equilibrium of any segment of the reinforcement of infinitesimal length ds as shown in Fig. 4 [4].

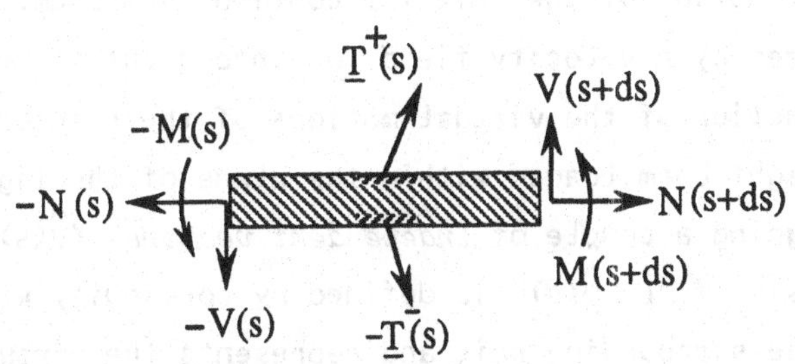

Fig. 4 : Equilibrium of an infinitesimal segment of reinforcement.

In a way, Eqs (6-a) and (6-b) may be interpreted as expressing the *mechanical interaction* between the soil and the reinforcement.

Finally, the static approach can be summarized as follows. The reinforced structure is (potentially) stable ($K \leq K^*$ or $\Gamma^* \geq 1$) if there exists at least one "mixed" stress field $\{\underline{\underline{\sigma}}; (N, V, M)\}$ verifying :

a) the equilibrium equations : (5) for $\underline{\underline{\sigma}}$ and (6) for (N, V, M) ;

b) the boundary conditions : $\underline{\underline{\sigma}}.\underline{n} = 0$ on ∂V (stress-free surfaces); $N = V = M = 0$ at point O $(s = 0)$;

c) the strength criteria : (1) for $\underline{\underline{\sigma}}$; (2) for (N, V, M) .

Anytime such a stress field can be exhibited, the value of $\gamma h/C$ it equilibrates constitutes a *lower bound estimate* of $K^* = (\gamma h/C)^*$. Unfortunately, the method remains quite difficult to implement, unless resorting to finite element numerical techniques [5]. By contrast, the *kinematic approach* will prove far easier to handle.

1.3. KINEMATIC APPROACH [6]

The yield design kinematic approach stems from the mathematical dualization of the previous static approach through the *principle of virtual work* ([2] [3]). As a preliminary, it relies upon the definition of the virtual motions to be considered in the mechanical description of the system. Within the framework of a mixed modelling approach, any such virtual motion is defined in the following way :

• The virtual motion of the soil (2D continuous medium) is classically characterized by a velocity field $\underline{\hat{U}}$ throughout its volume V .

• The construction of the virtual motions of the reinforcement, modelled as a straight beam loaded within the plane of the figure, is achieved by assigning a couple of *independent vectors* $(\underline{\hat{U}}(s)\ ,\ \underline{\hat{\Omega}}(s))$ to any point $P(s)$ of L . $\underline{\hat{U}}(s)$ is defined by continuity with the virtual motion in the surrounding soil and represents the *virtual velocity* of the "neutral axis" of the reinforcing inclusion at point $P(s)$, while $\underline{\hat{\Omega}}(s) = \hat{\Omega}(s)\ \underline{e}_z$ is the *virtual rate of rotation* of the "transverse section" at the same point (Fig. 5).

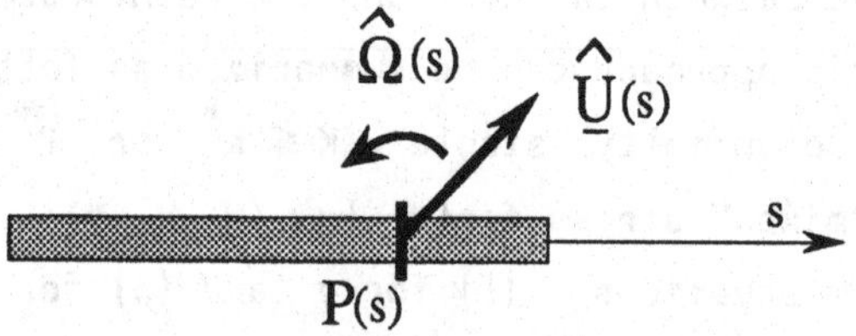

Fig. 5 : Virtual motions of the reinforcement.

The *principle of virtual work* states that for any statically admissible (s.a.) stress field $\{\underline{\underline{\sigma}}; (N, V, M)\}$ (i.e. satisfiying Eqs (5) and (6) along with boundary conditions) the following equality holds true :

$$W_e(\{\underline{\hat{U}}\ ;\ \hat{\Omega}\}) + W_i(\{\underline{\hat{U}}\ ;\ \hat{\Omega}\}) = 0 \tag{7}$$

whatever $\{\underline{\hat{U}}\ ;\ \hat{\Omega}\}$, which denotes a virtual motion as previously defined.

In this equality, W_e is the virtual work of the *external forces*, that is in the present case, the work of the gravity forces :

$$W_e(\{\underline{\hat{U}}\ ;\ \hat{\Omega}\}) = \int_V -\gamma\,\underline{\hat{U}}\,.\,\underline{e}_y\ dV \tag{8}$$

and W_i is the work of the *internal forces*, whose expression is :

$$W_i(\{\hat{\underline{U}}\ ;\ \hat{\Omega}\}) = -\int_V (\underline{\underline{\sigma}} : \hat{\underline{\underline{d}}})^{(\star)}\, dV - \int_L w_i(s)\, ds \qquad (9)$$

where $\hat{\underline{\underline{d}}} = 1/2\ (\underline{\text{grad}}\ \hat{\underline{U}} + {}^t\underline{\text{grad}}\ \hat{\underline{U}})$ is the virtual strain rate field associated with $\hat{\underline{U}}$, and [7]

$$w_i(s) = N(s)\frac{d\hat{\underline{U}}}{ds} \cdot \underline{e}_x + V(s) \cdot \left(\frac{d\hat{\underline{U}}}{ds}(s) \cdot \underline{e}_y - \hat{\Omega}(s)\right) + M(s)\frac{d\hat{\Omega}}{ds}(s) \quad . \qquad (10)$$

Remark : In order to account for possible discontinuities of $\hat{\underline{U}}$ and/or $\hat{\Omega}$ across surfaces (lines) Σ in the soil and/or at several points $P_k(s_k)$ along the reinforcement, the following additional contributions must be introduced in the expression of $W_i(\{\hat{\underline{U}}\ ;\ \hat{\Omega}\})$

- for the soil : $-\int_\Sigma (\underline{\underline{\sigma}} \cdot \underline{n}) \cdot [\![\hat{\underline{U}}]\!] d\Sigma$ (11)

- for the reinforcement :

$$\sum_k [\,(N(s_k)\,\underline{e}_x + V(s_k)\,\underline{e}_y) \cdot [\![\hat{\underline{U}}]\!](s_k) + M(s_k)\,[\![\hat{\Omega}]\!](s_k)\,] \qquad (12)$$

where $[\![\,.\,]\!]$ are the corresponding discontinuities.

Let then :

$$\pi(\hat{\underline{\underline{d}}}) = \sup_{\underline{\underline{\sigma}}} \{\underline{\underline{\sigma}} : \hat{\underline{\underline{d}}}\ ;\ f^s(\underline{\underline{\sigma}}) \leq 0\} \qquad (13)$$

and

$$\pi\left(\frac{d\hat{\underline{U}}}{ds}, \hat{\Omega}, \frac{d\hat{\Omega}}{ds}\right) = \sup_{(N, V, M)} \{w_i(s)\ ;\ f(N, V, M) \leq 0\} \qquad (14)$$

where functions f^s and f are defined by (1) and (2) respectively.

(⋆) $\underline{\underline{\sigma}} : \hat{\underline{\underline{d}}} = \sigma_{ij}\, \hat{d}_{ji}$ *with summation over repeated subscripts.*

According to the yield design reasoning and to the principle of virtual work which expresses the equilibrium requirement in a dual form, it appears that a *necessary* condition for the reinforced structure to be stable in the sense stated above, is :

$$W_e(\{\hat{\underline{U}} \; ; \; \hat{\Omega}\}) \leq W_r(\{\hat{\underline{U}} \; ; \; \hat{\Omega}\}) \qquad \text{whatever} \quad \{\hat{\underline{U}} \; ; \; \hat{\Omega}\} \tag{15}$$

$W_r(\{\hat{\underline{U}} \; ; \; \hat{\Omega}\})$ is called the *maximum resisting work* developed by both the soil and the reinforcement on account of their respective strength properties in the virtual motion under consideration. Its expression writes:

$$W_r(\{\hat{\underline{U}} \; ; \; \hat{\Omega}\}) = \int_V \pi(\hat{\underline{\underline{d}}}) \, dV + \int_L \pi\left(\frac{d\hat{\underline{U}}}{ds} \, , \, \hat{\Omega} \, , \, \frac{d\hat{\Omega}}{ds}\right) ds \tag{16}$$

to which one has to add up the following contributions in case of discontinuities

$$\int_\Sigma \pi(\underline{n} \; ; \; [\![\hat{\underline{U}}]\!]) \, d\Sigma + \sum_k \pi_k([\![\hat{\underline{U}}]\!](s_k) \; ; \; [\![\hat{\Omega}]\!](s_k)) \tag{17}$$

where the π-functions relating to discontinuities are defined as follows

$$\pi(\underline{n} \; ; \; [\![\hat{\underline{U}}]\!]) = \sup_{\underline{\underline{\sigma}}} \{(\underline{\underline{\sigma}} \, . \, \underline{n}) \, . \, [\![\hat{\underline{U}}]\!] \, , \, f^s(\underline{\underline{\sigma}}) \leq 0\} \tag{18}$$

$$\pi_k([\![\hat{\underline{U}}]\!](s_k \; ; \; [\![\hat{\Omega}]\!](s_k))$$

$$= \sup_{(N, V, M)} \{(N \, \underline{e}_x + V \, \underline{e}_y) \, . \, [\![\hat{\underline{U}}]\!](s_k) + M \, [\![\hat{\Omega}]\!](s_k) \; ; \; f(N, V, M) \leq 0\}. \tag{19}$$

The implementation of the yield design kinematic approach is based upon the fundamental inequality (15). The virtual motions $\{\hat{\underline{U}} \; ; \; \hat{\Omega}\}$ will be called "failure mechanisms" of the reinforced soil structure.

2. IMPLEMENTATION OF THE KINEMATIC APPROACH

2.1. KINEMATIC APPROACH USING ROTATIONAL FAILURE MECHANISMS

Like in the case of homogeneous soil structures, the kinematic approach can be implemented through the use of failure mechanisms such as that shown in Fig. 6 where a volume bounded by an arbitrary line AB is given a rigid body rotating motion of virtual velocity $\hat{\omega}$ about a pole F, thus inducing a velocity jump across AB between the rotating volume and the rest of the soil. Since such a mechanism does not involve any virtual deformation outside the discontinuity line AB in the soil and the discontinuity point P on the reinforcement, the *maximum resisting work* reduces to :

$$W_r(F, AB, \hat{\omega}) = \int_{AB} \pi(\underline{n} \,;\, [\![\underline{\hat{U}}]\!])\, d\ell + \pi_1\,([\![\underline{\hat{U}}]\!](s_1) \,;\, [\![\hat{\Omega}]\!](s_1))$$

with, $\underline{n}$ being the outwardly oriented normal to AB at current point P :

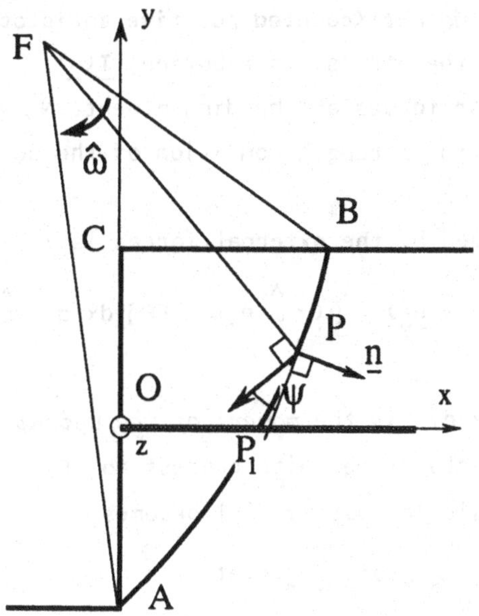

Fig. 6 : Rotational failure mechanism.

$$[\![\hat{\underline{U}}]\!] = \hat{\omega}\, \underline{e}_z \wedge \underline{FP} \; ; \; [\![\hat{\underline{U}}]\!](s_1) = \hat{\omega}\, \underline{e}_z \wedge \underline{FP_1}$$

$$\text{and} \qquad [\![\hat{\Omega}(s_1)]\!] = \hat{\omega}$$

where $\hat{\omega}$ is taken positive.

Putting those particular expressions of $[\![\hat{\underline{U}}]\!]$, $[\![\hat{\underline{U}}]\!](s_1)$ and $[\![\hat{\Omega}(s_1)]\!]$ into the general formulas *(18)* and *(19)* yields :

$$W_r(F, AB, \hat{\omega}) = \hat{\omega}(M_r^{soil} + M_r^{reinf}) \tag{20}$$

where :

$$M_r^{soil} = \int_{AB} \sup_{\underline{\underline{\sigma}}} . \{\underline{e}_z . [\underline{FP} \wedge (\underline{\underline{\sigma}} . \underline{n})] \; ; \; f^s(\underline{\underline{\sigma}}) \leq 0\}$$

and

$$M_r^{reinf} = \sup_{(N, V, M)} \{\underline{e}_z . [\underline{FP_1} \wedge (N\, \underline{e}_x + V\, \underline{e}_y)] + M \; ; \; f(N, V, M) \leq 0\} .$$

It is worth emphasizing that M_r^{soil} (resp. M_r^{reinf}) may be interpreted as the *maximum resisting moment* (counted positive anticlockwise) about point F developed by the stress distribution $\underline{T}(\underline{n}) = \underline{\underline{\sigma}} . \underline{n}$ along AB (resp. by the axial, shear forces and bending moment N, V and M at point P_1) on account of the strength condition of the soil (resp. of the reinforcement).

Likewise the work performed by the external forces is :

$$W_e(F, AB, \hat{\omega}) = \int_{ABC} (-\gamma\, \underline{e}_y) . [(-\hat{\omega}\, \underline{e}_z) \wedge \underline{FP}]\, dx\, dy = \hat{\omega} . M_w \tag{21}$$

where $M_w = \gamma \int_{ABC} \underline{FP} . \underline{e}_x \, dx\, dy$ is the *moment of the weight of the volume* ABC (counted positive clockwise) with respect to F .

As a consequence, the basic inequality *(15)* becomes

$$M_w \leq M_r^{soil} + M_r^{reinf} \tag{22}$$

which proves that performing the yield design kinematic approach with rotational failure mechanisms is pefectly equivalent to checking the *moment equilibrium* of volumes such as ABC (static approach from outside [2], [3]).

Let

$$\Gamma(F\,,AB) = (M_r^{soil} + M_r^{reinf}) / M_w \quad . \tag{23}$$

Since Ineq. (22) expresses a *necessary* condition for the structure to be stable (i.e. $\Gamma^* \geqslant 1$), it comes out that :

$$\Gamma^* \geqslant 1 \quad \Rightarrow \quad \Gamma(F\,,AB) \geqslant 1$$

and then

$$\Gamma^* \leqslant \Gamma(F\,,AB) \quad . \tag{24}$$

Given any volume ABC and point F, the ratio between the maximum resisting moment due to both the soil and the reinforcement, and the moment due to the weight of the soil ("driving moment") constitutes an *upper bound estimate* for the factor of confidence of the structure. The computation of M_r^{soil} (see for example [6] [8] [9]) shows that for its value to remain finite (and thus lead to a non-trivial upperbound estimate for Γ^*) it is necessary that the angle ψ between the velocity at any point of AB and the tangent at the same point (Fig. 6) be comprised between φ and $\pi - \varphi$. Furthermore in the case of a homogeneous soil structure it can be proved that for a given point F , the most critical volume ABC , associated with the minimum value of $\Gamma(F\,,AB)$, is such that $\psi = \varphi$. This implies that AB is an *arc of logspiral* of angle φ and focus F .

2.2. FAILURE MECHANISMS WITH CONTINUOUS DEFORMATIONS OF THE REINFORCEMENT

The implementation of the kinematic approach can be extented to failure mechanisms in which both the soil and the reinforcement undergo continuous deformations. As an example, the velocity discontinuity line of Fig. 6 may be replaced by a narrow "shear zone" across which the virtual velocity field remains continuous as sketched in Fig. 7-a. Such (virtual) deformation modes are suggested by the observation of (actual) failure patterns of reinforced soil structures occuring in full-scale experiments [10]. The main difference between such failure mechanisms and those with velocity discontinuities examined in the previous paragraph (2.1), lies in the quite different way the strength of the reinforcement is mobilized.

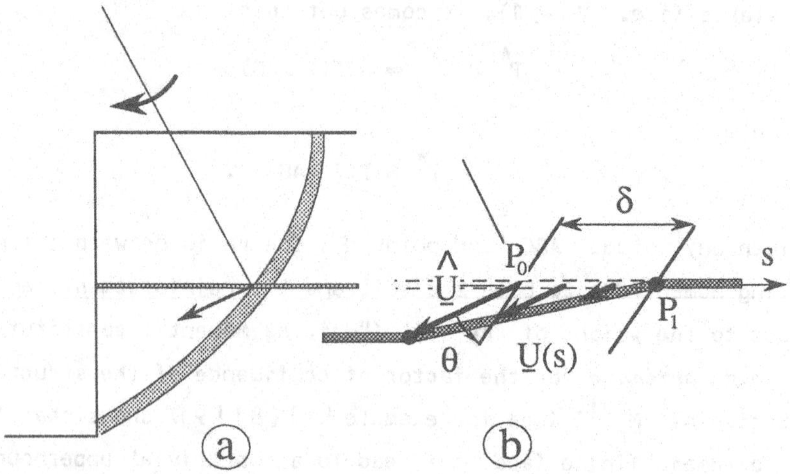

Fig. 7 : a) Failure mechanism with "shear zone"
b) "Double hinge" failure mechanism of the reinforcement.

In order to illustrate this point, the virtual motion of the reinforcement in the "shear zone" of the soil will be specified. Assuming for instance that the shear zone of soil intersects the reinforcement over a length δ so that the velocity along the reinforcement varies linearly between points $P_1(s_1)$ and $P_o(s_o)$ (Fig. 7-b), such a virtual motion may be defined as follows (for simplicity's sake the effect due to the rotation of the volume of reinforced soil is neglected) :

$$s < s_o \quad : \quad \underline{\hat{U}}(s) = \underline{\hat{U}} \quad , \quad \hat{\Omega}(s) = 0$$

$$s_o < s < s_1 = s_o + \delta \quad : \quad \underline{\hat{U}}(s) = \underline{\hat{U}}(s_1 - s)/\delta \quad , \quad \hat{\Omega}(s) = \hat{U}\cos\theta/\delta$$

$$s > s_1 \quad : \quad \underline{\hat{U}}(s) = 0 \quad , \quad \hat{\Omega}(s) = 0 \ .$$

In this virtual motion the "transverse sections" of the reinforcement remain normal to its "neutral axis" ($d\underline{\hat{U}}/ds \cdot \underline{e}_y = \hat{\Omega}(s)$: *Navier Bernoulli condition*) thereby inducing *hinges* (rotation discontinuities) at points P_o and P_1. The *maximum resisting work* in such a "double hinge" mechanism has been computed in [6]. Its expression may be put in the form :

$$W_r^{reinf}(\{\underline{\hat{U}} ; \hat{\Omega}\}) = N_o \, |\underline{\hat{U}}| \, (|\sin\theta| + \mu \, |\cos\theta|) \qquad (25)$$

where μ is a non dimensional factor whose definition is

$$\mu = 2 \, M_o / N_o \, \delta \; .$$

It is worth comparing formula (25) with the expression of the maximum resisting work developed by the reinforcement in the homologous mechanism with velocity jump [6] :

$$W_r^{reinf}(\{\underline{\hat{U}} ; \hat{\Omega}|) = N_o \, |\underline{\hat{U}}| \, [\sin^2\theta + \nu^2 \cos^2\theta]^{1/2} \qquad (26)$$

$$\text{with} \quad \nu = V_o / N_o \; .$$

While in the latter mechanism axial and *shear* strengths of the reinforcement (N_o and V_o) are both involved, Eq. (25) clearly shows that only its axial and *bending* strength characteristics (N_o and M_o) are mobilized in the "double hinge" mechanism.

2.3. A MODIFIED STRENGTH CONDITION FOR THE REINFORCEMENT

In accordance with the previous analysis, a simplified design method can be set up, based on the following idea. Instead of dealing with "shear zone" failure mechanisms directly, which proves a quite difficult, if not impossible task [11], it still remains possible to implement the kinematic approach using classical rotational failure mechanisms with velocity discontinuities, provided that the flexional deformation mode of the reinforcements be taken into account. This is achieved by adopting a modified strength condition for the reinforcement.

Indeed, it can be noticed that the maximum resisting work developed by the reinforcement in such a flexional deformation mode (Eq. (25)), amounts to what could be computed in the homologous velocity jump mechanism, given by Eq. (26), where a modified strength criterion for the reinforcement ($f^{\star}(N,V,M) \leq 0$) is to be substituted for the initial one ($f(N,V,M) \leq 0$). Such a condition is simply derived from the identification of formulas (25) and (26) expressed for any orientation θ of the velocity discontinuity with respect to the normal to the reinforcement.

Thus :

$$|\underline{U}| \sup_{(N,V,M)} \{N\sin\theta + V\cos\theta \; ; \; f^{\star}(N,V,M) \leq 0\} = |\underline{U}| \; N_o(|\sin\theta| + \mu\,|\cos\theta|) \quad .$$

This equation shows that the boundary surface defined by $f^{\star}(N,V,M) = 0$ in the space (N,V,M) may be constructed as the envelope of a family of planes, parallel to the M-axis, whose parametric equation writes :

$$N\sin\theta + V\cos\theta = N_o(|\sin\theta| + \mu\,|\cos\theta|)$$

thus defining the following domain :

$$f^{\star}(N,V,M) \leq 0 \quad \Leftrightarrow \quad \begin{cases} |N| \leq N_o \\ |V| \leq \mu\,N_o \end{cases} \quad .$$

Since the first condition is already fulfilled by the initial condition $f(N,V,M) \leq 0$ (see Ineq. (2)), the modified strength criterion finally writes :

$$f(N,V,M) \leq 0 \quad \text{with} \quad |V| \leq \mu\,N_o \quad . \tag{27}$$

This modification amounts to sharply reducing the shear strength available from the reinforcement, since it should be reminded that $\mu = 2\,M_o/N_o\,\delta$ is a dimensionless parameter which compares a transversal dimension of the reinforcement with the length of reinforcement δ involved by the shear zone. As an example, for a reinforcing inclusion of circular cross-section (used in the "soil nailing" technique), of diameter d, one gets :

$$\mu = \frac{4}{3\pi}\,d/\delta \quad .$$

Typical values of μ are of the order of 0.1, so **that** the shear strength capacities of the reinforcement can be cut back by as much as 80 % ($\mu N_0 = 0.1 N_0$ instead of $V_0 = 0.5 N_0$). The corresponding strength domain is represented in Fig. 8. It is simply obtained by truncating the initial strength domain by two symmetric planes of equations : $V = \pm \mu N_0$.

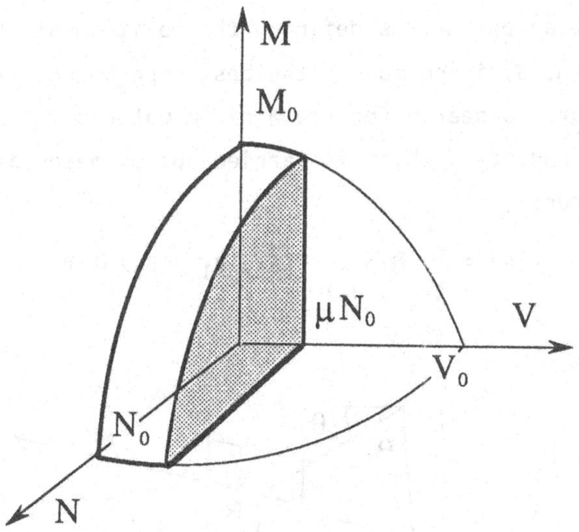

Fig. 8 : Truncated strength domain of the reinforcement.

3. Numerical implementation

The yield design kinematic method using rotational failure mechanisms is particularly well suited to a numerical implementation. As a matter of fact, a computer software based upon this method has been developed over the last few years ([8], [9], [12]). It enables any user of such a numerical code to perform the stability analysis of reinforced soil structures (slopes, retaining walls, embankments, ...) under various loading conditions.

3.1. GENERAL DESCRIPTION

Eq. (24) shows that for any volume ABC the ratio between the maximum resisting moment and the "driving" moment (or in other terms between the maximum resisting work and the work of the external forces in the associated failure mechanism) is an *upperbound estimate* for the factor of confidence of any structure. Considering logspiral failure mechanisms, this ratio appears as a function of A, θ_1 and θ_2 (Fig. 9), where A is the lower exit point of the logspiral on the outer surface, and θ_1, θ_2 are the angular parameters defining the position of the focus F, once A is known. Striving to get the best upperbound estimate of $\Gamma^\star$ requires therefore to search for the minimum value of $\Gamma(A, \theta_1, \theta_2)$ with respect to θ_1 and θ_2, which is carried out by means of a specific numerical procedure

$$\Gamma(A) = \underset{(\theta_1, \theta_2)}{\text{Min}} \; \Gamma(A, \theta_1, \theta_2) \geqslant \Gamma^\star \quad . \qquad (28)$$

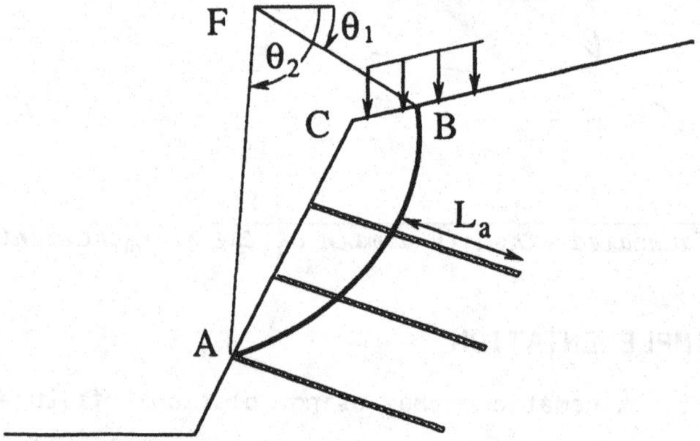

Fig. 9 : Logspiral failure mechanism.

The method still applies to structures comprising several layers of soil exhibiting different strength characteristics (i.e. different values of C and φ). In such a case, the rotating volume is bounded by a curve drawn by connecting several arcs of logspiral with the same focus, each of them being characterized by an angle equal to the friction angle of the intersected soil layer (Fig. 10) [9].

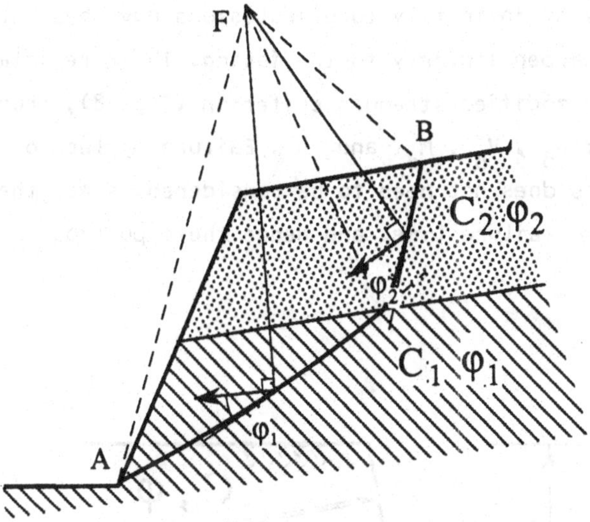

Fig. 10 : Failure mechanism in a multilayer soil structure.

Moreover, the possibility of failure by lack of adherence between the soil and the reinforcement, which up to now had been discarded in the analysis, can be taken into account in a very simple way, namely through the introduction of a "pull out" strength condition which may be put in the form :

$$N \leq f_\ell\, L_a \tag{29}$$

where L_a is the "anchoring" length of the reinforcement, that is, the distance between its intersecting point with the failure line and its free end (see Fig. 9), and f_ℓ is a lateral friction coefficient which characterizes the bonding between the soil and the reinforcement.

3.2. AN EXAMPLE OF PARAMETRIC STUDY ([12][13])

Owing to the high speed of computations carried out in the above mentioned programme, it becomes henceforth possible to undertake comprehensive parametric studies, such as the one presented in the sequel, which provides a clear assessment of the way shear and bending strengths of the reinforcements should be taken into account in design calculations.

The study refers to the structure displayed in Fig. 11. In order to ensure the stability of the initial homogeneous structure under its own weight, five rows of infinitely long inclusions have been placed at regular intervals perpendicularly to the facing. Their resistance is characterized by the modified strength criterion (Fig. 8), thus involving four parameters : N_o, V_o, M_o and μ. Failure by lack of adherence of the reinforcements does not need to be considered, since their anchoring length beyond any failure line, and hence their pull-out resistance, is infinite.

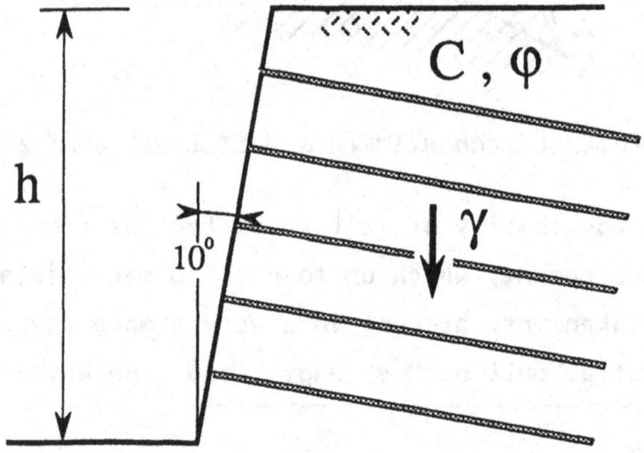

Fig. 11 : Structure investigated in the parametric study.

Several analytical expressions of the maximum resisting work developed by any reinforcement which obeys the strength condition (27), relating to different situations, have been calculated, then introduced in the computer code [13]. Varying the strength parameters of the reinforcement, while all the other parameters are being kept fixed, the best upper bound estimate of the factor of confidence of the structure, derived from a minimization over all logspiral failure mechanisms passing through the toe, appears as a function of the following non dimensional parameters :

$$r = N_o/Ch \quad , \quad \alpha = M_o/N_o h \quad \text{and} \quad \mu$$

the shear strength V_o being taken equal to 0.5 N_o .

Actually, those computations give evidence of a hardly noticeable influence of α , so that the estimate of $\Gamma^\star$, denoted by Γ , depends solely on parameters r and μ

$$\Gamma = \Gamma(r \, , \mu) \geqslant \Gamma^\star \quad .$$

The results of this parametric study have been put in the form of charts represented in Fig. 12-a, b and c, which correspond to $\varphi = 0°$, 20° and 40° respectively. They give the variations of the ratio $\Gamma(r \, , \mu) / \Gamma(r \, , \mu = 0)$ as a function of r and μ . This ratio accounts for the percentage of increase in stability which can be expected when adopting the reduced shear strength $\mu N_o = 2 \mu V_o$ instead of completely neglecting it $(\mu = 0)$. The following conclusions should be underlined.

• First of all, the results of such computations confirm what could be deduced from the theory of yield design (and from common sense !), namely, that, for the same class of failure mechanisms, ignoring the shear strength of the reinforcements *always* results in a *conservative estimate* for the stability of the structure as a whole. As a matter of fact it appears that for $\mu = \alpha = 0$ the strength domain defined by Ineq. $f(N \, , V \, , M) \leqslant 0$ in the $(N \, , V \, , M)$ space shrinks down to the segment $[- N_o \, , N_o]$ on the N-axis.

• The relative improvement of the factor of confidence (or more precisely of its upper bound estimate) due to the shear strength capacities increases with r and μ .

• By contrast, it undergoes a significant decline as the friction angle φ increases. As an example for $r = 3$ and $\mu = 0.1$, it may reach 15 % for a purely cohesive soil $(\varphi = 0°)$, whereas it amounts to no more than 11 % for $\varphi = 20°$, and reduces to less than 5 % for $\varphi = 40°$.

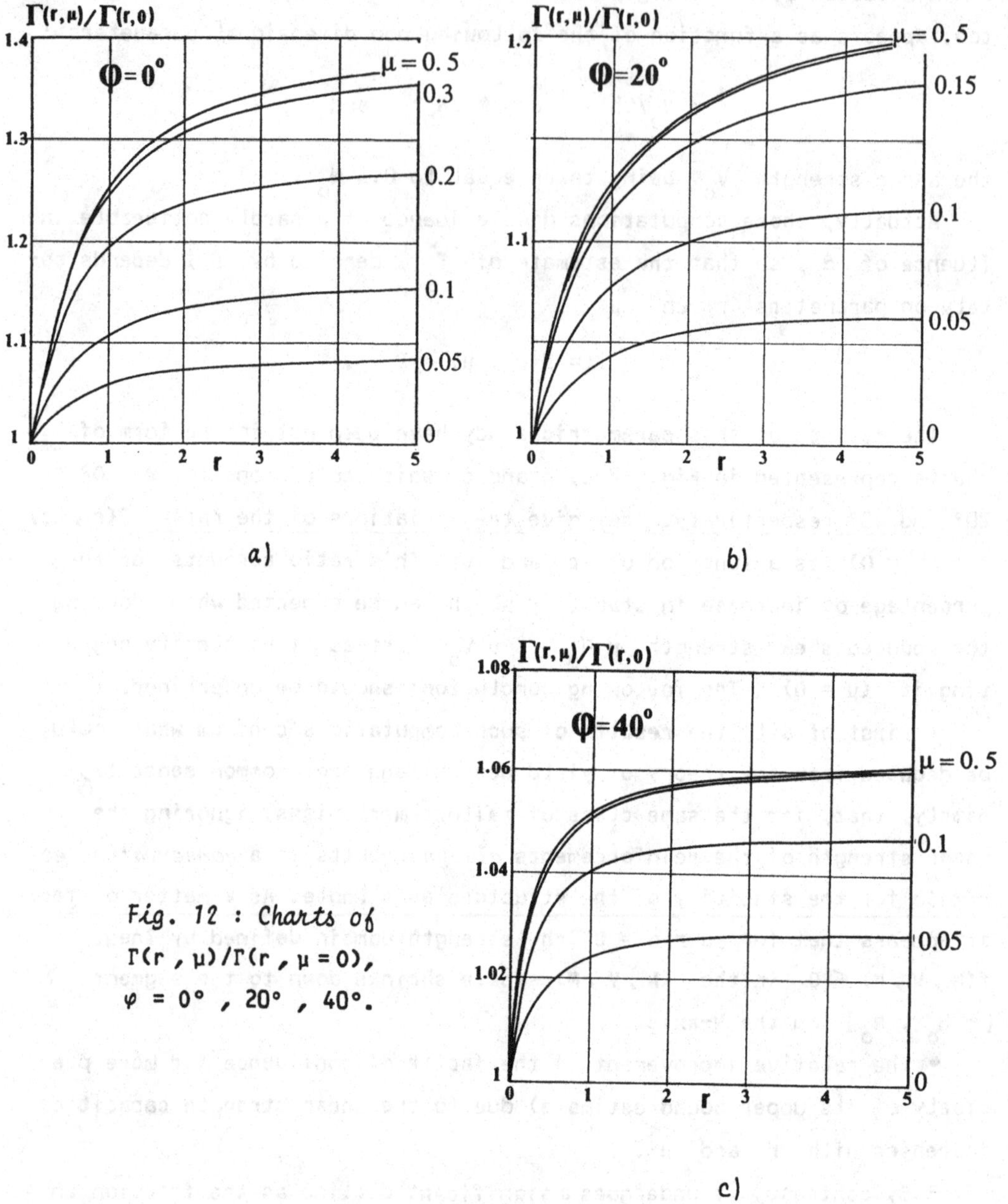

Fig. 12 : Charts of $\Gamma(r, \mu)/\Gamma(r, \mu = 0)$, $\varphi = 0°, 20°, 40°$.

Although obtained on a particular example, this last result is of paramount practical importance. Indeed, it provides a definite answer to the recently debated question ([14] [15]) to know whether and to what extent the contribution due to the shear strength of reinforcements such as nails should be taken into account. Leaving aside the particular case of reinforcements having large cross sectional dimensions (such as micropiles) which would require a specific investigation, it eventually turns out that this contribution remains in any case quite limited. It thus gives full legitimacy to the simplifying assumption according to which the reinforcements do not exhibit any resistance apart from that due to axial tensile forces.

RÉFÉRENCES

1. SALENÇON, J. : *Calcul à la Rupture et Analyse Limite*, Ed. Presses de l'E.N.P.C., Paris, 1983.
2. SALENÇON, J. : *An introduction to the yield design theory and its application to soil mechanics*, Eur. Jl Mech., A/Solids, 9, n° 5, 1990, 477-500.
3. SALENÇON, J. : *Yield design : a survey of the theory*, C.I.S.M. course "On the Evaluation of Global Bearing Capacities of Structures", 1992.
4. ANTHOINE, A. : *Mixed modelling of reinforced soils within the framework of the yield design theory*, Comp. Geotech., 7, 1989, 67-82.
5. PASTOR, J., TURGEMAN, S., CISS, A. : *Calculation of limit loads of structures in soils with metal reinforcement*, Eur. Conf. Num. Meth. Geomech., Stuttgart, Germany, 1986.
6. de BUHAN, P., SALENÇON, J. : *A comprehensive stability analysis of soil nailed structures*, To appear in Eur. Jl Mech., A/Solids, 1992.
7. SALENÇON, J. : *Cours de Mécanique des Milieux Continus*, T. II, Ed. Ellipses, Paris, 1988.
8. ANTHOINE, A. : *Une méthode pour le dimensionnement à la rupture des ouvrages en sols renforcés*, Rev. Fr. Géotech., 50, 1990, 5-21.
9. de BUHAN, P., DORMIEUX, L., SALENÇON, J. : *Stability analysis of reinforced soil retaining structures using the yield design theory*, Int. Conf. on retaining structures, Cambridge, 20-23 July, 1992.
10. PLUMELLE, C. : *Expérimentation en vraie grandeur d'une paroi en sol cloué*, Rev. Fr. Géotech., 40, 1987, 45-50.
11. ANTHOINE, A. : *Stabilité d'une fouille renforcée par clouage*, Proc. 4th French-Polish Coll. Appl. Soil Mech., Grenoble, France, 1987.
12. de BUHAN, P., DORMIEUX, L., SALENÇON, J. : *An interactive computer software for the yield design of reinforced soil structures*, "Computers and Geotechnics" Coll., Paris, 1992.
13. GARNIER, D. : *Analyse de la stabilité d'ouvrages en sols cloués avec prise en compte des résistances à la flexion et au cisaillement*, Graduation thesis, Lab. Mec. Sol., 1991.

14. JEWELL, R.A., PEDLEY, M.J. : *Soil nailing design : the role of bending stiffness*, Ground Engineering, March, 1990, 30-36.

15. SCHLOSSER, F. : *Discussion : the multicriteria theory in soil nailing*, Ground Engineering, November, 1991, 30-33.

PART II : HOMOGENIZATION APPROACH TO THE YIELD STRENGTH OF REINFORCED SOILS

1. THE YIELD DESIGN HOMOGENIZATION METHOD [2][3][4]

The principle of many soil improvement techniques consists in placing the reinforcing inclusions into the bulk of the soil, following a regular pattern : reinforced earth, high density soil nailing, reinforcement of foundation soils by networks of columns, etc. Even though the interval between the reinforcements is of the order of the meter, it may be considered small when compared with the overall dimensions of the structure. Making use of this essential property, the homogenization approach is based on the intuitive idea that, from a *macroscopic* point of view, that is, as regards the overall characteristics of the structure, the constituent composite soil can be perceived as a homogeneous medium, whose macroscopic strength criterion should be determined. On account of the existence of preferential orientations due to the reinforcements such a homogenized material will be expected to exhibit strongly marked anisotropic properties.

1.1. PRINCIPLE OF THE METHOD

Consider for illustrative purpose the example of a foundation soil subjected to a vertical load Q applied by means of a rigid strip footing of width B (Fig. 1). In order to increase its bearing capacity, the soil has been reinforced by a group of uniformly spaced columns placed in a zone located just beneath the footing.

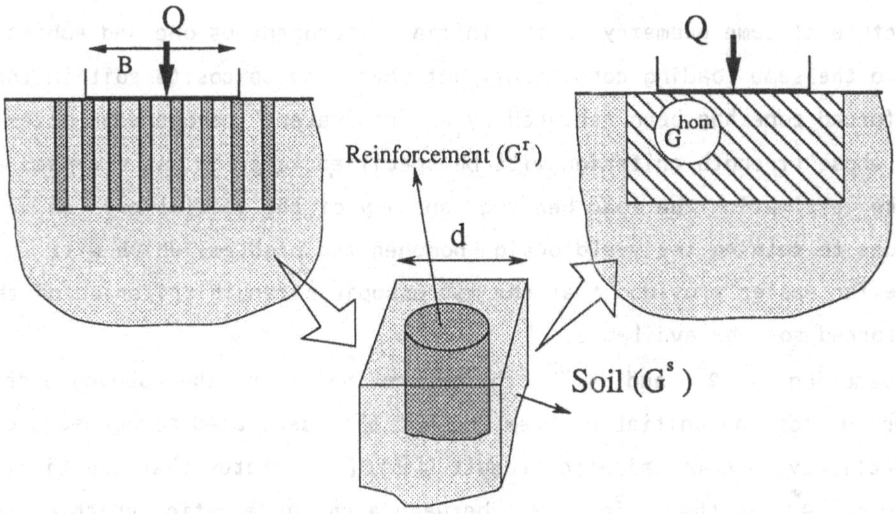

Fig. 1 : The yield design homogenization method.

Referring to the yield design theory, the ultimate load carrying capacity Q^* is defined as the maximum possible value of Q for which there exists at least one stress field $\underline{\underline{\sigma}}$ throughout the foundation soil which meets the following requirements :

a) $\underline{\underline{\sigma}}$ must be *statically admissible* with Q, i.e. satisfy the equilibrium equations and the stress boundary conditions of the problem as well.

b) It must also satisfy at any point the strength criterion of either the native soil or the constituent material of the columns. An additional strength condition relating to the interface between the soil and the reinforcements may possibly be introduced.

Due to the strong heterogeneity of the reinforced soil, many difficulties arise when attempting to implement either of the yield design static and kinematic approaches directly, unless resorting to sophisticated, and therefore of limited practical convenience, numerical techniques ([1]).

As an alternative approach, the homogenization method is based on the concept of *associated homogeneous structure* (Fig. 1), defined as a structure of same geometry as the initial heterogeneous one and subjected to the same loading conditions, but where the composite soil in the reinforced zone has been replaced by an "equivalent" homogeneous material whose strength criterion will be specified later on. As a consequence, estimating the load bearing capacity of the initial problem reduces to solving the yield design homogenized problem, which will prove far easier provided that the macroscopic strength criterion of the reinforced soil be available.

Denoting by $Q^\star$ and Q^{hom} the extreme values of the loading parameter Q for the initial problem and for the associated homogeneous one respectively, a homogenization result ([2], [3]) states that the limit value of $Q^\star$ as the ratio d/B, between a characteristic length of the reinforcement pattern d and the width of the footing B, tends to zero, remains below or equal to Q^{hom} :

$$\lim_{d/B \to 0} Q^\star \leq Q^{hom} \quad . \tag{1}$$

Actually, it can be proved from a theoretical point of view, that, on condition that "edge effects" occuring in the vicinity of the boundary surface of the structure be left aside, the two ultimate values do coincide, so that the implementation of the yield design homogenization method for predicting the load carrying capacity of reinforced soil structures is fully relevant.

1.2. MACROSCOPIC STRENGTH CRITERION OF THE REINFORCED SOIL

From the fundamental point of view of continuum mechanics, the reinforced soil may be perceived as a *periodic composite material*, so that determining its macroscopic strength criterion reduces to solving a yield design boundary value problem attached to a "unit cell", which may be regarded as the smallest representative volume of reinforced soil which contains all informations necessary to describe its structure completely (Fig. 2).

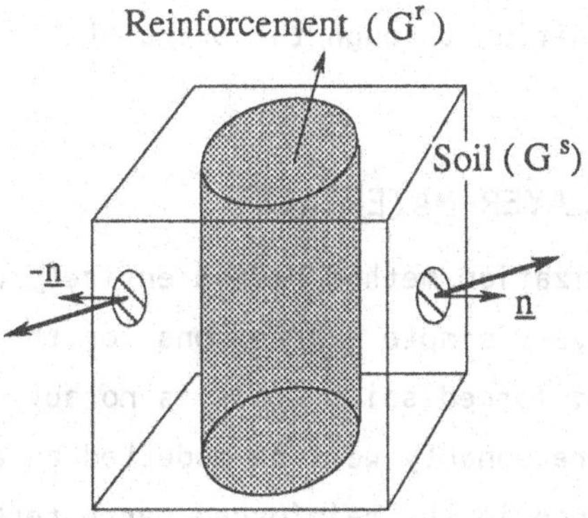

Fig. 2 : Unit representative cell C of the reinforced soil.

Denote by C the unit cell, and by C^s (resp. C^r) the subdomain occupied by the soil (resp. the inclusion). The strength criterion of each constituent is characterized by the set of allowable stress tensors $\underline{\underline{\sigma}}$, namely G^s for the soil and G^r for the reinforcement. Under these conditions the *macroscopic strength domain* G^{hom} of the reinforced soil may be defined as the set of "macroscopic" states of stress $\underline{\underline{\Sigma}}$, such that a stress field $\underline{\underline{\sigma}}$ can be exhibited over the unit cell which complies with all the following conditions ([2][3]).

a) $\underline{\underline{\Sigma}}$ is equal to the *average value* of $\underline{\underline{\sigma}}$ over C ;

b) div $\underline{\underline{\sigma}} = 0$ (along with $[\![\underline{\underline{\sigma}}]\!].\underline{n} = 0$ across possible stress discontinuity surfaces when piecewise continuous stress fields are considered);

c) $\underline{\underline{\sigma}}.\underline{n}$ *antiperiodic*, i.e. opposite on opposite sides of C , where $\underline{n}$ is the outer unit normal to C (Fig. 2) ;

d) and finally :

$$\underline{\underline{\sigma}}(\underline{x}) \in G^s \quad \text{(resp. } G^r\text{)} \qquad \text{whatever} \quad \underline{x} \in C^s \quad \text{(resp. } C^r\text{)}$$

(perfect bonding is assumed between the soil and the reinforcement).

It should be noted that despite the unusual conditions prescribed on the boundary surface of the unit cell, regarded as a *structure*, requirements a), b) and c) actually define a well posed boundary value problem, in which the components of $\underline{\underline{\Sigma}}$ may be considered as the *loading parameters*. Futhermore, a kinematic definition of G^{hom} ([2][3][4]), which involves *periodic* strain rate fields on the cell, may be obtained by dualizing the above static definition through the principle of virtual work.

2. Reinforced soils as multilayer materials

The feasibility of the homogenization method relies entirely upon the possibility of deriving relatively simple expressions for the macroscopic strength criterion of reinforced soils. This is notably the case when the reinforced soil can reasonably well be modelled as a multilayer material such as for instance in the reinforced earth technique.

2.1. STRENGTH CRITERION OF A MULTILAYER REINFORCED SOIL

In anticipation of the application of the yield design homogenization method to "plane strain" problems, the reinforced soil will be described as a *two-dimensional* multilayer material [3]. Referring to an Oxy coordinate system, a unit square may then be selected as representative cell of the reinforced soil (Fig. 3), the direction of the reinforcements being parallel to Ox . Combining the static and kinematic definitions of the macroscopic strength domain G^{hom} , it can be shown [3] that any allowable macroscopic stress $\underline{\underline{\Sigma}}$, i.e. belonging to G^{hom} , may be written in the form :

$$\begin{cases} \underline{\underline{\Sigma}} = (1-\eta)\,\underline{\underline{\sigma}}^s + \eta\,\underline{\underline{\sigma}}^r & \text{(2-a)} \\ \text{with} \quad \underline{\underline{\sigma}}^s \cdot \underline{e}_y = \underline{\underline{\sigma}}^r \cdot \underline{e}_y = \underline{\underline{\Sigma}} \cdot \underline{e}_y & \text{(2-b)} \\ \text{and} \quad \underline{\underline{\sigma}}^s \in G^s \ , \ \underline{\underline{\sigma}}^r \in G^r \ . & \text{(2-c)} \end{cases}$$

$\eta = e/\Delta H$ (Fig. 3) represents the voluminal fraction of the reinforcing material.

It can be easily seen from the general definition of G^{hom} given in para. 1.2., that (2) is obtained by using *piecewise constant* stress fields, i.e. stress fields which are homogeneous in each constituent. Condition (2-b) then expresses the continuity of the *stress vector* across the interface between soil and reinforcement.

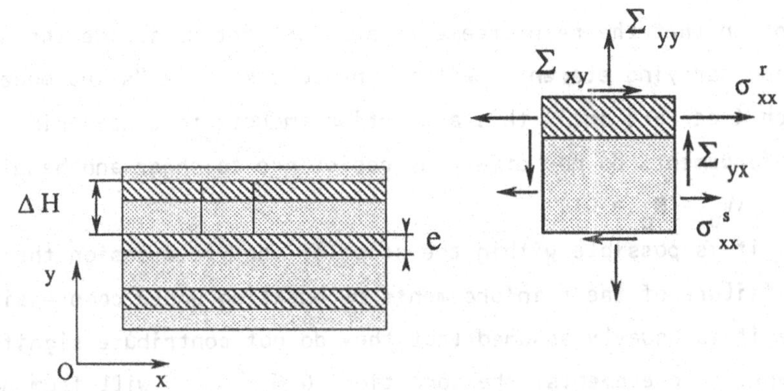

Fig. 3 : Multilayer model of reinforced soil.

In most cases the reinforcement layer is made of a material (such as a metal) whose strength characteristics are considerably greater than those of the native soil, whereas its thickness e is very small when compared with the spacing between two successive layers ($\eta = e/\Delta H$ is about 10^{-2} or 10^{-3}). Mathematically, this particular configuration (reinforcement by sheets, membranes or strips) may be derived from the above multilayer model by making η tend to zero, while the tensile strength R_t of the reinforcements per unit transverse length along Oz remains constant.

In such a case the definition of the macroscopic strength domain simplifies to ([3] [4])

$$\underline{\underline{\Sigma}} \in G^{hom} \quad \Updownarrow \quad \underline{\underline{\Sigma}} = \underline{\underline{\sigma}}^s + \sigma \, \underline{e}_x \otimes \underline{e}_x \tag{3}$$

with : $\underline{\underline{\sigma}}^s \in G^s$ and $|\sigma| \leq \sigma_o = R_t/\Delta H$.

The latter formulation is to be closely connected with those previously conjectured by several authors ([5] [6] [7]), based on the heuristic assumption that the reinforcements are just acting inside the soil as tensile load carrying elements. Within the context of a "mixed modelling" approach (see Part one), this assumption amounts to considering that the reinforcements do not offer any resistance to shear and bending solicitations ($V_o = M_o = 0$).

Moreover, it is possible within the scope of the yield design theory, to allow for failure of the reinforcements by buckling under compressive forces. Since it is usually assumed that they do not contribute significantly as compressive elements, the condition $0 \leq \sigma \leq \sigma_o$ will from now on be substituted for $|\sigma| \leq \sigma_o$ in (3).

2.2. GEOMETRICAL INTERPRETATION AND π-FUNCTIONS

Definition (3) of the macroscopic strength domain G^{hom} lends itself to a very simple graphical construction in the stress space. As a matter of fact, G^{hom} may be drawn as the convex envelope of the family of domains $G^s(\sigma)$ defined by

$$G^s(\sigma) = \{\underline{\underline{\Sigma}} \; ; \; \underline{\underline{\Sigma}} - \sigma \; \underline{e}_x \otimes \underline{e}_x \in G^s\} \tag{4}$$

with parameter σ, which represents the tensile stress in the reinforcements, varying from 0 to σ_o. $G^s(\sigma)$ is simply obtained by translating G^s along the Σ_{xx}-axis by the algebraic distance σ (Fig. 4). The increase in strength of the native soil due to the reinforcements is clearly evident from this geometrical construction, since G^s is always enclosed in G^{hom}.

Let f^s be a convex function associated with the strength domain of the soil :

$$f^s(\underline{\underline{\sigma}}^s) \leq 0 \iff \underline{\underline{\sigma}}^s \in G^s \; .$$

The macroscopic strength condition may be written in the following form

$$F^{hom}(\underline{\underline{\Sigma}}) \leq 0 \iff \underline{\underline{\Sigma}} \in G^{hom} \tag{5}$$

with for example :

$$F^{hom}(\underline{\underline{\Sigma}}) = \underset{\sigma}{\text{Min}} \; \{f^s(\underline{\underline{\Sigma}} - \sigma \; \underline{e}_x \otimes \underline{e}_x) \; ; \; 0 \leq \sigma \leq \sigma_o\} \; .$$

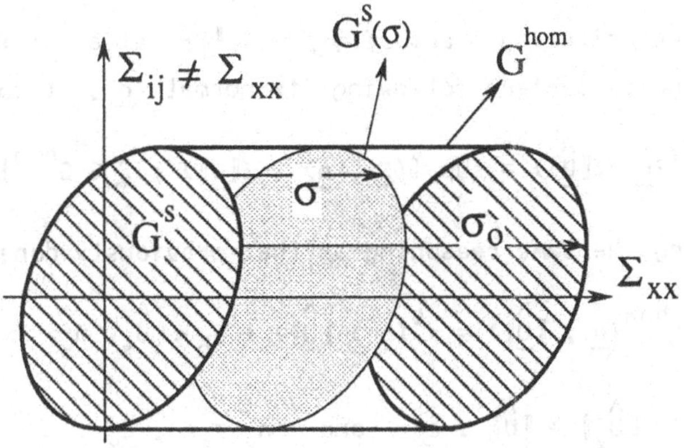

Fig. 4 : Geometrical construction of G^{hom} .

Looking forward to implementing the yield design homogenization method, and more particularly the kinematic approach, it is necessary to compute the π-functions (support functions) which define the macroscopic strength domain G^{hom} by duality.

• Given any *bidimensional* virtual strain rate tensor $\hat{\underline{\underline{D}}}$ in the Oxy plane, we get :

$$\pi^{hom}(\hat{\underline{\underline{D}}}) = \sup_{\underline{\underline{\Sigma}}} \{\underline{\underline{\Sigma}} : \hat{\underline{\underline{D}}} \ ; \ \underline{\underline{\Sigma}} \in G^{hom}\}$$

hence, from (3) :

$$\pi^{hom}(\hat{\underline{\underline{D}}}) = \sup_{(\underline{\underline{\sigma}}^s,\sigma)} \{(\underline{\underline{\sigma}}^s + \sigma\, \underline{e}_x \otimes \underline{e}_x) : \hat{\underline{\underline{D}}} \ ; \ \underline{\underline{\sigma}}^s \in G^s \ \text{ and } \ 0 \leq \sigma \leq \sigma_o\}$$

or, since $\underline{\underline{\sigma}}^s$ and σ vary independently

$$\pi^{hom}(\hat{\underline{\underline{D}}}) = \sup_{\underline{\underline{\sigma}}^s} \{\underline{\underline{\sigma}}^s : \hat{\underline{\underline{D}}} \ ; \ \underline{\underline{\sigma}}^s \in G^s\} + \sup_{\sigma} \{\sigma\, \hat{D}_{xx} \ ; \ 0 \leq \sigma \leq \sigma_o\}$$

that is, whatever $\underline{\underline{D}} \in \mathbb{R}^3$

$$\pi^{hom}(\hat{\underline{\underline{D}}}) = \pi^s(\hat{\underline{\underline{D}}}) + \sigma_o \langle \hat{D}_{xx} \rangle \tag{6}$$

where $\pi^s(\cdot)$ is the support function of G^s , defined in just the same way as $\pi^{hom}(\cdot)$ for G^{hom} , and $\langle\cdot\rangle$ denotes the positive part ($\langle x \rangle = x$ if $x \geq 0$, $= 0$ otherwise).

• Likewise, considering a *velocity jump* $[\![\underline{\hat{U}}]\!]$ at a point when crossing a discontinuity surface following its normal $\underline{n}$, it comes out :

$$\pi^{hom}(\underline{n} \; ; [\![\underline{\hat{U}}]\!]) = \sup \{(\underline{\underline{\Sigma}} \; . \; \underline{n}) \; . \; ([\![\underline{\hat{U}}]\!]) \; ; \; \underline{\underline{\Sigma}} \in G^{hom}\}$$

that is following the same reasoning as that previously done :

$$\pi^{hom}(\underline{n} \; ; [\![\underline{\hat{U}}]\!]) = \pi^{s}(\underline{n} \; ; [\![\underline{\hat{U}}]\!]) + \sigma_o < [\![\hat{U}_x]\!] \; n_x > \qquad (7)$$

with $[\![\hat{U}_x]\!] = [\![\underline{\hat{U}}]\!] \; . \; \underline{e}_x$ and $n_x = \underline{n} \; . \; \underline{e}_x$.

3. STRENGTH ANISOTROPY OF REINFORCED SOILS

This section is devoted to a more detailed study of the macroscopic strength criterion of a reinforced soil in the case when the original soil is isotropic, obeying either a Tresca's of a Coulomb's strength condition. Closed form analytical expressions for the corresponding strength condition of the reinforced soil are then available.

3.1. CASE OF A PURELY COHESIVE NATIVE SOIL ([3] [14])

The initially unreinforced soil is supposed to obey a Tresca's strength condition. Under plane strain conditions parallel to the Oxy plane, this condition can be expressed as a function of the three independent components of the stress in that plane :

$$f^s(\underline{\underline{\sigma}}) \leq 0 \iff (\sigma_{xx} - \sigma_{yy})^2 + (2\,\sigma_{xy})^2 \leq 4\,C^2 \qquad (8)$$

where C represents the *cohesion* (or "shear strength") of the soil.

Introducing the following notations :

$$p = (\sigma_{xx} + \sigma_{yy}) \,/\, \sqrt{2} \quad , \quad s = (\sigma_{yy} - \sigma_{xx}) \,/\, \sqrt{2} \quad , \quad t = \sqrt{2}\,\sigma_{xy} \qquad (9)$$

condition (8) may be rewritten in :

$$s^2 + t^2 \leq 2\,C^2$$

so that the strength domain G^s in the space of coordinates $(0, p, s, t)$ appears as a circular cylinder of radius $\sqrt{2}\,C$ and axis Op which represent the hydrostatic states of stress (Fig. 5-a).

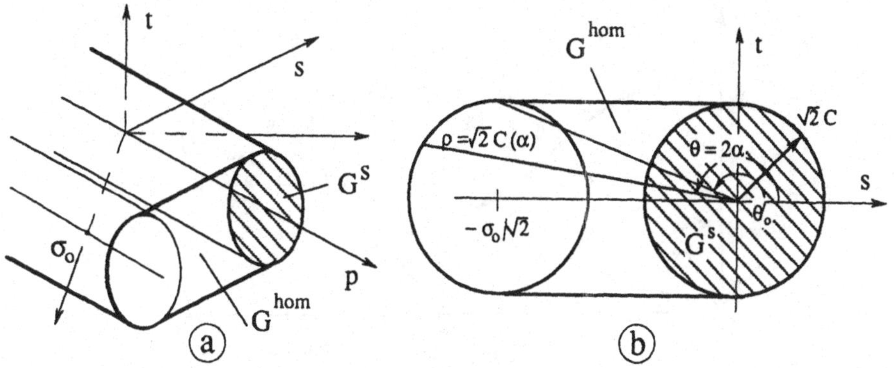

Fig. 5 : Representation of G^{hom}
a) in the stress space ; b) in any deviatoric plane $(p = ct)$.

Consequently, according to the geometrical interpretation given in para. 2.2., G^{hom} is obtained by drawing the envelope of this cylinder and of that deduced by a translation of components $(\sigma_o, 0, 0)$ in the $(\sigma_{xx}, \sigma_{yy}, \sqrt{2}\,\sigma_{xy})$ space or of components $(0, -\sigma_o/\sqrt{2}, 0)$ in the (p, s, t) coordinate system, since G^s remains unchanged by any translation parallel to the p-axis. It follows that G^{hom} is also a cylinder of axis Op, whose cross section by any *deviatoric plane*, defined by $p = ct$, is represented in Fig. 5-b.

This geometrical representation makes it possible to derive a closed form analytical expression for the macroscopic strength criterion of the reinforced soil. Denoting by Σ_I (resp. Σ_{II}) the major (resp. minor) principal component of $\underline{\underline{\Sigma}}$ in the Oxy plane, and by α the angle made by Σ_I with the Oy direction (Fig. 6) it comes out :

$$\Sigma_{yy} - \Sigma_{xx} = (\Sigma_I - \Sigma_{II}) \cos 2\alpha \qquad \Sigma_{xy} = \frac{\Sigma_I - \Sigma_{II}}{2} \sin 2\alpha$$

whence, adopting for $\underline{\underline{\Sigma}}$ the same notations as those introduced in (9) for $\underline{\underline{\sigma}}$

$$S = (\Sigma_I - \Sigma_{II}) \cos 2\alpha / \sqrt{2} \quad , \quad T = (\Sigma_I - \Sigma_{II}) \sin 2\alpha / \sqrt{2}$$

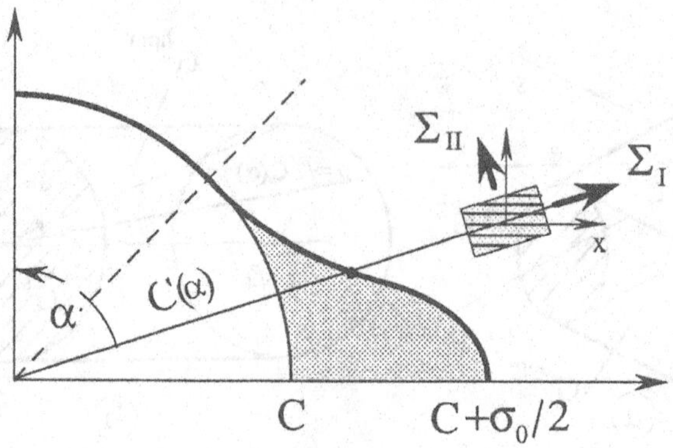

Fig. 6 : Polar diagram of cohesion for a reinforced soil.

or in terms of polar coordinates (ρ, θ) in the (s, t) plane (Fig. 5-b)

$$\rho = \sqrt{S^2 + T^2} = (\Sigma_I - \Sigma_{II}) / \sqrt{2} \quad , \quad \theta = 2\alpha \ .$$

The macroscopic strength condition may therefore be expressed in the following form :

$$\Sigma_I - \Sigma_{II} \leq 2\, C(\alpha) \tag{10}$$

where $C(\alpha)$, which can be interpreted as an *anisotropic cohesion* of the reinforced soil perceived as a homogeneous material, is calculated by noting that the equation in polar coordinates of the curve bounding the strength domain (Fig. 5-b) writes then :

$$\rho(\theta) = \sqrt{2}\, C(\theta/2) \ .$$

Thus for $\alpha \in [0, +\pi/2]$

$$C(\alpha) = \begin{cases} C & \text{if} \quad 0 \leqslant \alpha \leqslant \alpha/4 \\ C/\sin 2\alpha & \text{if} \quad \pi/4 \leqslant \alpha \leqslant \alpha_o \\ \dfrac{\sigma_o}{2}\,|\cos 2\alpha| + \sqrt{C^2 - \left(\dfrac{\sigma_o}{2}\sin 2\alpha\right)^2} & \text{if} \quad \alpha_o \leqslant \alpha \leqslant \pi/2 \end{cases} \qquad (11)$$

with $\tan 2\alpha_o = -2C/\sigma_o$ and of course $C(-\alpha) = C(\alpha)$.

The cohesion $C(\alpha)$ is plotted in Fig. 6 in the form of a *polar diagram,* quite similar to those experimentally drawn from triaxial tests performed on samples of naturally anisotropic clays ([8] [9]). It varies from the minimum value of C (cohesion of the soil prior to reinforcement) obtained when the major principal stress is inclined at less than $\pi/4$ with respect to Oy to a maximum value of $C + \sigma_o/2$ reached for $\alpha = \pi/2$. The shaded area in Fig. 6 thus visualizes the increase in strength due to the reinforcements.

3.2. A MODEL FOR "REINFORCED EARTH" [10] [15]

It corresponds to the case when the native soil is a dry cohesionless sand characterized by an *internal friction angle* φ , i.e. whose strength criterion is the classical Mohr-Coulomb's condition :

$$f^s(\underline{\underline{\sigma}}) \leqslant 0 \iff [(\sigma_{xx} - \sigma_{yy})^2 + 4\,\sigma_{xy}^2]^{1/2} \leqslant -(\sigma_{xx} + \sigma_{yy})\sin\varphi \qquad (12)$$

("plane strain" criterion)

or in terms of the (p, s, t) variables previously introduced

$$\rho = \sqrt{s^2 + t^2} \leqslant -p\sin\varphi \qquad (p \leqslant 0) \quad .$$

The related strength domain G^s is a circular cone with its axis coinciding with the p-axis and its vertex located at the origin 0 , while the macroscopic strength domain is once again constructed by taking the envelope of G^s and of the cone translated from G^s by the distance σ_o along the σ_{xx}-axis . The so-obtained "double cone" shaped strength domain is sketched in Fig. 7.

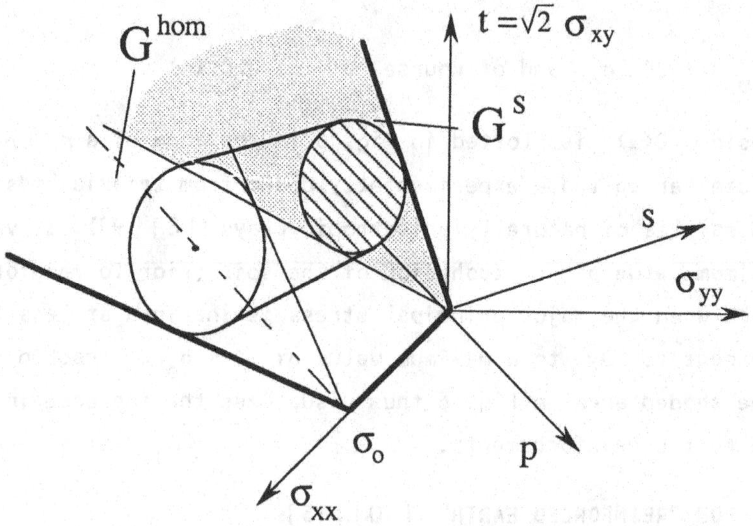

Fig. 7 : Macroscopic strength domain of reinforced earth.

Its cross section by a deviatoric plane of equation $p = ct$ (with $p < 0$) is represented in Fig. 8. It comprises two arcs of circle of respective radii $(- p \sin \varphi)$ and $(- p + \sigma_o/\sqrt{2}) \sin \varphi$ and respective centers $(0 , 0)$ and $(- \sigma_o/\sqrt{2} , 0)$ connected by two segments inclined at angle φ with the s-axis.

This geometrical representation gives a clear interpretation of three possible *failure modes* for the reinforced earth. They correspond to a subdivision of the boundary surface of G^{hom} into three complementary zones

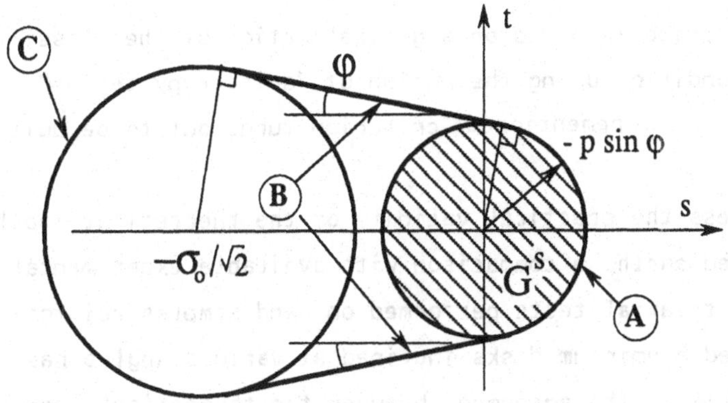

Fig. 8 : Section of G^{hom} by a deviatoric plane.

• Region A , which is common to the boundaries of G^s and G^{hom}, is the locus of points associated with states of stress $\underline{\underline{\Sigma}}$ for which the strength of the soil is fully mobilized ($f^s(\underline{\underline{\sigma}}^s) = 0$ in (3)) while the reinforcements have reached their ultimate compressive strength ($\sigma = 0$).

• Region B is made up of two portions of planes tangent to the cones (segments in Fig. 8). The corresponding stresses $\underline{\underline{\Sigma}}$ are such that only the soil is at failure, the tension in the reinforcements strictly remaining within its ultimate bounds ($0 < \sigma < \sigma_o$) .

• Finally region C of the boundary surface represents the macroscopic solicitations $\underline{\underline{\Sigma}}$ for which both constituents undergo failure, the reinforcements having reached their ultimate tensile strength ($\sigma = \sigma_o$) .

Some comments are worth being made

• The *anisotropy* of reinforced earth as a homogenized material is clearly ascertained by this theoretical model, since any isotropic strength criterion (such as the Coulomb criterion) would be represented in the (p , s , t) space by an *axisymmetrical* domain about the p-axis, which is obviously not the case here.

• The concepts of cohesion and friction angle, as classically introduced for isotropic or even purely cohesive anisotropic soils ([8] [9]), prove no longer relevant to account for the anisotropic characteristics of the above described criterion. Nor can such an anisotropy be conveniently dealt with by criteria based on a generalization of the classical Coulomb failure condition using the notion of "anisotropy tensor" ([11]). Consequently the aforementioned criterion turns out to be quite original.

• In order to assess the practical validity of the theoretical model proposed for reinforced earth, a comparison with available experimental data corresponding to triaxial tests performed on sand samples reinforced by regularly spaced aluminium disks inclined at various angles has been made ([12] [13] [14]). The agreement between the theoretical predictions and the observed results was shown to be excellent.

4. AN EXAMPLE OF APPLICATION

Making use of the above strength criterion of reinforced earth, the yield design homogenization method will now be applied to the design of a reinforced earth retaining structure.

4.1. STABILITY ANALYSIS OF A VERTICAL REINFORCED EARTH WALL ([15] [16])

The problem under consideration is that of a vertical retaining structure whose purely frictional backfill soil has been reinforced by regularly spaced horizontal reinforcements. The vertical spacing between two successive reinforcement layers is considered small when compared with the total height h of the structure, so that the homogenization method fully applies. The strength criterion of reinforced earth as a homogeneous material is therefore characterized by parameters σ_o and φ.

Designing that structure from the stability analysis point of view comes down to determining, or at least estimating, the *critical height* $h^\star$ of the wall, that is the value of h beyond which failure of the structure under its own weight γ will certainly occur. A simple dimensional analysis reasoning shows that $h^\star$ is necessarily of the form :

$$h^\star = \sigma_o / \gamma\, K^\star(\varphi)$$

where $K^\star(\varphi)$ is a dimensionless factor function of the sole friction angle φ.

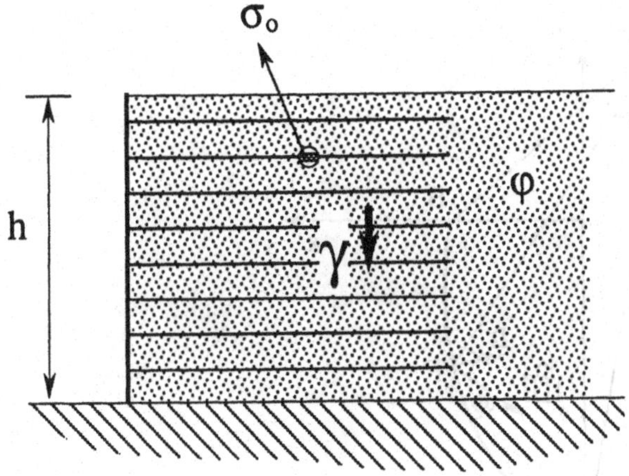

Fig. 9 : Vertical reinforced earth wall.

Considering failure mechanisms as that sketched in Fig. 10, where a volume of reinforced soil bounded by an arbitrary line AB is given a virtual rigid body rotating motion of velocity $\hat{\omega}$ about a pole F, the yield design kinematic approach will enable us to derive *upperbound estimates* for h^* (or K^*), since a necessary condition for the structure to remain stable (i.e. $h \leq h^*$) writes

$$W_e(F, AB, \hat{\omega}) \leq W_r(F, AB, \hat{\omega}) \quad . \qquad (13)$$

W_e is the work developed by the weight of the rotating volume, that is, referring to Oxy axes, where Oy is oriented upwards,

$$W_e(F, AB, \hat{\omega}) = \int_{OAB} - \gamma \hat{U}_y \, dx\, dy \qquad (\gamma > 0) \qquad (14)$$

where $\hat{U}_y$ is the vertical component of the velocity at any point (x, y).

On the other hand, W_r represents the *maximum resisting work* developed by the reinforced soil along the velocity jump line AB. Its expression is :

$$W_r(F, AB, \hat{\omega}) = \int_{AB} \pi^{hom}(n ; [\![\hat{\underline{U}}]\!])\, ds \qquad (15)$$

with $[\![\hat{\underline{U}}]\!]$ denoting the velocity jump across AB at current point P following its normal $\underline{n}$.

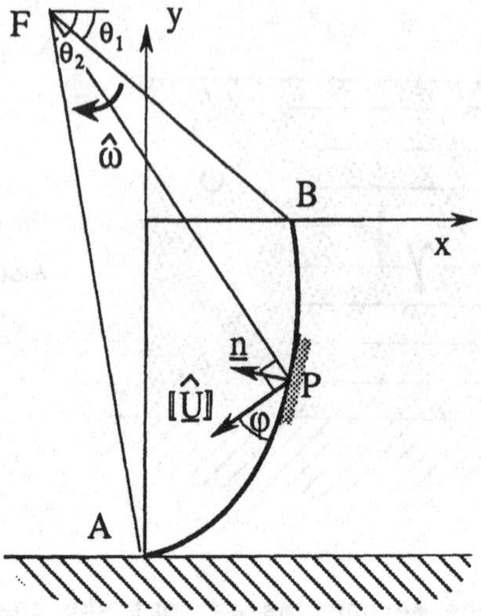

Fig. 10 : Stability analysis of the reinforced earth wall by means of rotational failure mechanisms.

Function π^{hom} is given by formula *(7)*, where π^s represents the support function of the Coulomb's strength condition with no cohesion. Therefore :

$$\pi^{hom}(\underline{n}\ ;\ [\![\underline{\hat{U}}]\!]) = \begin{cases} \sigma_o < [\![\hat{U}_x]\!]\ n_x > & \text{if} \quad \varphi \leq \psi \leq \pi - \varphi \\ + \infty & \text{otherwise} \end{cases}$$

where ψ is the angle made by the velocity jump $[\![\underline{\hat{U}}]\!]$ with the tangent to line AB (Fig. 10).

For $\psi = \varphi$, AB reduces to an *arc of logspiral of angle* φ *and focus* F , which may be completely defined through angular parameters θ_1 and θ_2 , once AB is assigned to pass through the toe of the wall.

Expliciting the expressions of *(14)* and *(15)* (more detailed calculations may be found in [15] and [16]), the basic inequality *(13)* reduces to :

$$h \leq (\sigma_o/\gamma)\ K(\theta_1\ ,\ \theta_2\ ;\ \varphi) \qquad \text{whatever} \quad h \leq h^\star$$

and then

$$h^{*} \leq (\sigma_o/\gamma)\, K(\theta_1\,,\, \theta_2\,;\, \varphi)\ .$$

The minimum value of $K(\theta_1, \theta_2 ; \varphi)$ with respect to θ_1 , θ_2 and hence the *best upperbound estimate* of h^{*} which can be expected from such an analysis, is obtained through a numerical procedure :

$$K^{m}(\varphi) = \min_{(\theta_1\,,\,\theta_2)} K(\theta_1\,,\,\theta_2\,;\,\varphi)\ .$$

4.2. COMPARISON WITH EXPERIMENTS

Fig. 11 displays a comparison between predicted failure heights derived from the previous theoretical analysis and experimental results [17] obtained on reduced scale walls using a backfill sand with a friction angle of $\varphi = 35°$, reinforced by metallic strips of variable width b , thus leading to a variable strength parameter σ_o , since σ_o is directly proportional to b .

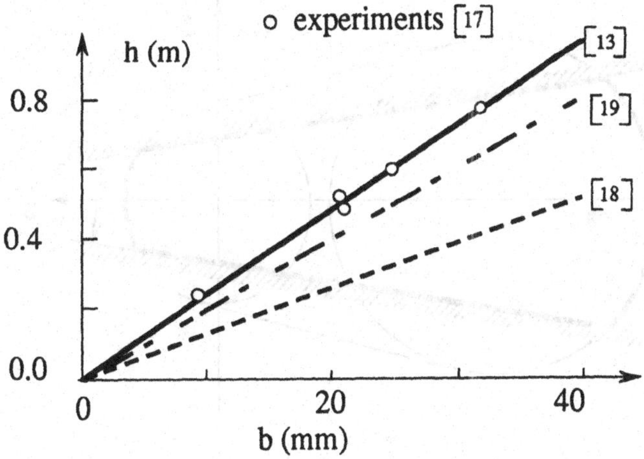

Fig. 11 : Predicted vs. experimental failure heights for model walls.

The good performance of the homogenization method is self evident, while other traditional design methods ([18] [19]) tend to significantly underestimate the actual failure heights.

5. CURRENT AND FUTURE DEVELOPMENTS OF THE THEORY

As shown by the previous application, the yield design theory combined with the concept of homogenization proves a fairly adequate approach for dealing with the stability analysis of reinforced soil structures, provided that the conditions allowing for the modelling of the reinforced soil as an anisotropic material be satisfied. The method still remains relevant when an additional failure condition relating to the interface between the reinforcements and the surrounding soil has to be taken into account ([3] [20]). By way of illustration, Fig. 12 shows the cross section by a deviatoric plane of the strength domain of reinforced earth, resulting from the introduction of a Coulomb's type interface failure condition characterized by a friction angle δ $(\delta < \varphi)$. It is simply obtained by truncating the initial cross section of G^{hom} corresponding to the case of perfect bonding (Fig. 8) with two straight lines inclined at $\pm\,\delta$ with the s-axis.

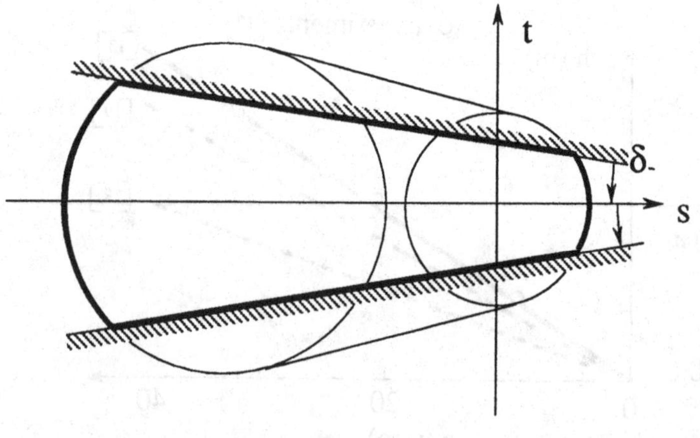

Fig. 12 : Truncated cross-section of G^{hom} *accounting for an interface failure condition.*

Two recent developments of the theory, among possible others, deserve to be more particularly mentioned.

• *Soils reinforced by continuous threads.* This highly innovative reinforcement technique consists in elaborating a composite material from the mixing of a granular soil (cohesionless sand) with randomly distributed synthetic threads [21]. Although no clear structural pattern can be distinguished from the intricate entanglement formed by the grains and the threads, experimental tests conducted on samples of this material have undoubtedly shown an anisotropy of its strength characteristics. A model has been recently proposed for ascertaining such an anisotropy [22]. It is simply derived through a generalization of the reinforced earth model described in section 3.2., by considering that the original soil is reinforced by two arrays of threads placed along two perpendicular directions : the major reinforcement direction, generally close to the horizontal plane, is characterized by the strength parameter σ_o^1 (tensile resistance of the threads per unit transverse area), while the minor reinforcement direction is associated with σ_o^2 ($\sigma_o^2 < \sigma_o^1$) (Fig. 13). The corresponding strength domain G^{hom} is then defined as

$$\underline{\underline{\Sigma}} \in G^{hom}$$
$$\Updownarrow$$
$$\underline{\underline{\Sigma}} = \underline{\underline{\sigma}}^s + \sigma^1 \, \underline{e}_x \otimes \underline{e}_x + \sigma^2 \, \underline{e}_y \otimes \underline{e}_y$$
$$\text{with} \quad 0 \leqslant \sigma^1 \leqslant \sigma_o^1 \quad \text{and} \quad 0 \leqslant \sigma^2 \leqslant \sigma_o^2 \ .$$

Its cross-section by the plane of equation $t = 0$ is sketched in Fig. 13.

Design methods based upon the kinematic approach making use of such a criterion are curently devised for analysing the stability of embankments.

• *Reinforcement by columns.* This technique is increasingly used today for improving the load carrying capacity of soft foundation soils (Fig. 1). Unlike other types of reinforcement, it turns out to be a truly *three-dimensional* reinforcement (Fig. 2), which cannot therefore be properly investigated by referring to a multilayer model. The macroscopic strength criterion of such reinforced soils has still to be determined starting from the general definition given in section 1.2.

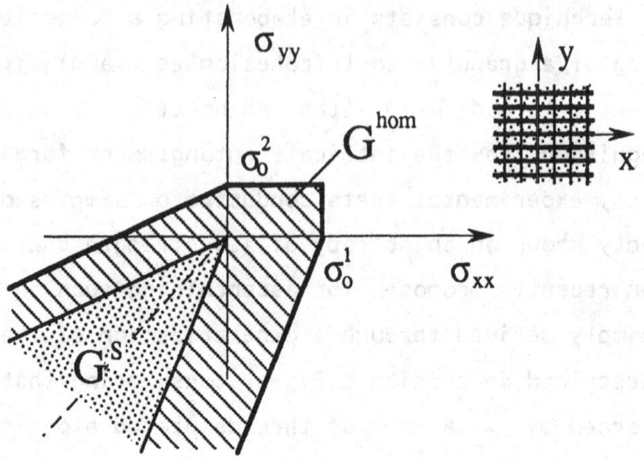

Fig. 13 : Strength domain of a purely frictional soil reinforced along two perpendicular directions.

As regards for example the case of a purely cohesive soil (clay) reinforced by columns of a purely frictional material ("stone columns" technique), it remains to be seen whether the strength properties to be expected for the homogenized reinforced soil, are actually improved with respect to those of the original soil, since the strength domain of the latter (G^s : Tresca criterion) is not included in that of the "reinforcing" inclusions (G^r : Coulomb criterion). A preliminary investigation of this question shows that the reinforcing effect is only apparent in the domain of compressive stresses.

REFERENCES

1. PASTOR, J., TURGEMAN, S., CISS, A. : *Calculation of limit loads of structures in soils with metal reinforcement*, Eur. Conf. Num. Meth. Geomech., Stuttgart, Germany, 1986.
2. SUQUET, P. : *Elements of homogenization in inelastic solid mechanics*, In "Homogenization techniques for composite media", C.I.S.M., Udine, Springer Verlag, 1985, 193-278.
3. de BUHAN, P. : *Approche fondamentale du calcul à la rupture des ouvrages en sols renforcés*, Th. Dr Sc., Paris, 1986.

4. de BUHAN, P., SALENÇON, J. : *Yield strength of reinforced soils as anisotropic media,* In "Yielding, Damage and Failure of Anisotropic Solids", Ed. J.P. Boehler, Mech. Eng. Publ., London, 1990, 791-803.

5. Mc LAUGHLIN, P.V. : *Plastic limit behaviour and failure of filament reinforced materials,* Int. Jl Solids Structures, 8, 1972, 1299-1318.

6. SAWICKI, A. : *Plastic limit behaviour of reinforced earth,* Jl Geotech. Eng., A.S.C.E., 109, n° 7, 1983, 1000-1005.

7. SAWICKI, A., LESNIEWSKA, D. : *Failure modes and bearing capacity of reinforced retaining walls,* Geotextiles and Geomembranes, 5, 1987, 29-44.

8. BISHOP, A.W. : *The strength of soils as engineering materials,* Géotechnique, 16, 1966, 89-130.

9. SALENÇON, J. : *Yield strength of anisotropic soils,* Invited lecture, Proc. 16th I.C.T.A.M., Lyngby, Denmark, North-Holland, 1984, 369-386.

10. de BUHAN, P., SALENÇON, J., SIAD, L. : *Critère de résistance pour le matériau "terre armée",* C.R. Acad. Sc., Paris, 302, série II, 1986, 377-381.

11. BOEHLER, J.P., SAWCZUK, A. : *Equilibre limite de sols anisotropes,* Jl Mécanique, 9, 1970, 5-33.

12. LONG, N.T., URSAT, P. : *Comportement du sol renforcé,* L.C.P.C. - L.R. Strasbourg, 1977.

13. MANGIAVACCHI, R., PELLEGRINI, G. : *Analisi teorica del comportamento della terra armata mediante una procedura di omogeneizazzione,* Tesi di Laurea, Politecnico di Milano, 1985.

14. de BUHAN, P., SALENÇON, J. : *Analyse de stabilité des ouvrages en sols renforcés par une méthode d'homogénéisation,* Rev. Fr. Géotech., n° 41, 1987, 29-43.

15. SIAD, L. : *Dimensionnement d'ouvrages en terre armée par une méthode d'homogénéisation,* Th. Dr E.N.P.C., Paris, 1987.

16. de BUHAN, P., MANGIAVACCHI, R., NOVA, R., PELLEGRINI, G., SALENÇON, J. : *Yield design of reinforced earth walls by a homogenization method,* Géotechnique, 39, n° 2, 1989, 189-201.

17. BEN ASSILA, A., EL AMRI, M. : *Comportement à la rupture des murs en terre armée et clouée,* Graduation thesis, E.N.P.C., Paris, 1984.

18. SCHLOSSER, F., VIDAL, H. : *La Terre Armée,* Bull. Liais. L.C.P.C., Paris, n° 41, 1969, 101-144.

19. JURAN, I., SCHLOSSER, F. : *Etude théorique des efforts de traction dans les armatures des ouvrages en terre armée,* Proc. Int. Conf. Soil Reinf., Paris, 1, 1979, 77-82.

20. de BUHAN, P., SIAD, L. : *Influence of a soil-strip interface failure condition on the yield-strength of reinforced earth,* Comp. and Geotech., 7, 1989, 3-18.

21. KHAY, M., GIGAN, J.P. : *Texsol : ouvrages de soutènement*, Guide technique L.C.P.C. - S.E.T.R.A., 1990.

22. de BUHAN, P., SALENÇON, J., VERZURA, L. : *Une modélisation de la résistance des sols renforcés par fils continus*, Scientific Report Greco "Géomatériaux", 1990.

APPLICATION OF HOMOGENIZATION THEORY AND LIMIT ANALYSIS TO THE EVALUATION OF THE MACROSCOPIC STRENGTH OF FIBER REINFORCED COMPOSITE MATERIALS

A. Taliercio
Politechnic of Milan, Milan, Italy

ABSTRACT

The macroscopic strength domain of a composite material reinforced by long, parallel fibers is, in general, unknown but for its theoretical definition. In this note it is shown how a homogenization technique applied to yield design theory allows the derivation of two domains (in the space of macroscopic stresses) which are a lower and an upper bound to the composite strength domain. The dependence of these domains on the fiber content and on the shape of the fiber array is pointed out. Analytical equations for the approximate uniaxial macroscopic strength of composites with Drucker–Prager or Von Mises type matrix are derived. For more complex stress conditions, the relevant strength domains are numerically evaluated as well. The discrepancy between the two bounds is in many cases relatively small. In particular, the two bounds yield the same value for the uniaxial strength of the composite along the fiber direction, which by consequence is exactly determined.

1. INTRODUCTION

The use of fiber reinforced composite materials, either with polymeric, metallic or ceramic matrix, is constantly expanding in a number of fields such as aerospace, motorcraft, biomechanic and civil engineering. The excellent elastic and ultimate properties of fiber composites often make of these materials the ones usable in applications where high stiffness–to–weight and strength–to–weight ratios are required.

As a consequence of their increasing use, a very large amount of theoretical and experimental work has been devoted, mainly in the last two decades, to the study of the mechanical behaviour of both composite materials and structures made of composite. Nevertheless, the problem of defining strength criteria supported not simply by a good comparison with experimental results, but also by a clear mechanical interpretation, does not seem to be fully solved yet. In fact, the most widely used strength criteria for these materials are 'phenomenological' criteria, such as those proposed by Azzi & Tsai [1], Hoffmann [2], Tsai & Wu [3]. These criteria are subsequent extensions of a criterion originally formulated by Hill [4] by extending, in turn, Von Mises strength criterion to anisotropic materials.

All of these criteria usually allow a good reproduction of the strength properties of composites subjected to different kinds of stress (see e.g. the comparisons shown by Wu [5]). On the other hand, these criteria are defined by a rather large number of parameters (Tsai and Wu criterion, for instance, requires six parameters in plane stress conditions), some of which are not easily obtained by the experiments (biaxial or pure shear tests may be required). This difficulty is a direct consequence of the fact that phenomenological criteria are not based on a micromechanical approach, accounting for the strength properties of the materials forming the composite; they merely aim at interpolating sets of experimental data, disregarding the actual heterogeneous nature of the composite which is considered as a homogeneous, orthotropic material.

In order to avoid the drawbacks related to a phenomenological approach, some authors have tried to formulate strength criteria directly taking into account the mechanical behaviour of fiber, matrix and, possibly, the fiber–matrix interface. This was done, for instance, by Boehler & Raclin [6], whose theoretical developments allow an excellent reproduction of experimental tests on composites subjected to uniaxial tension and triaxial compression. The criterion proposed by these authors allows for a distinction between failure modes, depending on the loading condition, and involves parameters related to the strength properties of the components. A limitation to the use of this criterion is the number of parameters required, which amount to eighteen. Similar considerations apply to the criterion proposed by Hashin [7], that was tested by the author under uniaxial conditions only.

Another interesting criterion based on a micromechanical approach was proposed by McLaughlin & Batterman [8] and Mc Laughlin [9], who used limit analysis theory to predict the ultimate load domain of the Representative Volume Element (RVE) featuring composite members and attributed to this domain the significance of strength domain for the composite material. Even though these authors do not operate within the framework of homogenization theory, which would provide a mathematical justification to their criterion, their theoretical results are well validated by comparisons with experimental tests on unidirectional and angle–ply metal composites subjected to uniaxial tension.

Criteria founded on a sound mathematical basis and defined by a reasonable number of parameters clearly related to the strength of the components can be derived through an approach based on the application of *limit analysis* to *homogenization theory*. Similarly to McLaughlin's approach, the basic idea is to consider RVEs as 'structures' and to determine their ultimate loads (which are identified with the 'macroscopic stress' in the composite structure). In limit analysis this is usually done by means of static and kinematic approaches, which yield lower and upper bounds, respectively, to the ultimate bearing capacity of the structure.

De Buhan & Taliercio [10,11] applied this approach to composites reinforced by one or more arrays of parallel fibers, embedded in a matrix conforming with Von Mises strength criterion. By defining suitable stress fields over the RVE, these authors obtained lower bounds to the macroscopic strength properties of the composite which, in many cases, were in excellent agreement with available experimental results. However, these lower bounds had to be proved not to be overconservative when compared with upper bounds. Also, the use of a Von Mises matrix does not permit to allow for the effects of isotropic pressure on the composite strength which, as shown by Boehler & Raclin [6], can be far from being negligible. This latter drawback was eliminated by Taliercio [12] by assuming Drucker–Prager strength criterion for polymeric matrices. Taliercio [12,13] also proposed upper bounds to the macroscopic strength properties of fiber composites, formulated by means of a limit analysis kinematic approach, requiring the definition of suitable failure mechanisms for the RVE. This was done by extending the results of de Buhan *et al.* [14], which actually were of interest only for composites in plane strain conditions.

After a brief review of the fundamental aspects of the theoretical approach used (Sec. 2), in Sec. 3 the lower bound to the macroscopic strength domain is derived and a simplified version of this bound, particularly useful in applications, is presented. Sec. 4 is devoted to the definition of upper bounds for composites with any type of components. Theoretical results are specialized in Sec. 5 to composites with Drucker–Prager and Von Mises type matrix. Both bounds are defined by only four parameters at most, for any stress state. These parameters are related to the uniaxial tensile and compressive strengths of the components and can be rather easily obtained by means of experiments. In Sec. 6 the problem is studied of determining the strength parameters on the basis of experimental failure data relevant to the composite as a whole. This is also done with reference to specific experimental results, which are compared to the theoretical predictions obtained through the model presented here. Finally, in Sec. 7 it is shown how a fiber–matrix interface strength criterion can be incorporated into the present model.

In spite of the simplicity of the stress fields used in the static approach and the failure mechanisms used in the kinematic approach, lower and upper bounds tend to be rather close for composites with different fiber arrangements and volume fractions, subjected to uni- and biaxial stress. In most cases, also the theoretical–experimental agreement is quite satisfactory.

2. PROBLEM FORMULATION

Consider a composite material consisting of an array of parallel, cylindrical fibers with circular cross section, embedded in a matrix of bonding material. Fibers are supposed to be 'long', in the sense that their length is comparable to the size of the structural member considered. If fibers form a regular array, the composite can be regarded as heterogeneous material with *periodic* internal structure. Periodic materials are featured by a 'representative volume element' (RVE–see e.g. Hashin [15]). The volume of the RVE will be indicated by $\mathcal{A}$; in subsequent analysis its measure will be supposed to be scaled to 1 (i.e. $|\mathcal{A}| = 1$). $\mathcal{A}_f$ and $\mathcal{A}_m$ will denote the parts of the RVE relevant to the fiber and the matrix material, respectively. Denoting by $\eta = |\mathcal{A}_f|/|\mathcal{A}|(= |\mathcal{A}_f|)$ the fiber volume fraction, the matrix volume fraction is then $1 - \eta = |\mathcal{A}_m|$. The points $\underline{x}$ of the RVE will be referred to an orthogonal reference frame $Oxyz$, with the x–axis collinear with the fibers. Let $\underline{e}_x$, $\underline{e}_y$, $\underline{e}_z$ be the base vectors. The length of the RVE along the x–axis is appearently arbitrary. As a consequence, subsequent analysis will be developed in any cross section of the composite perpendicular to the x–axis (thus staying in the (y, z) plane); for the sake of brevity, from here onwards 'RVE' will actually denote the intersection of the representative element with this plane.

Regular fiber arrays can be fundamentally of two kinds:

a) rectangular arrays, see Fig. 1a, for which RVEs are rectangles with sides parallel to the axes y and z, the fiber axes crossing the centers of any RVE. Let d_y, d_z be the length of the sides of any RVE.

b) hexagonal arrays, see Fig. 1b, such that in any section perpendicular to the fibers the fiber axes are located at the centers of more or less regular hexagons. In this case, z is defined by the centers of a row of fibers and y is perpendicular to z in that section. d_y and d_z will denote the distance between adjacent fibers along these axes.

Reinforcing arrays will be featured by an angle β, defined as $\beta = \mathrm{arctg} d_y/d_z$. The reference frame will be chosen in such a way that $d_y \geq d_z$ (so that $45° \leq \beta < 90°$). Set, for brevity, $s_\beta = \sin\beta$, $c_\beta = \cos\beta$. Depending on the value of β and the type of array, a certain limit value of the fiber volume fraction η exists which cannot be exceeded. Simple geometrical considerations lead to the conclusion that this limit value is $\bar{\eta}_r = \pi/4c_\beta/s_\beta$ for composites with rectangular fiber array ($\bar{\eta}_r = 0.79$ for square arrays), whereas it is $\bar{\eta}_e = \min\{\pi/(8s_\beta c_\beta); \pi/2c_\beta/s_\beta\}$ for hexagonal arrays ($\bar{\eta}_e = 0.91$ for regular hexagonal arrays).

In view of the definition of strength domains suitable for the description of the overall strength properties of periodic fiber composites subjected to any stress, it is customary to define global measures for the stress at any point of the composite structure. These are called *macroscopic stresses* and are the volume average of the microscopic stresses over the RVE (see e.g. Hashin & Rosen [16]). Thus, denoting by $\underline{\underline{\sigma}}(\underline{x})$ the microscopic stress field in the RVE and by $< \cdot >$ volume average over RVE, the macroscopic stress $\underline{\underline{\Sigma}}$ is

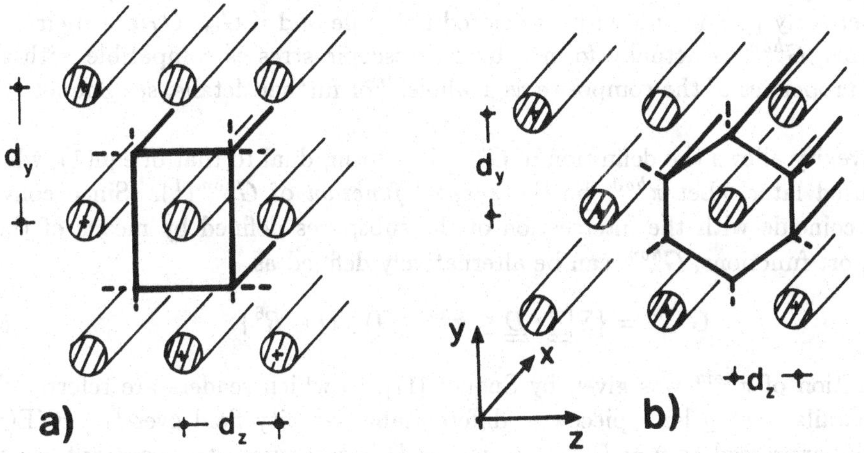

Fig. 1: Unidirectional fiber reinforced composite with (a) rectangular and (b) hexagonal reinforcing array. The relevant RVEs are also shown.

$$\underline{\underline{\Sigma}} = < \underline{\underline{\sigma}} > = \int_{\mathcal{A}} \underline{\underline{\sigma}}(\underline{x}) dA$$

Suppose that the strength domains of fibers and matrix are given convex domains in the stress space, denoted by G_f and G_m respectively. For the time being, the fiber-matrix interface will be supposed to be indefintely strong (i.e. there is 'perfect bond' between fibers and matrix); this restrictive assumption will be removed in Sec. 7. Suquet [17,18] proved that a convenient macroscopic strength domain can be defined as

$$\begin{aligned}\{G^{hom} = \underline{\underline{\Sigma}} | \underline{\underline{\Sigma}} = < \underline{\underline{\sigma}} >;\ \mathrm{div}\underline{\underline{\sigma}} = \underline{0}\ \forall \underline{x} \in \mathcal{A};\ [\![\underline{\underline{\sigma}}]\!] \cdot \underline{n}_S = \underline{0}\ \forall \underline{x} \in S; \\ \underline{\underline{\sigma}} \cdot \underline{n} \text{ anti-periodic over } \partial\mathcal{A};\ \underline{\underline{\sigma}} \in G_m\ \forall \underline{x} \in \mathcal{A}_m;\ \underline{\underline{\sigma}} \in G_f\ \forall \underline{x} \in \mathcal{A}_f\}\end{aligned} \tag{1}$$

and can be proved to be convex. Here S is any possible discontinuity surface for the microscopic stress field $\underline{\underline{\sigma}}$ in $\mathcal{A}$ and $\underline{n}_S$ is the normal unit vector at any point of S.

REMARK: The definition of G^{hom}, eqn(1), does not involve any assumption about the constitutive law of the component materials and requires that only the strength domains of matrix and fibers be known. Moreover, the boundary of these domains has not necessarily to correspond to the event of yielding or fracture of any component, but can be merely formed by stress states which are considered to be critical for some reason. As a consequence, the domain defined by eqn(1) contains *a priori* only 'potentially admissible macroscopic stresses', in the sense defined by Salençon [19]. This means that, if $\underline{\underline{\Sigma}} \notin G^{hom}$, $\underline{\underline{\Sigma}}$ cannot be associated with any microscopic stress field being statically admissible and, at the same time, compatible with the strength properties of fibers and matrix. On the other hand, if components are elastic–perfectly plastic and with associated flow rule and if G_m, G_f are their elastic domains, G^{hom} is actually formed by macroscopic stresses compatible with the strength properties of the composite as a whole. For further details, see Suquet [17].

Convexity allows the definition of G^{hom} in a form, dual to that of eqn(1), which will be used later. Let π^{hom} be the *support function* of G^{hom}([1]). Since convex domains coincide with the intersection of the subspaces defined by means of their own support functions, G^{hom} can be alternatively defined as

$$G^{hom} = \{\underline{\underline{\Sigma}} | \underline{\underline{\Sigma}} : \underline{\underline{D}} \leq \pi^{hom}(\underline{\underline{D}}) \; \forall \underline{\underline{D}} \in R^6\}. \tag{2}$$

The definition of π^{hom} was given by Suquet [17], to which readers are referred for further details. Let $\underline{v}$ be a piecewise differentiable velocity field over the RVE([2]). If $\underline{v}$ can be expressed as $\underline{v} = \underline{\underline{D}} \cdot \underline{x} + \underline{u}$, where $\underline{\underline{D}}$ is any symmetric constant second order tensor and $\underline{u}$ has the same kind of periodicity as the considered medium([3]), it is possible to show that the support function of G^{hom} is

$$\pi^{hom}(\underline{\underline{D}}) = \inf_{\underline{v}} \{ < \pi(\underline{\underline{d}}) > \; | \; \underline{\underline{D}} = < \underline{\underline{d}} >; \; \underline{\underline{d}} = \text{grad}^s \underline{v} \} \tag{3}$$

(superscript ^{s} denotes symmetric part of a rank two tensor), where $\underline{\underline{d}}$ is the strain rate associated to $\underline{v}$ at the points of $\mathcal{A}$ where this field is differentiable and $\pi(\underline{\underline{d}})$ is the support function of the strength domain of the material forming the RVE at any point $\underline{x}$. Because of the assumed periodicity, in eqn(3) $< \underline{\underline{d}} >$ can be computed as

$$< \underline{\underline{d}} > = \int_{\partial \mathcal{A}} \underline{v} \overset{s}{\otimes} \underline{n} \, dS,$$

where $\overset{s}{\otimes}$ denotes dyadic product and $\underline{n}$ is the outward unit normal to the boundary of $\mathcal{A}$. The volume average of $\pi(\underline{\underline{d}})$ is given by (see e.g. Salençon, [19])

$$< \pi(\underline{\underline{d}}) > = \int_{\mathcal{A}} \pi(\underline{\underline{d}}) d\mathcal{A} + \int_{S_v} \pi(\underline{n}_v; [\![\underline{v}]\!]) dS \tag{3'}$$

S_v being any possible discontinuity surface for $\underline{v}$, $\underline{n}_v$ the normal to this surface and $[\![\underline{v}]\!]$ the jump in $\underline{v}$ across this surface. The term computed on S_v is given by

$$\pi(\underline{n}_v; [\![\underline{v}]\!]) = \sup_{\underline{\underline{\sigma}}} \{ \underline{\underline{\sigma}} \cdot \underline{n}_v \cdot [\![\underline{v}]\!], \; \underline{\underline{\sigma}} \in G(\underline{x}) \}.$$

([1]) As for the definition of support function of a convex domain, refer e.g to Tyrrell Rockafellar [20])

([2]) The definition given by Suquet actually involves velocity fields more general than those used here.

([3]) i.e. $\underline{u}$ is $\mathcal{A}$–periodic.

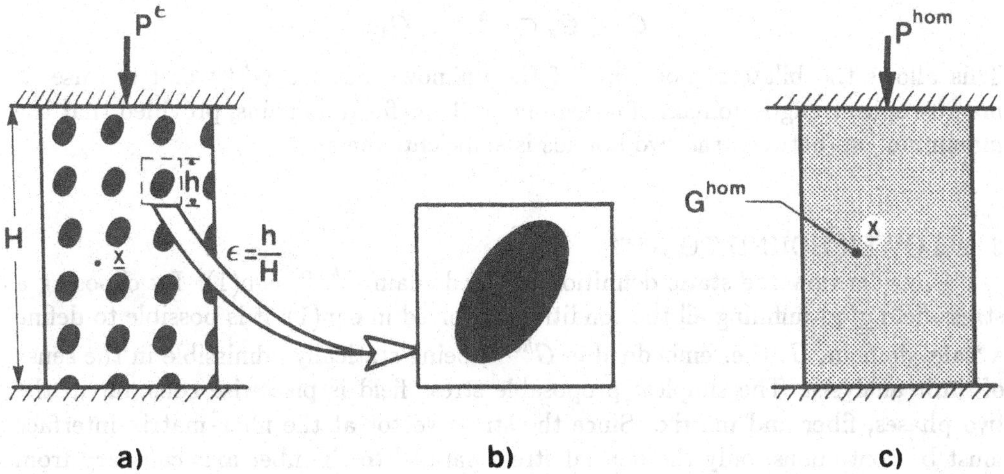

Fig. 2: (a) Generic periodic heterogeneous structure; (b) Representative Volume Element (RVE); (c) homogenized structure.

At this stage, it is worth emphasizing that generally G^{hom} is known only through its definition, eqn(1) or (2), but not explicitly.

Once the macroscopic failure criterion for the periodic material has been formulated, another important result established by Suquet [18] allows the correlation of the limit loads of a given heterogeneous structure, made of periodic material, and the limit load of a homogenized structure, supposed to be formed by a homogeneous material with G^{hom} as strength domain at each point (see Figs. 2a and 2c). These limit loads will be denoted by P^{ϵ} and P^{hom}, respectively. Here, the superscript ϵ denotes the ratio between two typical dimensions of the RVE and the entire structure (see Fig. 2a). Suquet's result reads:

$$(P^0 =) \lim_{\epsilon \to 0} P^{\epsilon} = P^{hom}.$$

This means that the limit load of the real periodic structure tends to that given by homogenization theory when inclusions tend to become indefinitely small if compared to the whole structure. Since this is always the case for structures made of fiber composites, the use of the macroscopic strength domain, G^{hom}, defined in the way recalled here, reveals effective in predicting the bearing capacity of these structures.

In the next two sections, two domains constituting a lower and an upper bound to G^{hom} will be defined. The lower bound is derived by means of a limit analysis static approach and is denoted by G_s. Furtherly, a simplified version of this domain is proposed (G_0) and it is shown to be a lower bound stricter than G_s for any fiber volume fraction. The upper bound is obtained by means of a limit analysis kinematic approach and is denoted by G_1. These domains are related one to each other and to the actual macroscopic strength domain, G^{hom}, through the set of inequalities

$$G_0 \subset G_s \subset G^{hom} \subset G_1.$$

This allows the bilateral bounding of the unknown domain G^{hom} and the use as macroscopic strength domain of either one of the defined domains, provided that the maximum gap between the two bounds is sufficiently small.

3. A LOWER BOUND TO G^{hom}

Consider now the static definition of the domain G^{hom}, eqn(1). By choosing a stress field $\underline{\underline{\sigma}}(\underline{x})$ fulfilling all the conditions involved in eqn(1), it is possible to define a 'safe' domain, G_s (i.e. embedded in G^{hom}) being statically admissible in the sense of limit analysis. The simplest proposable stress field is piecewise constant in the two phases, fiber and matrix. Since the stress vector at the fiber–matrix interface must be continuous, only the normal stress parallel to the fiber axis can vary from one phase to the other one. Hence, such stress field can be defined as:

$$\underline{\underline{\sigma}}(\underline{x}) = \begin{cases} \underline{\underline{\sigma}}_m & \forall \underline{x} \in \mathcal{A}_m; \\ \underline{\underline{\sigma}}_m + \sigma_f \underline{e}_x \otimes \underline{e}_x & \forall \underline{x} \in \mathcal{A}_f. \end{cases} \tag{4}$$

Assuming perfect bond between fiber and matrix, any stress field in the RVE is statically admissible if it is compatible with the strength criteria of fibers and matrix, whilst no interface criterion has to be fulfilled. The assumption of perfect bonding may turn out to be unrealistic in some cases and will be removed in Sec. 7; see also de Buhan & Taliercio [11]. On the basis of the stress field (4), the definition of the lower bound G_s is

$$G_s = \{\underline{\underline{\Sigma}} = \underline{\underline{\sigma}}_m + \eta\sigma_f \underline{e}_x \otimes \underline{e}_x;\ \underline{\underline{\sigma}}_m \in G_m;\ \underline{\underline{\sigma}}_m + \sigma_f \underline{e}_x \otimes \underline{e}_x \in G_f\}. \tag{5}$$

The domain defined by eqn(5) depends on the fiber volume fraction. However, it is possible to introduce a further domain (G_0) which, under some assumptions, can be proved to be embedded in G_s whatever the fiber volume fraction be. Let

$$G_0 = \{\underline{\underline{\Sigma}} = \underline{\underline{\sigma}}_m + \sigma \underline{e}_x \otimes \underline{e}_x;\ \underline{\underline{\sigma}}_m \in G_m; \quad -\bar{\sigma}^- \leq \sigma \leq \bar{\sigma}^+\} \tag{6}$$

where the lower and upper bound to parameter σ are $\bar{\sigma}^{\pm} = \eta(\sigma_f^{\pm} - \sigma_m^{\pm})$, $\sigma_m^{\pm}$ and $\sigma_f^{\pm}$ being the uniaxial tensile and compressive strengths along x of the matrix and the fibers, respectively. As for the physical meaning of G_0, refer to [11].

Without going into mathematical details, we just mention that the inequality

$$G_0 \subset G_s \tag{7}$$

holds, provided that the strength domains of matrix and fibers fulfil certain conditions, covering the case where the fiber strength domain is a homothetic expansion of G_m (as in the case of components undergoing Von Mises or Tresca criterion). The proof of eqn(7) was obtained by showing that $\pi_0(\underline{\underline{D}}) \leq \pi_s(\underline{\underline{D}})\ \forall \underline{\underline{D}} \in R^6$, π_0 and π_s

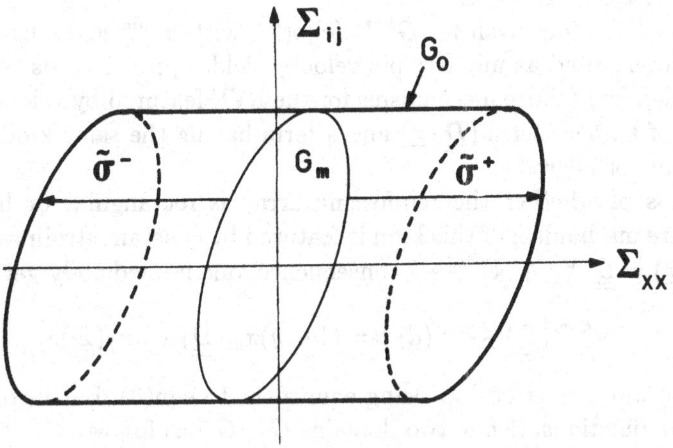

Fig. 3: Graphic construction of the simplified lower bound G_0 (see eqn(6)).

support functions of G_0 and G_s respectively, and was given by Taliercio [12] to which readers are referred for further details.

Eqn(7) makes G_0 more interesting to be used in comparison with G_s; in fact, in the definition of G_s, eqn(5), the *entire* strength domain of the fibers is involved, whereas only the *uniaxial* strength of the fibers is required to define G_0 (eqn(6)).

It is worth noting that in the definition of G_0 the possibility of buckling failure for the compressed fibers can be easily accounted for. In fact, it is sufficient to reduce the fiber compressive strength by a suitable coefficient $\mu \in [0,1]$ and to use $-\mu\bar{\sigma}^-$ as lower bound to σ.

The domain G_0 lends itself to a simple graphic construction. In fact, eqn(6) shows that the surface of G_0 is the convex envelope of the set of domains obtained by translating along the Σ_{xx} axis (in the stress space) the matrix strength domain between $-\bar{\sigma}^-$ and $\bar{\sigma}^+$. This graphic construction is shown in Fig. 3. This figure clearly shows how the presence of fibers enhances the strength properties of the composite in comparison with the matrix ones. This strengthening effect is particularly marked if the composite is subjected to almost uniaxial stress along the fiber direction.

REMARKS:

i) The definition of the strength domain G_0 coincides with that given by Mc Laughlin [9]. However, this author formulated his strength criterion for composites with infinitely thin fibers and did not operate within the framework of homogenization theory, which actually shows that this is only a lower bound to the actual macroscopic strength domain of a composite with any fiber volume fraction.

ii) The definition of G_0 can be rather easily extended to composites reinforced by N arrays of parallel fibers with different orientations (see [12]).

iii) No role is played by the geometric angle β, featuring the fiber array, in the definition of the lower bounds G_s and G_0.

4. AN UPPER BOUND TO G^{hom}

The dual definition given for G^{hom} (eqn(2), with π^{hom} given by eqn(3)) shows that, if $\underline{v}$ is interpreted as microscopic velocity field, upper bounds to G^{hom} can be obtained by defining failure mechanisms for the RVE featured by velocity fields being combination of a linear term ($\underline{\underline{D}} \cdot \underline{x}$) and a term having the same kind of periodicity as the medium considered ($\underline{u}$).

Regardless of whether the reinforcing array is rectangular or hexagonal, the simplest failure mechanism of this kind is featured by constant strain rate throughout the RVE ($\underline{\underline{d}}(\underline{x}) = \underline{\underline{D}}\ \ \forall \underline{x} \in \mathcal{A}$). As a consequence, one immediately gets (see eqn(3))

$$\pi^{hom}(\underline{\underline{D}}) \leq <\pi(\underline{\underline{d}})>= (1-\eta)\pi_m(\underline{\underline{D}}) + \eta\pi_f(\underline{\underline{D}}). \tag{8}$$

In order to obtain a relationship being equivalent to eqn(8), but involving domains and not scalar functions, define two domains G_f^+, G_f^- as follows:

$$\begin{aligned} G_f^+ =&\{\underline{\underline{\Sigma}}|\underline{\underline{\Sigma}} : \underline{\underline{D}} \leq \rho^+\pi_f(\underline{\underline{D}})\ \forall \underline{\underline{D}}|D^{xx} \geq 0; \\ &\underline{\underline{\Sigma}} : \underline{\underline{D}} \leq \rho^+\pi_f(\underline{\underline{D}} - D^{xx}\underline{e}_x \otimes \underline{e}_x)\ \forall \underline{\underline{D}}|D^{xx} \leq 0\}, \\ G_f^- =&\{\underline{\underline{\Sigma}}|\underline{\underline{\Sigma}} : \underline{\underline{D}} \leq \rho^-\pi_f(\underline{\underline{D}})\ \forall \underline{\underline{D}}|D^{xx} \leq 0; \\ &\underline{\underline{\Sigma}} : \underline{\underline{D}} \leq \rho^-\pi_f(\underline{\underline{D}} - D^{xx}\underline{e}_x \otimes \underline{e}_x)\ \forall \underline{\underline{D}}|D^{xx} \geq 0\} \end{aligned}$$

where $\rho^\pm = \eta + (1-\eta)/r^\pm$, ($\rho^\pm \leq 1$), with $r^\pm = \sigma_f^\pm/\sigma_m^\pm$ (≥ 1). For details about the geometrical meaning of these domains, readers are referred to [12]. It can be shown that, if the strength domains of fiber and matrix fulfil the same conditions that make eqn(7) hold true, the scalar inequality eqn(8) is equivalent to the inclusion

$$G^{hom} \subset (G_f^+ \cap G_f^-). \tag{8'}$$

In order to improve the upper bound just defined, periodic failure mechanisms featured by relative movement of rigid parts of RVE separated by slip planes will be considered. These planes are supposed not to cross any fiber, their intersection with RVE will be denoted by S_v and their normal by $\underline{n}_v$. Recalling eqn(3') and denoting by $\underline{V}(= [\![\ \underline{v}\ ]\!])$ the jump in the velocity field across the plane of failure, $<\pi(\underline{\underline{d}})>$ is then given by

$$<\pi(\underline{\underline{d}})>= |S_v|\pi_m(\underline{n}_v; \underline{V}).$$

For further developments, it is convenient to associate any of these mechanisms with a domain in the macroscopic stress space defined as follows:

$$G_m^*(\underline{n}_v) = \{\underline{\underline{\Sigma}}|\underline{\underline{\Sigma}} : \underline{\underline{D}} \leq \pi_m^*(\underline{\underline{D}})\},$$

where

$$\begin{aligned} \pi_m^*(\underline{\underline{D}}) &= \pi_m(\underline{n}_v; \underline{V}) \text{ if } \underline{\underline{D}} = \underline{V} \overset{s}{\otimes} \underline{n}_v; \\ &= +\infty \text{ otherwise.} \end{aligned}$$

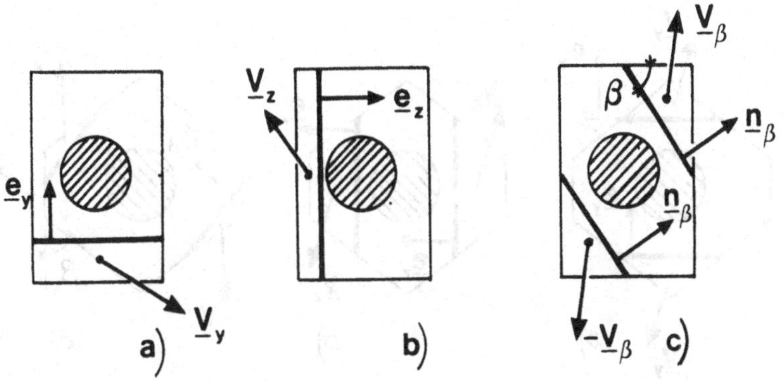

Fig. 4: Periodic failure mechanisms for rectangular RVEs, featured by relative rigid movement of blocks.

4.1 Rectangular array

If the fiber array is rectangular, a mechanism of the kind described above with failure plane perpendicular to $\underline{e}_y$ (or $\underline{e}_z$) can be defined for any percentage of reinforcement–see Figs. 4a,b. Let $\underline{V}_y$ (resp. $\underline{V}_z$) be the relative velocity between the two rigid blocks separated by the failure plane and let S_y (resp. S_z) be the trace of the failure plane in the (y, z) plane. Since $\underline{\underline{D}} = \underline{e}_y \overset{s}{\otimes} \underline{V}_y |S_y|$ (resp. $\underline{\underline{D}} = \underline{e}_z \overset{s}{\otimes} \underline{V}_z |S_z|$), by virtue of the definition of π^{hom}, eqn(3), the two following inequalities hold for any η ($\leq \bar{\eta}_r$):

$$\pi^{hom}(\underline{\underline{D}} = \underline{e}_y \overset{s}{\otimes} \underline{V}_y) \leq \pi_m(\underline{e}_y; \underline{V}_y); \qquad \pi^{hom}(\underline{\underline{D}} = \underline{e}_z \overset{s}{\otimes} \underline{V}_z) \leq \pi_m(\underline{e}_z; \underline{V}_z),$$

and amount at the inclusion between domains in the macroscopic stress space $G^{hom} \subset G^*_m(\underline{e}_y) \cap G^*_m(\underline{e}_z)$.

Periodic are also mechanisms featured by two parallel slip planes crossing each RVE, located at opposite sides with respect to the fiber and with traces in the (y, z) plane intersecting the edges of any RVE at their midpoints (see Fig. 4c). The unit vector perpendicular to these planes has the form $\underline{n}_\beta = \pm \underline{e}_y c_\beta \pm \underline{e}_z s_\beta$. The triangular wedge of RVE with $\underline{n}_\beta$ as inward normal is supposed to have relative velocity $\underline{V}_\beta$ with respect to the central block embedding the fiber, so that, by periodicity, the velocity of the triangular wedge with $\underline{n}_\beta$ as outward normal with respect to the central block must be $-\underline{V}_\beta$. If the failure planes have to cross only the matrix, the fiber volume fraction η must not exceed $\eta^*_r = \pi/4 s_\beta c_\beta$ (i.e. $\eta \leq \pi/8 = 0.39$ for square RVEs). If this restraint is fulfilled,

$$\pi^{hom}(\underline{\underline{D}} = \underline{V}_\beta \overset{s}{\otimes} \underline{n}_\beta) \leq \pi_m(\underline{V}_\beta; \underline{n}_\beta)$$

that is, in terms of domains $G^{hom} \subset G^*_m(\underline{n}_\beta)$.

As a conclusion, an upper bound to G^{hom} for composites with rectangular fiber arrays is given by the domain $G_1 = G^+_f \cap G^-_f \cap G^*_m$, where

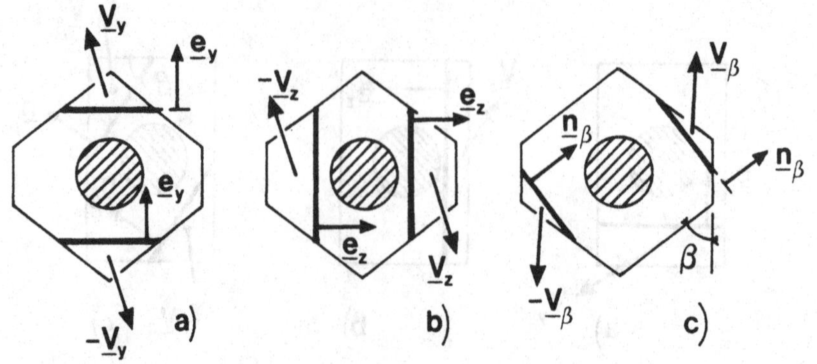

Fig. 5: Periodic failure mechanisms for hexagonal RVEs, featured by relative rigid movement of blocks.

$$G^*_m = G^*_m(\underline{e}_y) \cap G^*_m(\underline{e}_z) \cap G^*_m(\underline{n}_\beta) \text{ if } \eta \leq \eta^*_r \tag{9a}$$
$$= G^*_m(\underline{e}_y) \cap G^*_m(\underline{e}_z) \text{ otherwise.} \tag{9b}$$

4.2 Hexagonal array

For hexagonal arrays, periodic failure mechanisms are featured by two parallel slip planes per RVE, perpendicular to $\underline{e}_y$, $\underline{e}_z$ or $\underline{n}_\beta$, symmetric with respect to the fiber and fulfilling all the periodicity requirements (see Fig. 5). Let $\underline{n}_v$ denote the unit normal to the failure planes; these planes do not intersect any fiber if:

$\eta \leq \pi/8s_\beta/c_\beta$ (i.e. if $\eta \leq \pi\sqrt{3}/8 = 0.68$ for regular hexagonal arrays) for $\underline{n}_v = \underline{e}_y$;

$\eta \leq \pi/8c_\beta/s_\beta$ (i.e. if $\eta \leq \pi\sqrt{3}/24 = 0.23$ for regular hexagonal arrays) for $\underline{n}_v = \underline{e}_z$;

$\eta \leq \pi/2s_\beta c_\beta$ (i.e. if $\eta \leq 0.68$ for regular hexagonal arrays) for $\underline{n}_v = \underline{n}_\beta$.

After having defined domains $G^*_m(\underline{e}_y)$, $G^*_m(\underline{e}_z)$ and $G^*_m(\underline{n}_\beta)$, similar to those introduced in Sec. 4.1 for rectangular arrays, it is possible to state that $G^{hom} \subset G_1$, where G_1 can be expressed as $G_1 = G^+_f \cap G^-_f \cap G^*_m$ and

$$\begin{aligned}
G^*_m &= G^*_m(\underline{e}_y) \cap G^*_m(\underline{e}_z) \cap G^*_m(\underline{n}_\beta) \text{ if } \eta < \pi/8c_\beta/s_\beta \\
&= G^*_m(\underline{e}_y) \cap G^*_m(\underline{n}_\beta) \text{ if } \pi/8c_\beta/s_\beta < \eta < \min\{\pi/8s_\beta/c_\beta; \pi/2s_\beta c_\beta\} \\
&= G^*_m(\underline{n}_\beta) \text{ if } \pi/8c_\beta/s_\beta < \eta < \pi/8s_\beta/c_\beta \text{ and } \beta \leq 60° \\
&= G^*_m(\underline{e}_y) \text{ if } \pi/2s_\beta c_\beta < \eta < \min\{\pi/8s_\beta/c_\beta; \bar{\eta}_e\} \text{ and } \beta \geq 60° \\
&= R^6 \text{ otherwise.}
\end{aligned}$$

5. APPLICATIONS

The theoretical results presented up to here will be now applied to the approximate evaluation of the maximum solicitation that fiber reinforced composites subjected to particular states of stress can sustain. The material forming the matrix will be supposed to undergo Drucker–Prager strength criterion. This two–parameter criterion was chosen for the reasons discussed in Sec. 1 and because there are experimental results showing that the ultimate strength of some polymeric matrices is well described by Drucker–Prager criterion under some stress conditions (see e.g. the results reported by Hull [21]).

As is well known, denoting by k_m the shear strength of the matrix and by α_m a parameter related to the 'internal angle of friction' of the matrix, Drucker–Prager criterion reads (see e.g. Salençon [19]):

$$\alpha_m \,\mathrm{tr}\underline{\underline{\sigma}}_m + \sqrt{\frac{1}{2}\left(\underline{\underline{\sigma}}_m : \underline{\underline{\sigma}}_m - \frac{1}{3}\mathrm{tr}^2\underline{\underline{\sigma}}_m\right)} \leq k_m$$

and the relevant support function is

$$\begin{aligned}\pi_m(\underline{\underline{D}}) &= \frac{k_m}{3\alpha_m}\mathrm{tr}\underline{\underline{D}} \quad \text{if } \mathrm{tr}\underline{\underline{D}} \geq \frac{3}{2}\alpha_m\sqrt{(\underline{\underline{D}} : \underline{\underline{D}} - \frac{1}{3}\mathrm{tr}^2\underline{\underline{D}})} \\ &= +\infty \quad \text{otherwise.}\end{aligned} \tag{10}$$

The parameters α_m, k_m defining Drucker–Prager criterion are related to the uniaxial strength values of the matrix by the relationships

$$\alpha_m = \frac{1}{\sqrt{3}}\frac{\sigma_m^- - \sigma_m^+}{\sigma_m^- + \sigma_m^+}, \qquad k_m = \frac{2}{\sqrt{3}}\frac{\sigma_m^-\sigma_m^+}{\sigma_m^- + \sigma_m^+}$$

Only the case $\sigma_m^- \geq \sigma_m^+$ (i.e. $0 \leq \alpha_m \leq 1/\sqrt{3}$) will be treated, for its greater interest in practical applications.

When failure mechanisms with slip planes perpendicular to $\underline{n}_v$ are considered, setting $\underline{V} \overset{s}{\otimes} \underline{n}_v = \underline{\underline{D}}$ the condition $\mathrm{tr}\underline{\underline{D}} \geq \frac{3}{2}\alpha_m\sqrt{(\underline{\underline{D}} : \underline{\underline{D}} - \frac{1}{3}\mathrm{tr}^2\underline{\underline{D}})}$ can be easily shown to be equivalent to

$$\underline{V} \cdot \underline{n}_v \geq \frac{3\alpha_m}{\sqrt{1 - 3\alpha_m^2}}|\underline{V}|.$$

In other words, finite values of π_m (and, consequently, significant upper bounds to π^{hom}) are obtained if the relative velocity vector $\underline{V}$ between rigid blocks is inclined at an angle not exceeding $\arccos 3\alpha_m/\sqrt{1-3\alpha_m^2}$ to the slip plane. This implies $\alpha_m > 0.5/\sqrt{3}$; thus, for Drucker–Prager matrices with extremely high compressive–to–tensile strength ratios (i.e. $0.5/\sqrt{3} < \alpha_m \leq 1/\sqrt{3}$), no bound is obtained by this kind of mechanisms.

Drucker–Prager criterion reduces to Von Mises criterion if the matrix has equal strength in uniaxial compression and tension ($\sigma_m^+ = \sigma_m^- = \sigma_m = k_m\sqrt{3}$ and $\alpha_m = 0$). The support function of Von Mises strength criterion can be deduced from eqn(10) noting that $\alpha_m \to 0$ implies $\mathrm{tr}\underline{\underline{D}} \to 0$ if finite values for π_m (and consequently significant upper bounds to the macroscopic strength) are to be obtained. Thus (see also Salençon [19]):

$$\begin{aligned}\pi_m(\underline{\underline{D}}) &= \sigma_m\sqrt{2/3\underline{\underline{D}}:\underline{\underline{D}}} \text{ if } \mathrm{tr}\underline{\underline{D}}=0\\ &= +\infty \text{ if } \mathrm{tr}\underline{\underline{D}}\neq 0\end{aligned} \qquad (10')$$

Since $\underline{\underline{D}} = \underline{V} \overset{s}{\otimes} \underline{n}_v$, the condition $\mathrm{tr}\underline{\underline{D}} = 0$ is equivalent to $\underline{n}_v \perp \underline{V}$. In other words, for composites with Von Mises type matrix, the relative velocity between rigid blocks in movement has to be tangent to the failure plane.

Also the material forming the reinforcing fibers will be assumed to undergo a Drucker–Prager type criterion, defined by parameters α_f and k_f. For the sake of illustration, suppose that the ratios between the uniaxial strengths of fibers and matrix are such that $\sigma_f^+/\sigma_m^+ = \sigma_f^-/\sigma_m^- = r(\geq 1)$: in this case, the conditions under which the bilateral bounding of G^{hom} was obtained are fulfilled. α_f and k_f can be expressed in terms of α_m, k_m as

$$\alpha_f = \alpha_m; \quad k_f = rk_m(\, r \geq 1).$$

For composites with Drucker–Prager matrix, the general definition of the domain G_0, eqn(6), specializes to:

$$\begin{aligned}G_0 = \{\underline{\underline{\Sigma}} | \alpha_m(\mathrm{tr}\underline{\underline{\Sigma}} - \sigma) + \sqrt{\frac{1}{2}(\underline{\underline{\Sigma}}:\underline{\underline{\Sigma}} - \frac{1}{3}\mathrm{tr}^2\underline{\underline{\Sigma}}) + \sigma(\frac{1}{3}\mathrm{tr}\underline{\underline{\Sigma}} - \Sigma_{xx}) + \frac{1}{3}\sigma^2} - k_m \leq 0,\\ -\bar{\sigma}^- \leq \sigma \leq \bar{\sigma}^+\},\end{aligned} \qquad (11)$$

whereas the equations of the domains defined in Sec. 4 are

$$G_m^*(\underline{e}_y) = \{\underline{\underline{\Sigma}} | \frac{\alpha_m\Sigma_{yy} - k_m/3}{1/3 - 4\alpha_m^2} + \sqrt{12\alpha_m^2\left(\frac{\alpha_m\Sigma_{yy} - k_m/3}{1/3 - 4\alpha_m^2}\right)^2 + \Sigma_{xy}^2 + \Sigma_{yz}^2} \leq 0\}; \qquad (12a)$$

$$G_m^*(\underline{e}_z) = \{\underline{\underline{\Sigma}} | \frac{\alpha_m\Sigma_{zz} - k_m/3}{1/3 - 4\alpha_m^2} + \sqrt{12\alpha_m^2\left(\frac{\alpha_m\Sigma_{zz} - k_m/3}{1/3 - 4\alpha_m^2}\right)^2 + \Sigma_{yz}^2 + \Sigma_{zx}^2} \leq 0\}; \qquad (12b)$$

$$\begin{aligned}G_m^*(\underline{n}_\beta) = \{\underline{\underline{\Sigma}} | &\frac{\alpha_m\Sigma_{nn} - k_m/3}{1/3 - 4\alpha_m^2} + \sqrt{12\alpha_m^2\left(\frac{\alpha_m\Sigma_{nn} - k_m/3}{1/3 - 4\alpha_m^2}\right)^2 + \Sigma_{nt}^2 + \Sigma_{nx}^2} \leq 0;\\ &\Sigma_{nn} = \Sigma_{yy}c_\beta^2 \pm 2\Sigma_{yz}c_\beta s_\beta + \Sigma_{zz}s_\beta^2;\ \Sigma_{nx} = \pm\Sigma_{xy}c_\beta \pm \Sigma_{zx}s_\beta;\\ &\Sigma_{nt} = \pm(\Sigma_{zz} - \Sigma_{yy})c_\beta s_\beta + \Sigma_{yz}(c_\beta^2 - s_\beta^2)\};\end{aligned} \qquad (12c)$$

$$G_f^+ \cap G_f^- = \{\underline{\underline{\Sigma}} | \alpha_m\mathrm{tr}\underline{\underline{\Sigma}} + \sqrt{\frac{1}{2}(\underline{\underline{\Sigma}}:\underline{\underline{\Sigma}} - \frac{1}{3}\mathrm{tr}^2\underline{\underline{\Sigma}})} \leq [1 + \eta(r-1)]k_m\}. \qquad (12d)$$

5.1 Bilateral bounding of the macroscopic uniaxial strength

Suppose that the composite is subjected to uniaxial tension (resp. uniaxial compression) acting in the (x, y) plane. Let θ $(0 \leq \theta \leq 90°)$ be the orientation of the uniaxial stress, Σ, to the fibers. Set, for brevity, $s_\theta = \sin\theta$ and $c_\theta = \cos\theta$. The aim now is computing the functions $\Sigma_0^+(\theta)$ and $\Sigma_1^+(\theta)$ (resp. $\Sigma_0^-(\theta)$ and $\Sigma_1^-(\theta)$) constituting a lower and an upper bound, respectively, to the anisotropic macroscopic tensile strength $\Sigma^+(\theta)$ (resp. compressive strength $\Sigma^-(\theta)$) of the composite.

The simplified lower bounds to Σ^+ and Σ^- are given by eqn(11) and read:

$$\Sigma_0^\pm(\theta) = \sup_\sigma\{\Sigma | \alpha_m(\Sigma - \sigma) + \sqrt{\frac{1}{3}(\Sigma - \sigma)^2 + \Sigma\sigma s_\theta^2} - k_m \leq 0; -\bar{\sigma}^- \leq \sigma \leq \bar{\sigma}^+\}$$

that is

$$\Sigma_0^\pm(\theta) = \pm\bar{\sigma}^+ + \frac{\sqrt{3}}{1-3\alpha_m^2}\sqrt{\left(\frac{3}{2}\alpha_m\bar{\sigma}^+ s_\theta^2 + k_m\right)^2 - (\bar{\sigma}^+)^2(1-\frac{3}{4}s_\theta^2)(1-3\alpha_m^2)s_\theta^2}$$
$$\mp 3\frac{\frac{1}{2}\bar{\sigma}^+ s_\theta^2 + \alpha_m k_m}{1-3\alpha_m^2} \quad \text{if } \sigma \geq \bar{\sigma}^+ \tag{13a}$$
$$= \left[\frac{\mp 3/2\alpha_m + \sqrt{(1-3/4s_\theta^2)(1-3\alpha_m^2)}/s_\theta}{(1-3\alpha_m^2-3/4s_\theta^2)}\right]k_m \quad \text{if } -\bar{\sigma} \leq \sigma \leq \bar{\sigma}^+ \tag{13b}$$
$$= \mp\bar{\sigma}^- + \frac{\sqrt{3}}{1-3\alpha_m^2}\sqrt{\left(\frac{3}{2}\alpha_m\bar{\sigma}^- s_\theta^2 + k_m\right)^2 - (\bar{\sigma}^-)^2(1-\frac{3}{4}s_\theta^2)(1-3\alpha_m^2)s_\theta^2}$$
$$\mp 3\frac{-\frac{1}{2}\bar{\sigma}^- s_\theta^2 + \alpha_m k_m}{1-3\alpha_m^2} \quad \text{if } \sigma \leq -\bar{\sigma}^- \tag{13c}$$

where

$$\frac{\sigma}{k_m} = \frac{\pm 3/2\alpha_m}{1-3\alpha_m^2-3/4s_\theta^2} + \frac{\sqrt{1-3/4s_\theta^2}}{\sqrt{1-3\alpha_m^2}s_\theta}\frac{1-3\alpha_m^2-3/2s_\theta^2}{1-3\alpha_m^2-3/4s_\theta^2}.$$

As for the upper bound, note first of all that $G_m^*(\underline{e}_z)$ does not impose any bound to the uniaxial strength in the (x, y) plane, being $\Sigma_{zz} = \Sigma_{yz} = \Sigma_{zx} = 0$. Setting

$$\Sigma_y^\pm = \frac{k_m}{(\sqrt{1-12\alpha_m^2}c_\theta \pm 3\alpha_m s_\theta)s_\theta};$$

$$\Sigma_\beta^\pm = \frac{k_m}{(\sqrt{1-12\alpha_m^2}\sqrt{1-s_\theta^2 c_\beta^2} \pm 3\alpha_m s_\theta c_\beta)s_\theta c_\beta}; \quad \Sigma_f^\pm = k_m\frac{1+\eta(r-1)}{\pm\alpha_m + 1/\sqrt{3}},$$

the remaing domains, eqns(12a,c,d), yield:

- for rectangular RVEs with $\eta < \pi/4s_\beta c_\beta$ and for hexagonal RVEs with $\eta < \min\{\pi/8s_\beta/c_\beta;\ \pi/2s_\beta c_\beta\}$:

$$\Sigma_1^\pm(\theta) = \min\{\Sigma_y^\pm; \Sigma_\beta^\pm; \Sigma_f^\pm\}; \tag{14a}$$

– for rectangular RVEs with $\eta > \pi/4s_\beta c_\beta$ and for hexagonal RVEs with $\pi/2s_\beta c_\beta < \eta < \min\{\pi/8s_\beta/c_\beta; \bar{\eta}_e\}$ and $\beta \geq 60°$:

$$\Sigma_1^{\pm}(\theta) = \min\{\Sigma_y^{\pm}; \Sigma_f^{\pm}\}; \qquad (14b)$$

– for hexagonal RVEs with $\pi/8s_\beta/c_\beta < \eta < \pi/2s_\beta c_\beta$ and $\beta \leq 60°$:

$$\Sigma_1^{\pm}(\theta) = \min\{\Sigma_\beta^{\pm}; \Sigma_f^{\pm}\}; \qquad (14c)$$

– for hexagonal RVEs with fiber volume fraction different from those specified above, it is only possible to state that:

$$\Sigma_1^{\pm}(\theta) = \Sigma_f^{\pm}. \qquad (14d)$$

In Fig. 6 plots of $\Sigma_0^{\pm}$ and $\Sigma_1^{\pm}$ versus θ for composites with Drucker–Prager type fibers and matrix (with $r = 5$) and regular hexagonal reinforcing array are shown. The gap between the two bounds is appearently small, apart from the cases with greater α_m values. However, note that in Fig. 6 the fiber volume fraction was implicitely assumed to be sufficiently lower than the maximum one compatible with each kind of reinforcing array, so that the strictest upper bound was used (see eqn(14a)). The gap between bounds may tend to increase at some θ for larger volume fractions.

Anyway, for any α_m and whatever the geometry (β, η) of the composite be, the lower and the upper bound coincide at $\theta = 0$, so that

$$\Sigma_0^{\pm}(0) = \Sigma_1^{\pm}(0) = k_m \frac{1 + \eta(r-1)}{\pm\alpha_m + 1/\sqrt{3}} (= \Sigma^{\pm}(0)) \qquad (15)$$

is the actual macroscopic strength of the composite along the fiber direction. Note that $\Sigma^{\pm}(0) = (1-\eta)\sigma_m^{\pm} + \eta\sigma_f^{\pm}$ is appearently the weighted average of the uniaxial strengths of fiber and matrix, the weights being the relevant volume fractions. Thus, eqn(15) is a rigourous validation of a well known semi–empirical formula widely used in practice, usually called 'rule of mixtures' (see e.g. Hashin [15]).

5.2 Bilateral bounding of the macroscopic biaxial strength

If the strength properties of the composite have to be estimated under stress conditions not as simple as the uniaxial ones considered in Sec. 5.1, the solution of the problem is likely to be obtained numerically. As a rule, this can be done by choosing a radial path in the stress space and by determining the maximum norm of the macroscopic stress tensor compatible with G_0 or G_1. The lower and upper bound to the macroscopic strength domains for the state of stress considered are thus derived point by point.

For the sake of illustration, the lower and upper bounds to the biaxial strength of composites with Drucker–Prager type matrix will be now derived. Denote by Σ_I,

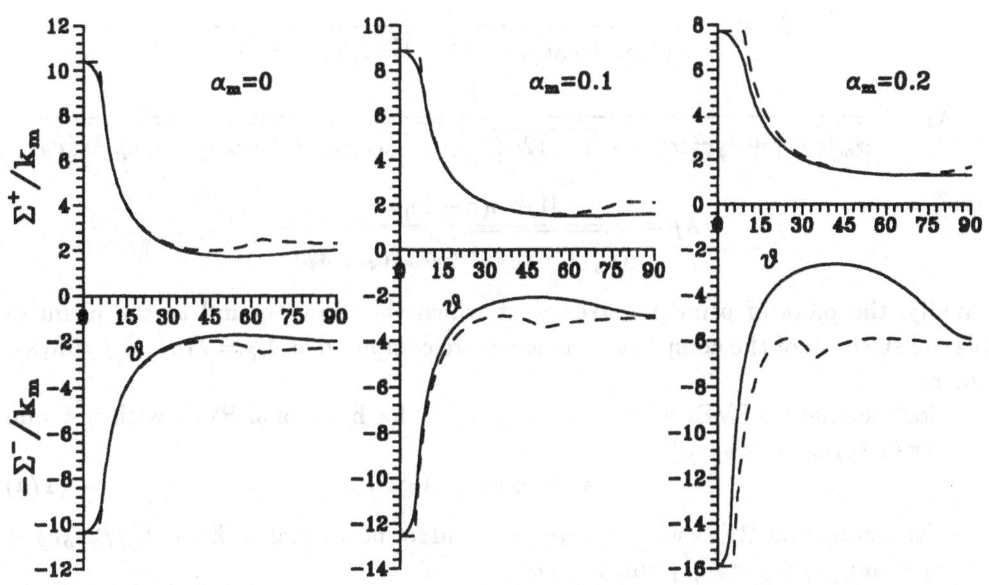

Fig. 6: Bounds to the uniaxial tensile (Σ^+) and compressive (Σ^-) strength for composites with regular hexagonal reinforcing array, $\eta < 0.68$, $\sigma_f^+/\sigma_m^+ = \sigma_f^-/\sigma_m^- = 5$ and Von Mises ($\alpha_m = 0$) or Drucker–Prager ($\alpha_m = 0.1, 0.2$) matrix. Solid lines are lower bounds; dashed lines are upper bounds.

Σ_{II} the two nonvanishing principal stresses and by θ the orientation of Σ_I to the fibers. Suppose that the principal stresses act in the (x, y) plane.

Let λ be a positive parameter and let ϕ be an angle varying between 0° and 360° in the plane (Σ_I, Σ_{II}). Setting $\Sigma_I = \lambda \cos\phi (= \lambda c_\phi)$, $\Sigma_{II} = \lambda \sin\phi (= \lambda s_\phi)$, for any prescribed orientation θ of the principal stresses to the fibers, the lower bound to the biaxial strength is obtained by computing the maximum value of $\lambda(\phi)$ compatible with the definition of G_0, eqn(11), with $\Sigma^{xx} = \Sigma_I c_\theta^2 + \Sigma_{II} s_\theta^2$, which is given by:

$$\lambda_0(\phi) = \sup_{\sigma}\{\lambda \mid \frac{1}{\sqrt{3}}\sqrt{\lambda^2(1 - c_\phi s_\phi) - \frac{1}{2}\lambda\sigma[c_\phi + s_\phi + 3(c_\phi - s_\phi)(c_\theta^2 - s_\theta^2)] + \sigma^2} + \lambda\alpha_m(c_\phi + s_\phi) - \alpha_m\sigma - k_m \leq 0;\ -\bar{\sigma}^- \leq \sigma \leq \bar{\sigma}^+\}. \tag{16}$$

Numerical solution of problem (16) yields the lower bound sought.

The upper bound is formed by the intersection of the domains defined by eqns (12). In the present case $G_m^*(\underline{e}_z)$ does not furnish any significant bound, since $\Sigma_{zx} = \Sigma_{yz} = \Sigma_{zz} = 0$. Following the same procedure as for the lower bound, for any prescribed orientation θ three values for the parameter $\lambda(\geq 0)$, denoted by λ_y, λ_β and λ_f, are obtained by finding the maximum value of λ such that Σ_I and Σ_{II} are compatible with $G_m^*(\underline{e}_y)$, $G_m^*(\underline{n}_\beta)$ and $G_f^+ \cap G_f^-$, respectively. In this case, unconstrained maximization problems have to be solved, yielding:

$$\lambda_y = \frac{k_m}{3\alpha_m(c_\phi s_\theta^2 + s_\phi c_\theta^2) + \sqrt{1-12\alpha_m^2}|c_\phi - s_\phi| s_\theta c_\theta}$$

$$\lambda_\beta = \frac{k_m}{3\alpha_m(c_\phi s_\theta^2 + s_\phi c_\theta^2)c_\beta^2 + \sqrt{1-12\alpha_m^2}\sqrt{(c_\phi - s_\phi)^2 c_\theta^2 s_\theta^2 + (c_\phi s_\theta^2 + s_\phi c_\theta^2)^2 s_\beta^2} c_\beta}$$

$$\lambda_f = \frac{[1+\eta(r-1)]k_m}{\sqrt{\frac{1}{3}(1 - c_\phi s_\phi)} + \alpha_m(c_\phi + s_\phi)}$$

Finally, the pairs of principal stresses at which the upper bound to the ultimate biaxial strength of the composite is reached are computed as $\Sigma_I = \lambda_1 c_\phi$, $\Sigma_{II} = \lambda_1 s_\phi$, where

- for rectangular RVEs with $\eta < \pi/4 s_\beta c_\beta$ and for hexagonal RVEs with $\eta <$ min $\{\pi/8\ s_\beta/c_\beta;\ \pi/2\ s_\beta c_\beta\}$:

$$\lambda_1 = \min\{\lambda_y; \lambda_\beta; \lambda_f\}; \tag{17a}$$

- for rectangular RVEs with $\pi/4 s_\beta c_\beta < \eta$ and for hexagonal RVEs with $\pi/2 s_\beta c_\beta < \eta < \min\{\pi/8 s_\beta/c_\beta; \bar{\eta}_e\}$ and $\beta \geq 60°$:

$$\lambda_1 = \min\{\lambda_y; \lambda_f\}; \tag{17b}$$

- for hexagonal RVEs with $\pi/8 s_\beta/c_\beta < \eta < \pi/2 s_\beta c_\beta$ and $\beta \leq 60°$:

$$\lambda_1 = \min\{\lambda_\beta; \lambda_f\}; \tag{17c}$$

- for other types of hexagonal arrays:

$$\lambda_1 = \lambda_f. \tag{17d}$$

Fig. 7 shows the bounds computed for composites with regular hexagonal RVEs ($\beta = 60°$) according to the procedure just described, at three different orientations (i.e. $\theta = 0°, 22.5°$ and $45°$). As a rule, the two bounds are very close. Again note that the strictest upper bound (computed according to eqn(17a)) was used, so that at greater volume fractions a larger gap might be obtained.

6. DETERMINATION OF THE MODEL PARAMETERS. COMPARISON BETWEEN THEORETICAL PREDICTIONS AND EXPERIMENTAL RESULTS

In order to check the reliability of the theoretical model presented here, it would be advisable to have strength data for fibers and matrix separately available. The bounds to the macroscopic strength criterion could then be directly computed through the presented equations and compared with the results of experimental tests on composite samples. Actually, the authors of experimental tests on composites only seldom report the strength data of fibers and matrix separately; hence, the strength

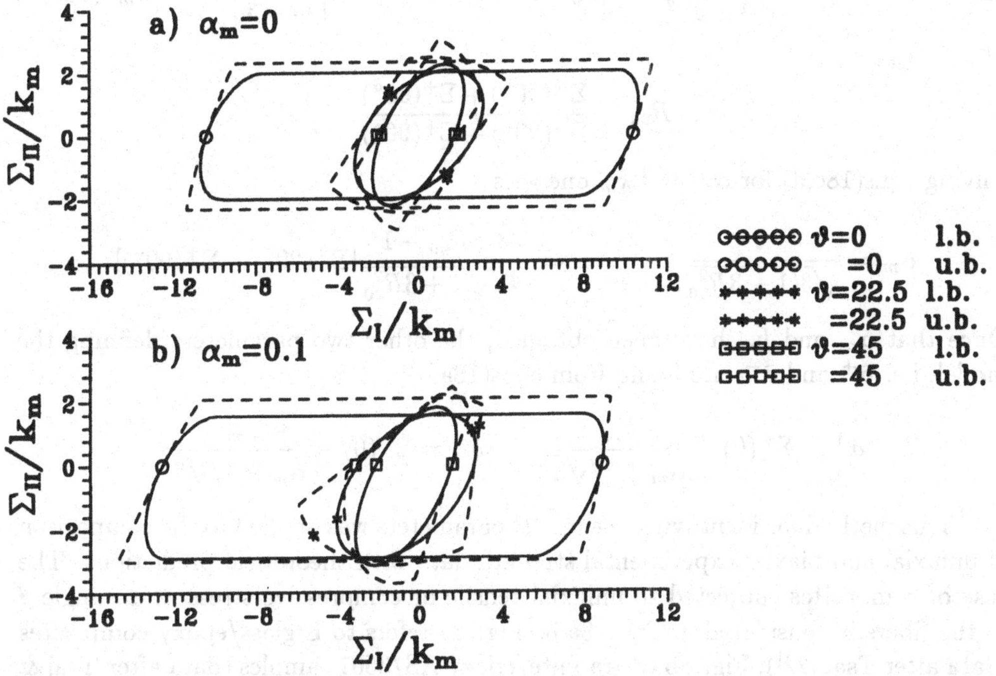

Fig. 7: Lower and upper bounds to the biaxial strength domains of composites with hexagonal reinforcing array (with $\beta = 60°$ and $\eta < 0.68$), at different orientations θ of Σ_I to the fibers. (a) Von Mises matrix; (b) Drucker–Prager matrix.

parameters to be employed in the model have to be determined on the basis of results relative to the entire composite, after having made some reasonable assumption about the behaviour of the component materials. We briefly examine here how these strength parameters can be deduced on the basis of simple uniaxial tests.

Since in most cases lower and upper bounds have turned out to be very close one to each other, either one of the bounds can be used as strength criterion. For convenience, only the simplified lower bound G_0 will be considered.

Equating the uniaxial tensile and compressive strength values yielded by the present model along the fiber direction (eqn (15)) to the experimental values $\Sigma^{\pm}(0)$, one gets

$$\Sigma^+(0) = k_m \frac{1+\eta(r-1)}{\alpha_m + 1/\sqrt{3}} = \bar{\sigma}^+ + \frac{k_m}{\alpha_m + 1/\sqrt{3}}, \tag{18a}$$

$$\Sigma^-(0) = k_m \frac{1+\eta(r-1)}{-\alpha_m + 1/\sqrt{3}} = \bar{\sigma}^- + \frac{k_m}{-\alpha_m + 1/\sqrt{3}}. \tag{18b}$$

Similarly, the experimental uniaxial strength values measured transverse to the fibers, $\Sigma^{\pm}(90°)$, are related to the theoretical predictions (see eqn(13b)) by

$$\Sigma^+(90°) = \frac{1}{2}\left[\frac{\sqrt{1-3\alpha_m^2} - 3\alpha_m}{1/4 - 3\alpha_m^2}\right]k_m, \; \Sigma^-(90°) = \frac{1}{2}\left[\frac{\sqrt{1-3\alpha_m^2} + 3\alpha_m}{1/4 - 3\alpha_m^2}\right]k_m \quad (18c,d)$$

([4]). Setting

$$R_{90} = \frac{\Sigma^-(90°) + \Sigma^+(90°)}{\Sigma^-(90°) - \Sigma^+(90°)},$$

solving eqns(18c,d) for α_m and k_m one gets

$$\alpha_m = \frac{1}{\sqrt{3(1+3R_{90}^2)}}; \qquad k_m = \frac{\sqrt{3}}{4}\frac{R_{90}^2 - 1}{\sqrt{1+3R_{90}^2}}[\Sigma^-(90°) - \Sigma^+(90°)].$$

Once that α_m and k_m have been obtained, the other two parameters defining the model, i.e. $\bar{\sigma}^+$ and $\bar{\sigma}^-$, are found from eqns(18a,b):

$$\bar{\sigma}^+ = \Sigma^+(0) - \frac{k_m}{\alpha_m + 1/\sqrt{3}}; \qquad \bar{\sigma}^- = \Sigma^-(0) - \frac{k_m}{\alpha_m - 1/\sqrt{3}}.$$

This method for identifying the model parameters was applied to the comparison of uniaxial and biaxial experimental strength data with theoretical predictions. The case of composites subjected to uniaxial tensile or compressive stress at an angle θ to the fibers is considered in Figs. 8a,b: Fig. 8a refers to E-glass/epoxy composites (data after Tsai [22]), Fig. 8b to graphite/epoxy AS/3501 samples (data after Tsai & Hahn [23] - see Fig. 7.6 in that reference). Plots were obtained through eqns(13). In both examples, a satisfactory agreement between model and experiments is observed.

Consider now combined shear and normal stress tests, performed at different orientations of the normal stress to the fibers. In Fig. 9 experimental data obtained by Wu [5] on graphite/epoxy unidirectional composite samples are compared with theoretical predictions. In all the cases shown, the experimental points are well matched by the theoretical curves.

It is worth emphasizing that the quite satisfactory results presented here are given by a model requiring at most four strength parameters only for its complete definition and that these parameters can be obtained through simple uniaxial tests.

7. ACCOUNTING FOR AN INTERFACE STRENGTH CRITERION

Up to here, the fiber–matrix interface was assumed to be of unlimited strength. As a matter of fact, there are experimental results that show that interface is the failure surface for composites subjected to particular load conditions. In this section the possibility of improving the theoretical model presented accounting for an interface strength criterion will be described.

([4]) Note that it is assumed here that, at 90°, the parameter σ involved in eqns(13) does not attain any limit value, so that the macroscopic strength is given by eqn(13b) and not by either one of eqns(13a,c).

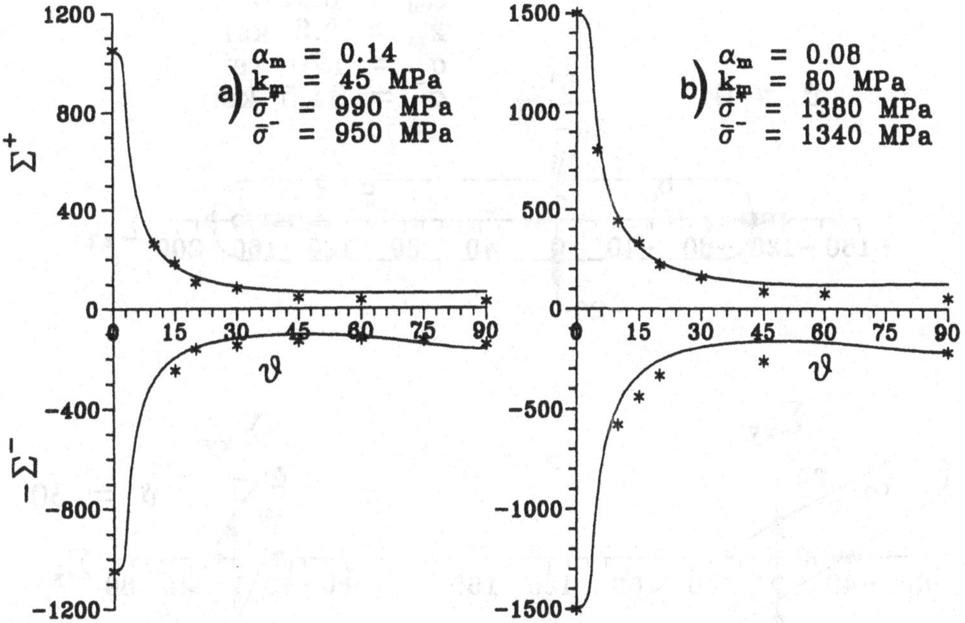

Fig. 8: Theoretical curves and experimental points for fiber reinforced composite samples submitted to uniaxial tests at different angles θ to the fibers. Data are in MPa units. Σ^+ = tensile strength; Σ^- = compressive strength. (a) E-glass/epoxy; (b) graphite/epoxy AS/3501.

First, by means of the static approach, the lower bounds obtained in Sec. 3 will be extended for composites with interface of limited strength (Sec. 7.1). Secondly, failure mechanisms for the RVE partially involving the fiber–matrix interface will be defined and upper bounds to the macroscopic strength will be obtained (Sec. 7.2). Note that the failure mechanisms used in Sec. 4 were defined by planes cutting the matrix material only, which could be done for composites with volume fraction not exceeding certain limit values. This restriction is removed considering failure surfaces partially involving the interface. Thus, the extension presented in this section allows obtaining a double advantage in terms of upper bounds: defining bounds capable of accounting for an interface strength criterion and valid for any percentage of reinforcement.

Theoretical results will be specialized to composites with Von Mises matrix and purely cohesive (Tresca) interface. Analytical equations of bounds to the uniaxial and pure shear strength of the composite will be derived (Sec. 7.3). The gap between bounds is small for composites with certain geometrical features at some orientations of the applied stress to the fibers, but it can turn out to be rather large in other cases. The reason of this event will be discussed and the fundamental role played by the strength criteria adopted will be outlined.

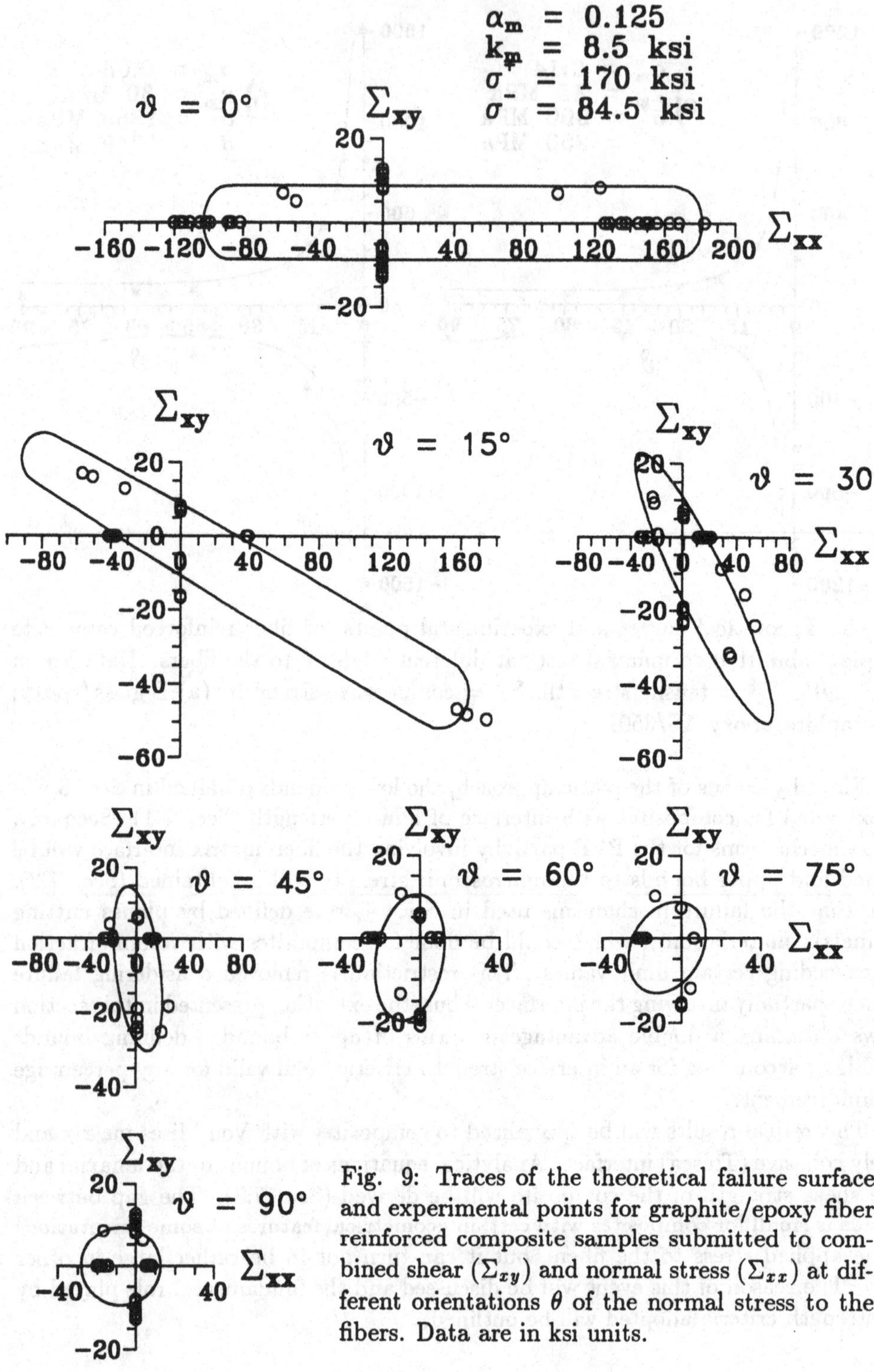

Fig. 9: Traces of the theoretical failure surface and experimental points for graphite/epoxy fiber reinforced composite samples submitted to combined shear (Σ_{xy}) and normal stress (Σ_{xx}) at different orientations θ of the normal stress to the fibers. Data are in ksi units.

It is worth recalling that an attempt to formulating strength criteria for composites accounting for the interface strength properties was also made by Aboudi (see e.g. [24]). This author assumes that slippage between fiber and matrix occurs when the interface stress reaches a limit value according to Coulomb's condition. Fibers are supposed to have square cross section. Explicit equations for the stress and displacement fields in fiber and matrix are obtained supposing these components to behave elastically up to a certain threshold. The approach used here is rather different, since the actual constitutive law of the components is disregarded and only their strength criteria are required for the definition of macroscopic criteria. This leads to a micromechanical approach quite simpler than Aboudi's one. Of course, the counterpart of this simplification is the lack of information about the actual stresses and displacements in the RVE.

7.1 Lower bound – Static approach.

Accounting for an interface criterion, the static definition given by Suquet [17,18] for the macroscopic strength domain of a fiber composite with periodic reinforcing array (eqn(1)) reads:

$$\begin{aligned}\{G^{hom} = & \underline{\underline{\Sigma}} | \underline{\underline{\Sigma}} = < \underline{\underline{\sigma}} >;\ \text{div}\underline{\underline{\sigma}} = \underline{0}\ \forall \underline{x} \in \mathcal{A};\ [\![\underline{\underline{\sigma}}]\!] \cdot \underline{n}_S = \underline{0}\ \forall \underline{x} \in S; \\ & \underline{\underline{\sigma}} \cdot \underline{n} \text{ anti-periodic over } \partial\mathcal{A};\ \underline{\underline{\sigma}} \in G_m\ \forall \underline{x} \in \mathcal{A}_m;\ \underline{\underline{\sigma}} \in G_f\ \forall \underline{x} \in \mathcal{A}_f; \\ & \underline{\underline{\sigma}}(\underline{x}) \cdot \underline{n}(\underline{x}) \in g_{int}\ \forall \underline{x} \in S_{int}\}\end{aligned} \qquad (19)$$

where S_{int} is the fiber–matrix interface and $g_{int}(\in R^3)$ its strength domain.

By making use of the piecewise constant stress field given by eqn(4), the following lower bound to G^{hom} is obtained:

$$G_s^{int} = G_s \cap G^{int} \qquad (20a)$$

where G_s was obtained in Sec. 3 in the case of perfect bond between fiber and matrix (eqn(5)) and

$$G^{int} = \{\underline{\underline{\Sigma}} |\ \underline{T}(\underline{x}) = \underline{\underline{\Sigma}} \cdot \underline{n}(\underline{x}) \in g_{int}\ \forall \underline{x} \in S_{int}\}. \qquad (20b)$$

Of course, a lower bound stricter than G_s^{int} is the domain

$$G_0^{int} = G_0 \cap G^{int}, \qquad (20c)$$

G_0 being the simplified lower bound for perfectly bonded fiber and matrix, defined by eqn(6). Note that in both lower bounds (20a) and (20c) only the domain G^{int} accounts for the interface strength properties, whose influence is uncoupled from those of fiber and matrix.

For metal composites, it is reasonable to assume that matrix fulfils Von Mises strength criterion (with k_m as pure shear strength); then, eqn(6) takes the form:

$$G_0 = \{\underline{\underline{\Sigma}} | \frac{1}{2}[\underline{\underline{\Sigma}} : \underline{\underline{\Sigma}} - \frac{1}{3}\mathrm{tr}^2\underline{\underline{\Sigma}} - 2\sigma(\Sigma_{xx} - \frac{1}{3}\mathrm{tr}\underline{\underline{\Sigma}}) + \frac{2}{3}\sigma^2] \leq k_m^2;\ -\bar{\sigma}^- \leq \sigma \leq \bar{\sigma}^+\}. \quad (5')$$

The interface strength criterion will be supposed to limit only the shear the stress acting at any point $\underline{x}$ of this surface, $\tau(\underline{x})$ (Tresca criterion). Denoting by k_{int} the interface shear strength, eqn(20b) reads:

$$G^{int} = \{\underline{\underline{\Sigma}} |\ |\tau(\underline{x})| \leq k_{int}\ \forall \underline{x} \in S_{int}\}.$$

Let (x, n, t) be a local orthogonal reference frame with origin at any point of the interface and with n perpendicular to the fiber. Denoting by ψ the angle between n and the $y-$axis of the global reference frame, one gets:

$$|\tau|^2 = (\Sigma^{nt})^2 + (\Sigma^{nx})^2 = \left[\frac{1}{2}(\Sigma_{zz} - \Sigma_{yy})\sin 2\psi + \Sigma_{yz}\cos 2\psi\right]^2 + \left[\Sigma_{xy}\cos\psi + \Sigma_{zx}\sin\psi\right]^2. \quad (21)$$

The condition in eqn(21) is fulfilled at any point of the interface if $\max_\psi |\tau| \leq k_{int}$. Setting $d|\tau|^2/d\psi = 0$, the following equation for ψ is obtained:

$$a_0 + a_1 \mathrm{tg}\psi + 6a_2 \mathrm{tg}^2\psi + a_3 \mathrm{tg}^3\psi + a_4 \mathrm{tg}^4\psi = 0, \quad (22)$$

where

$$a_0 = a_2 + \Sigma_{xy}\Sigma_{zx},\ a_1 = \Sigma_{zx}^2 - \Sigma_{xy}^2 + (\Sigma_{zz} - \Sigma_{yy})^2 - 4\Sigma_{yz}^2,\ a_2 = (\Sigma_{zz} - \Sigma_{yy})\Sigma_{yz},$$

$$a_3 = \Sigma_{zx}^2 - \Sigma_{xy}^2 - (\Sigma_{zz} - \Sigma_{yy})^2 + 4\Sigma_{yz}^2, \quad a_4 = a_2 - \Sigma_{xy}\Sigma_{zx}.$$

The maximum value of $|\tau|$ has to be sought amongst the values obtained by substitution in eqn(21) of the different solutions of eqn(22).

7.2 Upper bound – Kinematic approach.

For the sake of simplicity, only rectangular reinforcing arrays will be dealt with in this section.

First of all, note that the upper bound eqn(8'), obtained considering constant strain rate fields, holds regardless of whether the bond between fiber and matrix is perfect or not.

Then, consider failure mechanisms for the RVE featured by slip surfaces including part of the interface. Since fibers are here supposed to have circular cross section (with radius R), the trace of the curvilinear part of the failure surface will be a circular arc; let 2ψ be the angle subtended by this arc and set, for brevity, $s_\psi = \sin\psi$, $c_\psi = \cos\psi$. Denote by π^{int} the support function of the interface strength domain.

If the flat part of the failure surface is perpendicular to $\underline{e}_z$ (Fig. 10a), eqns(3),(3') yield:

$$\pi^{hom}(\underline{\underline{D}}) \leq <\pi(\underline{\underline{d}})> = (1 - \frac{2R}{d_y}s_\psi)\pi_m(\underline{V}_z \overset{s}{\otimes} \underline{e}_z) + \frac{1}{d_y}\int_{-\psi}^{+\psi} \pi^{int}(\underline{n}_v(\theta); \underline{V}_z) R d\theta \quad (23)$$

$\underline{V}_z$ being the relative velocity between blocks and $\underline{\underline{D}} = (\int_{\partial \mathcal{A}} \underline{V}_z \overset{s}{\otimes} \underline{n})/d_y = \underline{V}_z \overset{s}{\otimes} \underline{e}_z$. Since matrix and interface are respectively supposed to fulfil Von Matrix criterion and Tresca criterion, π_m is given by eqn(10') and π^{int} reads (see e.g. Salençon, [19])

$$\pi^{int}(\underline{n}_v; \underline{V}_z) = k_{int} \underline{V}_z \text{ if } \underline{V}_z \perp \underline{n}_v; \qquad = +\infty \quad \text{otherwise.}$$

Thus, the function in the integral of eqn(23) takes finite values for any $\theta \in [-\psi, +\psi]$ only if the relative velocity is parallel to the fiber at any point, so that it can be expressed as $\underline{V}_z = V_z^x \underline{e}_x$. In this case,

$$< \pi(\underline{\underline{d}}) >= \frac{1}{d_y} \left[(d_y - 2Rs_\psi) k_m |V_z^x| + 2k_{int} |V_z^x| R\psi \right].$$

The best upper bound to π^{hom} that can be obtained by this kind of failure mechanism is obtained by finding $\min_\psi < \pi(\underline{\underline{d}}) >$. This value is is attained at $\cos\psi = k_{int}/k_m \equiv \rho_k$ and is

$$\pi^{hom}(\underline{\underline{D}}) \leq \pi_m(\underline{\underline{D}}) \rho_z \ \forall \underline{\underline{D}} = V_z^x \underline{e}_x \overset{s}{\otimes} \underline{e}_z. \tag{24}$$

The ratio ρ_z is given by

$$\rho_z = 1 - \sqrt{\frac{\eta}{\bar{\eta}_r}} \text{cotg}\beta (\sqrt{1 - \rho_k^2} - \rho_k \arccos \rho_k)$$

and accounts for the fact that the composite strength properties are reduced with respect to the matrix ones by the presence of a weakening interface. In fact $\rho_z < 1$ $\forall k_{int} < k_m$ and $\rho_z = 1$ only if $k_m = k_{int}$.

Eqn(24) shows that, if matrix and interface are of equal strength, the failure surface does not include any part of interface ($\psi = 0$), since it reduces to a plane cutting the matrix only in order to be of the least possible extension. The weaker the interface is in comparison with the matrix, the larger the part of interface involved in the failure surface is. In the limit, if $k_{int} \to 0$, one halfth of the interface is part of the failure surface ($2\psi = \pi$).

The scalar inequality eqn(24) is equivalent to:

$$G^{hom} \subset \{ \underline{\underline{\Sigma}} | \ |\Sigma_{zx}| \leq k_m \rho_z \}. \tag{25a}$$

If the flat part of the failure surface is perpendicular to $\underline{e}_y$, similar considerations lead to the following inequality:

$$G^{hom} \subset \{ \underline{\underline{\Sigma}} | \ |\Sigma_{xy}| \leq k_m \rho_y \}, \tag{25b}$$

where

$$\rho_y = 1 - \sqrt{\frac{\eta}{\bar{\eta}_r}} (\sqrt{1 - \rho_k^2} - \rho_k \arccos \rho_k) \quad (\leq 1).$$

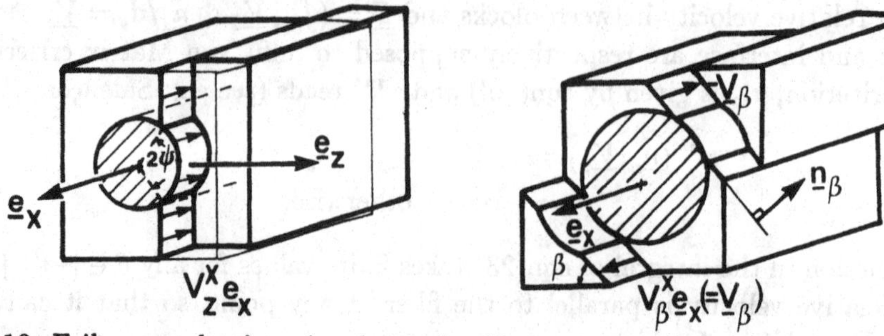

Fig. 10: Failure mechanisms involving the fiber–matrix interface.

Finally, consider failure mechanisms with the flat part of the failure surfaces is skew and perpendicular to $\underline{n}_\beta$ (Fig. 10b). Let $\underline{V}_\beta$, $-\underline{V}_\beta$ be the velocities of the outer blocks respect to the inner one. If the fiber volume fraction is smaller than η_r^* (see Sec. 4), the failure surfaces cut the matrix only and the case of Sec. 4.1 is recovered (see eqn(9a)). Thus, $\eta \geq \eta_r^*$ will be assumed. Also note that, unlike the previously treated case, here the amplitude of angle ψ is imposed by the composite geometry. Simple geometrical considerations yield $\cos\psi = \sqrt{\pi/8 \sin 2\beta/\eta} = \sqrt{\eta_r^*/\eta}$. Since $\underline{V}_\beta$ must be parallel to the fibers (for Von Mises matrices and purely cohesive interfaces), it can be expressed as $\underline{V}_\beta = V_\beta^x \underline{e}_x$. As a consequence, computation of $< \pi(\underline{\underline{d}}) >$ for the mechanism at hand leads to

$$\pi^{hom}(\underline{\underline{D}}) \leq < \pi(\underline{\underline{d}}) > = \pi_m(\underline{\underline{D}})\rho_\beta \qquad \forall \underline{\underline{D}} = V_\beta^x \underline{e}_x \overset{s}{\otimes} \underline{n}_\beta \tag{26}$$

where

$$\rho_\beta = 1 - 2\sqrt{\frac{\eta}{\bar{\eta}_r}} c_\beta \Big(\sqrt{1 - \frac{\eta_r^*}{\eta}} - \rho_k \text{arccos}\sqrt{\frac{\eta_r^*}{\eta}}\Big).$$

Note that $\rho_\beta = 1$ if $\eta = \eta_r^*$, in which case the failure surface is a plane tangent to the fiber. In some cases ρ_β may be greater than 1, which means that the failure mechanism just considered can give upper bounds less significant than those derived in Sec. 4.1 (at equal macroscopic strain, $\underline{\underline{D}}$), even though the interface is weaker than the matrix.

In terms of strength domains, eqn(26) is equivalent to

$$G^{hom} \subset \{\underline{\underline{\Sigma}} | \, |\Sigma_{nx}| = |\Sigma_{xy} c_\beta + \Sigma_{zx} s_\beta| \leq k_m \rho_\beta\}. \tag{25c}$$

7.3 Applications

The results obtained in this section will be now applied to the evaluation of lower and upper bounds to the strength of fiber composites subjected to uniaxial tension, Σ, in the (x, y) plane, at an angle θ to the fibers. The lower and upper bounds for composites with perfectly bonded fibers and matrix were obtained in the previous sections. These bounds will be here denoted by $\tilde{\Sigma}_0^+(\theta)$ and by $\tilde{\Sigma}_1^+(\theta)$ respectively and, from eqns(13), (14) specialized to Von Mises matrices ($\alpha_m = 0$), read

$$\tilde{\Sigma}_0^+(\theta) = \bar{\sigma}^+(1 - 3/2s_\theta^2) + \sqrt{3}\sqrt{k_m^2 - (\bar{\sigma}^+)^2(1 - 3/4s_\theta^2)s_\theta^2} \quad \text{if } \sigma \geq \bar{\sigma}^+ \tag{27a}$$

$$= \frac{k_m}{s_\theta\sqrt{1 - 3/4s_\theta^2}} \quad \text{if } \sigma < \bar{\sigma}^+, \tag{27b}$$

where $\sigma = k_m(1 - 3/2s_\theta^2)/(s_\theta\sqrt{1 - 3/4s_\theta^2})$, and

$$\tilde{\Sigma}_1^+(\theta) = k_m \min\{\frac{2}{\sin 2\theta}; \frac{1}{\sqrt{1 - s_\theta^2 c_\beta^2}} s_\theta c_\beta; \sqrt{3}(1 + \eta(r-1))\} \text{ if } \eta \leq \eta_r^* \tag{28a}$$

$$= k_m \min\{\frac{2}{\sin 2\theta}; \sqrt{3}(1 + \eta(r-1))\} \text{ if } \eta > \eta_r^* \tag{28b}$$

In the lower bound, the limited interface strength is accounted for solving eqn(22) for ψ and getting from eqn(21) the maximum values of the interface shear stress. By imposing that these values have not to exceed k_{int}, one gets:

$$\Sigma_0^+(\theta) = \begin{cases} \min\{\tilde{\Sigma}_0^+(\theta); \dfrac{2k_{int}}{\sin 2\theta}\} & \text{at } 0° \leq \theta \leq 45° \\ \min\{\tilde{\Sigma}_0^+(\theta); 2k_{int}\} & \text{at } 45° \leq \theta \leq 90° \end{cases}$$

In the upper bound, the limited interface strength is accounted for through the domains (25a,b,c), yielding:

$$\Sigma_1^+(\theta) = \begin{cases} \min\{\tilde{\Sigma}_1^+(\theta); \dfrac{2k_m}{\sin 2\theta}\rho_y; \dfrac{2k_m}{c_\beta \sin 2\theta}\rho_\beta\} & \text{if } \eta \leq \eta_r^* \\ \min\{\tilde{\Sigma}_1^+(\theta); \dfrac{2k_m}{\sin 2\theta}\rho_y\} & \text{if } \eta_r^* \leq \eta \end{cases}$$

Fig. 11 shows the just computed bounds in the case of square reinforcing arrays ($\beta = 45°$) with two different volume fractions. If $\eta = 0.35(< \eta_r^*)$, the agreement between the two bounds is rather good (Fig. 11a). If $\eta = 0.7(>> \eta_r^*)$, the gap between the two bounds tends to become large at $\theta > 60°$ (i.e. if tension is almost perpendicular to fibers, see Fig. 11b). The reason for this is that in mechanisms with failure velocity parallel to the fibers, like the ones considered for composites with high percentages of reinforcement, the part of Σ that 'does work' reduces with increasing θ.

8. CONCLUDING REMARKS

Homogenization theory applied to limit analysis has proved to be an effective tool for obtaining lower and upper bounds to the strength of composites under any stress condition in a relatively easy manner. In particular, analytical equations for lower and upper bounds to the macroscopic strength exhibited by the composite under uniaxial stress were obtained (Sec. 5). Provided that certain conditions regarding

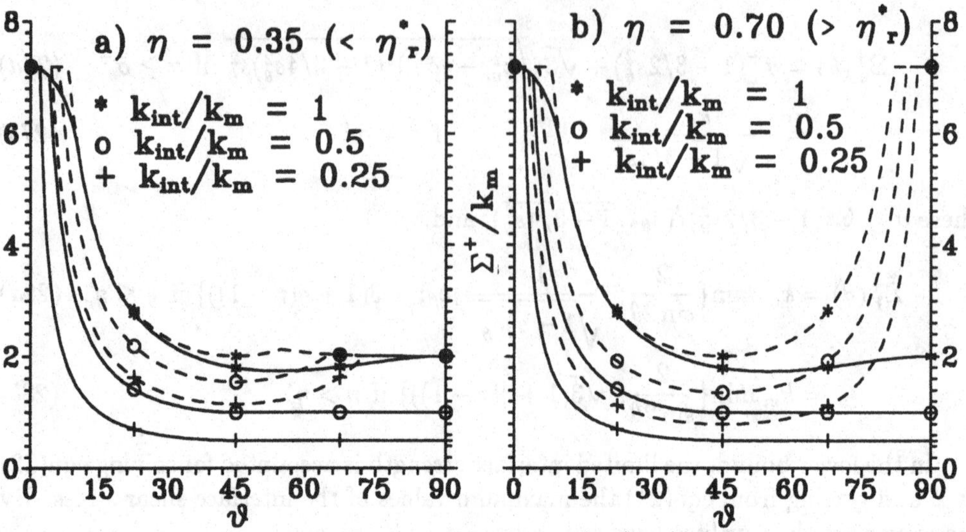

Fig. 11: Bounds to the uniaxial tensile strength in the (x, y) plane versus orientation of the applied stress to the fibers. $\beta = 45°$.

the geometry of the composite and the strength properties of the components are fulfilled, the upper bounds do not excessively overestimate the simplified lower bounds obtained by making use of the domain G_0 defined in Sec. 3 (and, in some cases, coincide with the latter ones).

If the geometry of the composite (defined by the percentage of reinforcement η and by the angle β) is such that the discrepancy between the two bounds is excessive, an approach more complex than the one presented here might become necessary. Upper bounds could be improved by taking into account periodic failure mechanisms for the RVE not as simple as the ones considered in Sec. 4. Secondly, provided that a suitable computer program is available, a structural analysis of the RVE could be performed in order to find with fair approximation the actual macroscopic strength of the composite: however, this approach turns out to be quite cumbersome in terms of computational cost (as shown by similar numerical studies performed by Turgeman & Pastor [26] for heterogeneous layered media and by Marigo *et al.* [27] for perforated plates).

Note that in principle the approximations defined in Sec. 3 and 4 can be used for composites with components undergoing any kind of strength criteria. If suitable experimental results are available, the definitions of G_0 and G_1 can be specialized in order to properly take into account the actual strength properties of matrix and fibers. However, the results shown in Sec. 6 show that, assuming that the matrix undergoes Drucker–Prager strength criterion, the lower bound to the strength of the composite both under uniaxial tension and under combined normal stress and shear is, in many cases, a close estimate of that measured by other authors on composite samples.

Finally, in Sec. 7 an extension of the theoretical model was presented with the aim of accounting for the limited strength properties of a purely cohesive fiber–matrix interface. Actually, the aim here was mainly to show the flexibility of homogenization theory, which allows accounting for any interface criterion, rather than using a strength criterion applicable to polymeric or ceramic composites – the most widely used ones. The obtained upper bounds improve those obtained by Taliercio [24], which were usable only for composites of specific geometry. On the other hand, the simple failure mechanisms used do not yield significant upper bounds under certain stress conditions. It was remarked that the smaller the stress components acting along the collapse velocity vector is, the poorer upper bounds are. In particular, if part of the interface is involved in the failure surface and the collapse velocity is parallel to the fibers, no upper bound is obtained to a tension perpendicular to the fibers. Appearently, this does not occur if matrix and interface undergo friction type strength criteria, such as Mohr–Coulomb criterion; in this case, failure mechanisms could be defined featured by simultaneous sliding and separation of the regions into which RVEs are divided by the failure surfaces. The derivation of homogenized strength criteria for composites with frictional matrix and interface is still to be examined and will probably lead to a further improvement in the understanding of the ultimate behaviour of polymeric and ceramic composites.

REFERENCES

1. Azzi, V.D. and Tsai, S.W.: Anisotropic strength of composites, Experimental Mechanics, 5 (1965), 286–288.
2. Hoffmann, O.: The brittle strength of orthotropic materials, J. Compos. Mater, 1 (1967), 200–206.
3. Tsai, S.W. and Wu, E.M.: A general theory of strength for anisotropic materials, J. Compos. Mater., 5 (1968), 58–80.
4. Hill, R.: The Mathematical Theory of Plasticity, Clarendon Press, Oxford, 1950.
5. Wu, E.M.: Phenomenological anisotropic failure criterion, in: Composite Materials, Vol. II (Eds. L.J. Broutman and R.H. Krock), Academic Press, New York, 1974.
6. Boehler, J.P. and Raclin, J.: Failure criteria for glass–fiber reinforced composites under confining pressure, J. Struct. Mech., 13 (1984), 371–393.
7. Hashin, Z.: Failure criteria for unidirectional fiber composites, ASME Trans., J. Appl. Mech., 47 (1980), 329–334.
8. McLaughlin, P.V. Jr. and Batterman, S.C.: Limit behaviour of fibrous materials, Int. J. Solids Structures, 6 (1970), 1357–1376.
9. McLaughlin, P.V.: Plastic limit behaviour and failure of filament reinforced materials, Int. J. Solids Structures, 8 (1972), 1299–1318.

10. de Buhan, P. and Taliercio, A.: Critère de résistance macroscopique pour les matériaux composites à fibres, C.R. Acad. Sci., Paris, 307 (1988), Série II, 227–232.
11. de Buhan, P. and Taliercio, A.: A homogenization approach to the yield strength of composite materials, Eur. J. Mech., A/Solids, 10 (1991), 129–154.
12. Taliercio, A.: Lower and upper bounds to the macroscopic strength domain of fiber reinforced composite materials, Int. J. Plasticity (1992 - to appear).
13. Taliercio, A.: Influenza dell'interfaccia fibre–matrice nella formulazione di modelli di resistenza macroscopici per materiali compositi, Proc. Nat. Congr. on *Problems of Material and Structure Mechanics*, Amalfi (Italy), June 3–5, 1991 (to appear).
14. de Buhan, P. Salençon, J. and Taliercio, A.: Lower and upper bounds estimates for the macroscopic strength criterion of fiber composite materials, in: Inelastic Deformation of Composite Materials (Ed. G.J. Dvorak), Springer–Verlag, New York, 1991, 563–579.
15. Hashin, Z.: Analysis of composite materials–a survey, ASME Trans., J. Appl. Mech., 50 (1983), 481–505.
16. Hashin, Z. and Rosen, B.W.: The elastic moduli of fiber–reinforced materials, ASME Trans., J. Appl. Mech., 31 (1964), 223–232.
17. Suquet, P.: Plasticité et homogénéisation, Thèse d'Etat, Université Pierre et Marie Curie, Paris, 1983.
18. Suquet, P.: Analyse limite et homogénéisation, C.R. Acad. Sci., Paris, 296 (1983), Série II, 1355–1358.
19. Salençon, J.: Calcul à la rupture et analyse limite, Presses de l'E.N.P.C., Paris, 1983.
20. Tyrrell Rockafellar, R.: Convex Analysis, Princeton Univ. Press, Princeton, N.J. (USA), 1970.
21. Hull, D.: An introduction to composite materials, Cambridge Univ. Press, 1981.
22. Tsai, S.W.: Strength theories of of filamentary structures, in: Fundamental Aspects of Fiber Reinforced Plastic Composites (Eds. R.T. Schwartz and H.S. Schwartz), Wiley Interscience, New York, 1968.
23. Tsai, S.W. and Hahn, H.T.: Introduction to Composite Materials, Technomic Pub. Co., Westport, Conn. (USA), 1980.
24. Aboudi, J.: Micromechanical analysis of fibrous composites with Coulomb frictional slippage between the phases, Mech. Mater., 8 (1989), 103–115.
25. Taliercio, A.: Delimitazione bilaterale del dominio di resistenza macroscopico di un materiale fibrorinforzato, Proc. X Nat. Congr. AIMETA, Pisa (Italy), October 2–5, 1990, 141–146.
26. Turgeman, S. and Pastor, J.: Comparaison des charges limites d'une structure hétérogène et homogénéisée, J. Méc. Th. Appl., 6 (1987), 121–143.
27. Marigo, J.P., Mialon, P., Michel, J.P. and Suquet, P.: Plasticité et homogénéisation: un example de prévision des charges limites d'une structure hétérogène périodique, J. Méc. Th. Appl., 6 (1987), 47–75.

RIGID-PLASTIC ANALYSIS AND DESIGN

M. Save
Faculté Polytechnique, Mons, Belgium

ABSTRACT

The paper first states the basic theorems of rigid-perfectly plastic limit analysis, proofs and details being given in appendices 1 and 2. It then explains how to base the description of structures on the theorem of virtual powers. Application to beams in bending follows, again with appendices 3 and 4 for details. Plates, shells and disks are briefly considered, the general solutions process being illustrated on circular plates examples. Multiple loading, optimal design and post-yield behaviour are then examined very rapidly. Experimental verification of the theory is discussed and the paper ends with remarks on the development of the numerical approach.

1. INTRODUCTION

We are interested in perfectly plastic solids, for which there exists a *yield condition* indicating when the stresses are such that plastic behaviour is possible. Perfect plasticity implies :

1) that this yield condition is independent of previous plastic deformations and
2) that plastic strains may be extremely large without causing fracture.

For this particular solids, when the extension of plastified regions become sufficient, *untrestricted plastic flow* can occur, under constant load as long as the change of geometry due to this flow remains negligible. This incipient unrestricted plastic flow is called limit state, the associated load limit load and the velocity field at this state is often called collapse mechanism.

The following text is concerned with the fundamentals of limit load evaluation.

The rigid-perfectly plastic model is the simplest possible model for the analysis and design of plastic structures, when interest is concentrated on the evaluation of the *limit load for proportional loading*. For purely mechanical loading, the elastic-perfectly plastic model introduces no improvement. Indeed, at incipient collapse, the structure deforms purely plastically under constant load and state of stress. As the undeformed geometry is referred to in describing the limit state, the latter is therefore identical with that of the rigid-perfectly plastic structure. With both models, the limit load will be obtained only if the elastic-plastic deformations for smaller loads do not substantially alter the geometry of the real structure. This can be verified only through a step-by-step analysis of the elastic-plastic deformations, or at least by an evaluation of these deformations just prior to theoretical collapse. In the following we shall assume that such a verification can successfully be made, and that no instability phenomenon of any kind will interfere with the formation of the unrestricted plastic flow mechanism.

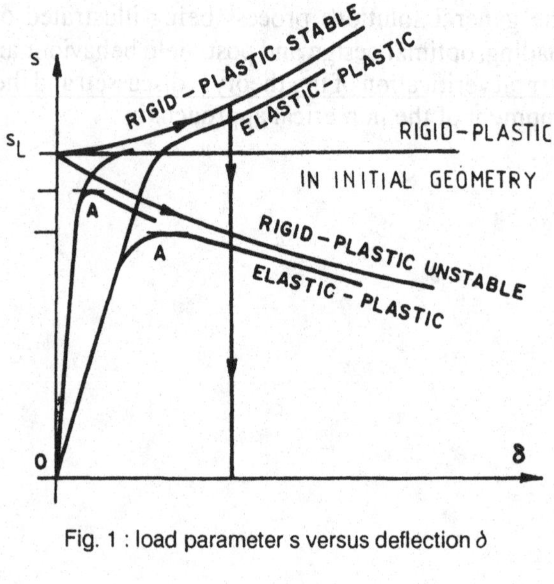

Fig. 1 : load parameter s versus deflection δ

On the other hand, the rigid-perfectly plastic model is able to indicate the real physical significance of the limit load. Indeed, a post-yield behaviour analysis will give the relation between the load parameter s and a relevant deflection δ as shown in fig. 1. A rising curve indicates favourable changes of geometry (so-called geometrical work-hardening). The structure remains stable under the limit load and higher loads, but unloading will result in large permanent deflection that most often will render the structure unusable. A downward sloping curve indicates instability of the structure at the limit load, resulting most often in catastrophic, complete collapse (and even fracture), except if important material workhardening has been neglected. Also, the theoretical limit load is likely not to be reached because prior elastic-plastic deformations will accelerate the onset of the global instability phenomenon (point A in fig. 1).

In rigid-perfectly plastic structures, no deformation occurs before the limit state is reached. Therefore, the limit load also is the "yield point load" of the structure. The basic concepts used in limit load evaluation are as follows :

- a stress-field is called *statically admissible* for the given load $\lambda \mathbf{F}_\alpha$ (*) if it satisfies all the equilibrium and static boundary conditions for these loads ;
- a stress-field is called *plastically admissible* if it does not violate the yield condition at any point ;
- a stress-field is called *licit* (**) if it is simultaneously statically and plastically admissible. The load factor corresponding to such a field is called a *licit static multiplier* and is denoted by s' :
- a strain-rate field is called *kinematically admissible* if it is derived from a field of velocities $\dot{\mathbf{u}}_\alpha$ that satisfies the kinematic boundary conditions ;
- a strain-rate field is called *plastically admissible* if (i) the velocity field from which it is derived is such that the corresponding exterior power $\sum_\alpha \mathbf{F}_\alpha \cdot \dot{\mathbf{u}}_\alpha$ is non-negative, and (ii) belongs to the set on which the interior dissipation function of a typical structural element is defined :
- a strain-rate field is called *licit* if it is simultaneously kinematically and plastically admissible. Its "power equation", equating the exterior power to the total interior rate of dissipation, yields a load factor s" called *licit kinematic multiplier.*

(*) Bold-face letters denote vectors.

(**) The terms safe and unsafe were avoided in the general theory because, (1) for structures, a safe multiplier must be *not larger* than the real collapse factor ; (2) for metal forming processes, a safe estimate of the necessary load factor must be *not smaller* than its real value.

(***) See appendix 1.

The basic theorems of limit analysis establish that the load factor s_L at collapse, any licit static multiplier s', and any licit kinematic multiplier s" satisfy the continued fundamental inequality :

$$s' \leq s_L \leq s" \tag{1}$$

Relation (1) applies to *perfectly-plastic "standard"* structures for every element of which :

(i) there exists a convex yield locus, fixed in the stress space ;

(ii) the internal dissipation is obtained by application of the normality rule to this yield locus. (***)

2. STRUCTURAL LIMIT ANALYSIS AND DESIGN

2.1. STRUCTURES

A structure is composed of solids of particular shape, (beams, plates, shells, disks) some dimensions of which are small with respect to others. Also, these small dimensions (cross-sectional dimensions of a beam, thickness of a plate, shell or disk) may vary with position in a continuous, "sufficiently smooth" manner. As a consequence of this definition, some a priori assumptions may be made on the deformations of the structures : conservation of plane sections in beams, or conservation of normals in plates and shells. Hence, the collapse mechanism of a rigid-perfectly plastic structure can conveniently be described by the generalized strain rate $\dot{\mathbf{q}}$ of its "layout" (axes, mid-surfaces), hence reducing a three dimensional problem to two or one dimension. The continuity of the deformation requires that these strain rates satisfy *compatibility relations* because they are derived from a velocity field $\dot{\mathbf{u}}$, in which the number of components of $\dot{\mathbf{u}}$ is in general smaller that that of $\dot{\mathbf{q}}$.

The generalized stresses $\mathbf{Q}$ associated with $\dot{\mathbf{q}}$ are then obtained from the basic assumption that the *principle of virtual power* is valid :

$$\int \left(\mathbf{F} \cdot \dot{\mathbf{u}} - \mathbf{Q} \cdot \dot{\mathbf{q}}\right) dx = 0 \tag{2}$$

where the integral extends over the layout of the structure and dx denotes an element of layout. The field $\mathbf{Q}$ must be statically admissible for the loads $\mathbf{F}$, whereas $\dot{q}$ is a kinematically admissible field that is derived from $\dot{\mathbf{u}}$.

(*) See appendix 2 for details.

We now use the relations expressing $\dot{\mathbf{q}}$ as derivative of $\dot{\mathbf{u}}$, substitute in (2) and integrate by parts. The resulting integral contains only the components of $\dot{\mathbf{u}}$ with factors in terms of **Q** and **F**. These factors must vanish because, in the vanishing integral, the kinematically admissible field $\dot{\mathbf{u}}$ is arbitrary. We obtain in this manner the equilibrium equations defining **S** as a function of the applied loads. The physical nature of the components of **Q** can then be recognized by comparison with the usual equilibrium equations. We can also use the equations of equilibrium to express the loads in terms of the generalized stresses and their derivatives and then remove the latter by integration by parts. In the resulting integral, the coefficients of the components of **Q** must vanish because **Q** is an *arbitrary* statically admissible field. We obtain, in this manner, the equations defining $\dot{\mathbf{q}}$ as a function of $\dot{\mathbf{u}}$ and its derivatives. We conclude that equation (2) unambiguously defines **Q** when $\dot{\mathbf{q}}$ is chosen, *and conversely*; this is particularly useful when the assumptions of conservation of plane sections or normals are abandoned (*). The components of **Q** and $\dot{\mathbf{q}}$ are known most often. from the physical character of the problem (as will be seen in the next section). It is sufficient to verify that the specific virtual internal power is given by their scalar product. The applied forces associated with vanishing displacements are called *external reaction*. Similarly, stresses associated with vanishing strain rates are called internal reactions. Some strain rates are seen to vanish either due to assumptions made about the structure (conservation of plane sections or normals) or of particular symmetry conditions. Obviously the reactions do not appear in the power equation of a licit mechanism. Moreover, it is possible to eliminate the internal reaction from the yield condition by application of the normality rule. Finally, if considered suitable, equilibrium conditions can also be expressed by the principle of virtual powers, in which the reactions do not appear. Hence, a limit analysis or design problem may be treated in terms of generalized stresses **Q** and strain rates $\dot{\mathbf{q}}$ only, without recourse to the internal reactions, even if these reactions exhibit non-vanishing values.

2.2. BEAMS AND FRAMES.

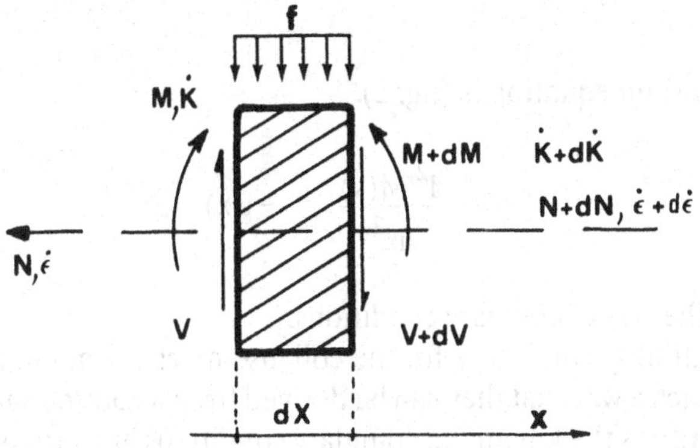

Fig. 2 : Combined bending and axial forces

According to the basic assumption that plane cross-sections remain plane and normal to the deformed axis, the generalized strain rates are the rates of longitudinal extensions $\dot{\varepsilon}$ and the rates of curvatures $\dot{\kappa}$ .The associated generalized stresses are the axial force N and the bending moment M (fig. 2). The shear force V is an internal reaction. For simplicity, we assume that the normal yiel stress σ_0 is the same in tension and compression. Extension of the theory to different yield stresses is straightforward.

2.2.1. Simple Plastic Theory.

In the so-called simple plastic theory for beams and frames, we assume the influence of N to be negligible. Hence, the only generalized variables are M and $\dot{\kappa}$. The (M, κ) diagram is that of fig. 3, which represents the stress-strain diagram of the material. The plastic moment M_0 is given by $M_0 = Z\sigma_0$ can be attained at m relative extrema of bending moment diagram, called *critical sections.* For further simplification, we assume that the plastified regions will be concentrated at the corresponding cross-sections, i.e., in beam elements of infinitely small length dx. Then the adjacent rigid beam segments exhibit relative rotation rates $\dot{\theta}_i$, evaluated with the same sign convention as for the bending moment M_i acting at section i, that is called a *plastic hinge*. The *constitutive equations* of perfect plasticity are then :

$$|M_i| \leq M_{oi} \text{ (yield condition)} \tag{3}$$

$$\dot{\theta}_i = 0 \text{ if } |M_i| < M_{oi} \quad \text{(a)}$$

$$\text{sign } \dot{\theta}_i = \text{sign } M_i \text{ if } |M_i| = M_{oi} \quad \text{(b)} \qquad \text{(4) (flow-law)}$$

The equilibrium equation is (fig. 2)

$$\frac{d^2 M(x)}{dx^2} = -f(x) \tag{5}$$

to be used with the static boundary conditions.

The compatibility conditions for the collapse mechanism relate the rotation rates θ (i = 1,2,...) in such a way that they can be derived from a continuous field of transversal velocity $\dot{u}$ that satisfies the kinematic boundary conditions and articulates the rigid beam segments together. The minimum number of plastic hinge is, in general, equal to r + 1, where r is the degree of redundancy of the system. A corresponding licit one degree of freedom collapse mechanism is defined except for an arbitrary scalar factor. Its licit kinematic multiplier is given by its power equation

$$s'' \sum_{\alpha} \mathbf{F}_{\alpha} \cdot \dot{\mathbf{u}}_{\alpha} = \sum_{i} M_{pi} \left| \dot{\theta}_i \right| \tag{6}$$

In constructing a licit bending moment diagram, equation (5) can suitably be replaced by a set of (m-r) independent equilibrium equations relating the moments M_i at the m critical sections. These equations are best obtained by applying the theorem of virtual powers to (m-r) linearly independent collapse mechanisms. Another possibility is to express any moment M_i by its statically determinate and indeterminate parts :

$$M_i = M^d_{\ i} + m_{i1} X_1 + ... + m_{ir} X_r .$$

In this relation, m_{i1} and $M^d_{\ i}$ are the bending moments at section i, in a statically determinate system (obtained from the given frame by r so-called simple cuts) under the action of a unit redundant X_j (j = 1,2,...r) and of the given loads, respectively. Equilibrium is then automatically satisfied by the very nature of $M^d_{\ i}$ and m_{ij}.

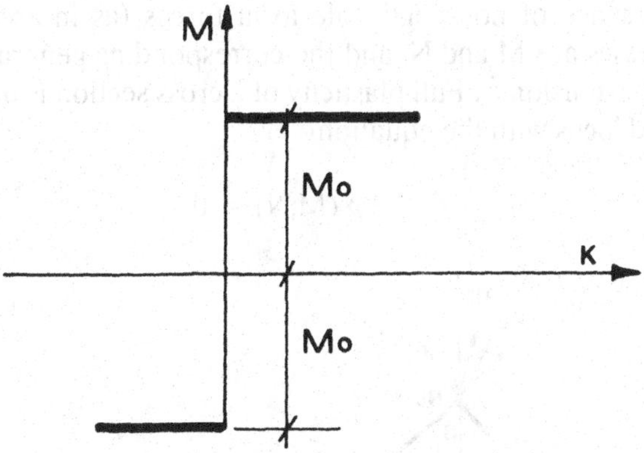

Fig. 3 : Moment-curvature diagram

According to the relation (1), s can be regarded as (i) the upper bound of the set of s' (purely statical approach) ; (ii) the lower bound of the set of s" (purely kinematical approach) ; and (iii) the only common element of the two sets (complete solution).

The two first approaches give separately the exact value of s only when the corresponding set can be explored completely, otherwise they result in approximate values as shown by (1). On the other hand, a complete solution, with its exact value of s, is obtained when the licit moment diagram associated to s' corresponds to the licit mechanism related to s" by the flow law (4). Indeed, if we label with ' and " the licit stress field and the licit mechanism, and express the statical admissibility of field' by equation (2) (only concentrated forces $\mathbf{F}_{\alpha}$ are considered for simplicity) :

$$s'\sum_{\alpha} \mathbf{F}_{\alpha} \left(\dot{\mathbf{u}}_{\alpha}\right)'' = \sum_{i} M_i \left(\theta_i\right)'' \tag{8}$$

Applying (6) to the mechanism, we have :

$$s''\sum_{\alpha} \mathbf{F}_{\alpha} \left(\dot{\mathbf{u}}_{\alpha}\right)'' = \sum_{i} M_{pi} \left|\theta_i\right|'' . \tag{9}$$

The correspondence of the two fields by the flow law (4) implies that the right-hand sides of (8) and (9) are identical. Hence s' = s" and, from (1) s' = s" = s_L.

For small-size problems, et is easy to begin with the kinematic approach, using the method of combination of fundamental mechanisms (see appendix 3), and verify that a licit bending moment diagram exists that corresponds to the obtained mechanism by the flow law.

For large-size problems with several loading cases, limit load evaluation is formulated as a mathematical (here, linear) programming problem that can be solved by suitable computer codes (see appendix 4).

2.2.2. Influence of the Axial force.

In the presence of non-negligable axial forces (as in arches for example), the generalized stresses are M and N, and the corresponding generalized strains are $\dot{\kappa}$ and the rate of axial extension $\dot{\varepsilon}$. Full plasticity of a cross section is obtained by stress point (M,N) on a yield locus with the equation

$$\varphi(M,N) = 0 \tag{10}$$

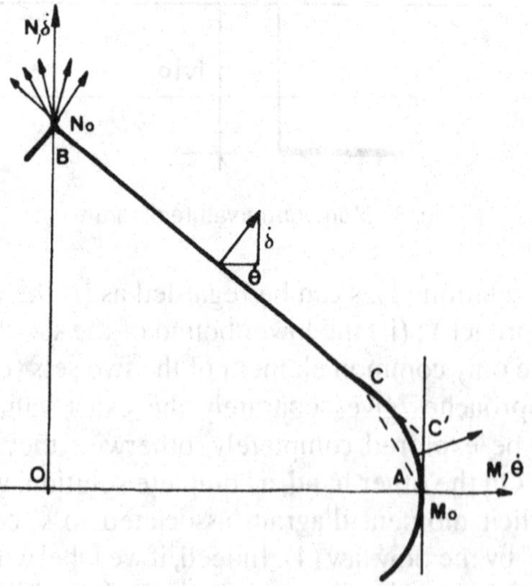

Fig. 4 : M-N yield locus for an I-beam

A typical yield Focus for an I beam is shown in fig. 4. As in simple plastic theory, yielding regions can be described as generalized plastic hinges concentrated at some of the potentially critical cross sections. At each of these hinges, the state of stress (M,N) is represented by a point on the yield locus, and the adjacent rigid parts are allowed to exhibit a relative rotation rate $\dot{\theta}$ *and a relative axial displacement rate* $\dot{\delta}$. The specific dissipation is

$$D = M\dot{\theta} + N\delta \tag{11}$$

where the strain rate vector $\dot{\mathbf{q}}$ with components $\dot{\theta}$ and $\dot{\delta}$ must have the direction of the outward normal to the yield locus at the stress point (ML,N) located on the locus. At vertices of the yield locus, all the outward normals in the cone of normals at infinitely nearby points are acceptable. From this normality or flow rule, the specific dissipation is seen to be a single valued function of $\dot{\mathbf{q}}$, for any shape of the convex yield locus.

It may prove appropriate to approximate the yield locus by an inscribed (i) or circumscribed(c) polygon as shown in fig. 4, where BCA and BCC'A are the inscribed and circumscribed approximations, respectively. The part BC of the parabola is nearly straight. If we denote as :

s'_i an allowable static multiplier for polygon i ;
s_i the (exact) limit multiplier for polygon i ;
s_L the (exact) limit multiplier for the yield locus ;
s_c the (exact) limit multiplier for polygon c ;
s''_c an allowable kinematic multiplier for polygon c,

from (1) it is easily proved that :

$$s'_i \leq s_i \leq s_L \leq s_c \leq s''_c \tag{12}$$

if we note that s_i is a licit static multiplier for the real structure and s_L is a licit static multiplier for the same structure that would have polygon c as its yield locus.

2.3. PLATES, SHELLS AND DISKS

2.3.1. Circular Plates

We develop the theory on the example of an axisymmetrically loaded and supported circular plate, which is the simplest case that retains all the essential features of a general shell problem.

We assume that, in a mechanism at incipient collapse, the mid-plane is not strained and the conservation of normals is satisfied. Hence, with axial symmetry, and in polar coordinates (fig. 5a), the mechanism is completely describe by the two principal rates of curvature $\dot{\kappa}_r$ (r) and $\dot{\kappa}_\theta$ (r). The dissipation per unit area of the mid-plane (fig. 5) is :

$$D = M_r \dot{\kappa}_r + M_\theta \dot{\kappa}_\theta. \tag{13}$$

Accordingly, the generalized stresses are the principal moments per unit of length, M_r and M_θ. Under the action of a distributed load f(r), the equilibrium equation is :

$$\frac{d}{dr}\left(r\,M_r\right) = M_\theta - \int_0^r f(r)r\,dr, \tag{14}$$

to be used with the static boundary conditions.

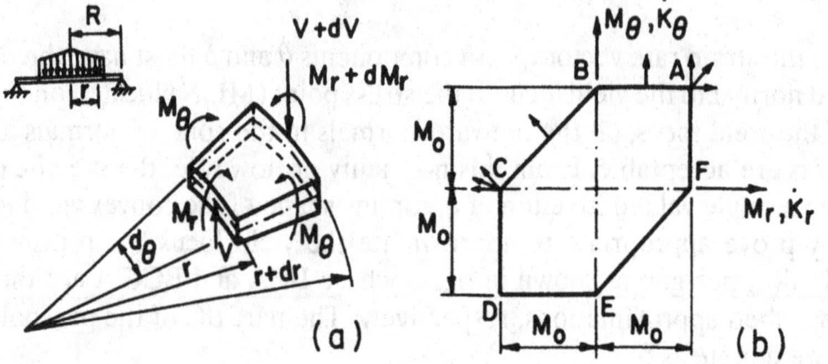

Fig. 5 : Axisymmetric circular plate (a) polar coordinates ; (b) Tresca yiel hexagon

A collapse mechanism will be described by its transversal velocity field $\dot{u}$ (r), from which the principal curvature rates are derived as :

$$\dot{\kappa}_r = -\frac{d^2\dot{u}}{dr^2} \quad \text{(a)}$$

$$\dot{\kappa}_\theta = -\frac{1}{r}\frac{d^2\dot{u}}{dr^2} \quad \text{(b)} \tag{15}$$

The particular nature of the otherwise "standard" plate will enter only by its yield locus. For metal plates, we can use the Tresca yield hexagon of fig. 5b. The application of (13) and of the normality rule to the Tresca hexagon give : for stress points (M_r, M_θ) on

$$\text{side AB}\left(\dot{\kappa}_r = 0\right),\ D = M_o\,\dot{\kappa}_\theta \quad \text{(a)}$$

$$\text{side BC}\left(\dot{\kappa}_\theta = -\dot{\kappa}_r > 0\right),\ D = M_o\,\dot{\kappa}_\theta \quad \text{(b)} \tag{16}$$

$$\text{corner A}\left(\dot{\kappa}_r \le 0\,,\dot{\kappa}_\theta \ge 0\right),\ D = M_o\left(\dot{\kappa}_\theta + \dot{\kappa}_r\right) \quad \text{(c)}$$

and similar relations for other sides and corners. Unlike the case of beams and frames, where collapse mechanisms are formed with concentrated plasticity at cross sections that exhibit discontinuities in slope $\frac{d\dot{u}}{dx}$ and axial elongation $\dot{\delta}$, in plate and shell problems the collapse mechanisms often contain finite plastified regions, sometimes extending over the whole structure (except in reinforced concrete structures). These plastified regions may be accompanied by lines of discontinuities where a concentrated dissipation occurs. To illustrate this fact, let us consider successively the following examples.

a) Simply supported Tresca plate subjected to uniformly distributed load f.

It is physically reasonable to assume that the bending moments will be positive throughout the plate, together with positive circumferential rates of curvature $\dot{\kappa}_\theta$. Hence, the *stress profile* AB is chosen in fig. 5b, with A and B corresponding to r = 0 and r = R, respectively. If, in the equilibrium equation (14), we use

$$\int_0^r f(r)\, r\, dr = f\,\frac{r^2}{2},$$

let $M_\theta = M_o$, integrate and use the boundary conditions

$$M_r = M_o \ \text{ in } r = 0 \ \text{ and } \ M_r = 0 \ \text{ in } \ r = R,$$

we obtain

$$M_r = M_o - \frac{r^2}{6} \tag{17}$$

and

$$f = \frac{6\,M_o}{R^2} \tag{18}$$

It is readily found that $\frac{6\,M_o}{R^2}$ is the exact limit load by showing that there exists a licit mechanism associated with the moment field given by (17) and $M_\theta = M_o$. Indeed, the stress profile AB is associated with $\dot{\kappa}_r = 0$ and $\dot{\kappa}_\theta \geq 0$. Using these conditions in (15), we obtain :

$$\dot{u}(r) = \dot{u}_c\left(1 - \frac{r}{R}\right), \tag{19}$$

where $\dot{u}_c$ denotes the velocity of the centre point. We can verify that the licit kinematical load parameter given by the mechanism (19) is $\frac{6 M_o}{R^2}$. Indeed, from (15b),

$$\dot{\kappa}_\theta = \frac{\dot{u}_c}{rR} . \tag{20}$$

From (16a),

$$D = \frac{M_o}{rR} . \dot{u}_c . \tag{21}$$

The total dissipation is

$$D = \int_0^{2\pi} \left\{ \int_0^R \frac{M_o \dot{u}_c}{rR} r\, dr \right\} d\theta = 2\pi M_o . \dot{u}_c \quad . \tag{22}$$

The power of the applied load is

$$P = \int_0^{2\pi} \left\{ \int_0^R f . \dot{u}_c \left(1 - \frac{r}{R}\right) r\, dr \right\} d\theta = f \dot{u}_c \pi \frac{R^2}{3} . \tag{23}$$

The power equation $P = D$ effectively gives $f = \frac{6 M_o}{R^2}$. We also verify that another licit mechanism gives a higher load : for example,

$$\dot{u} = \dot{u}_c \left(1 - \frac{r^2}{R^2} \right)$$

gives

$$D = 4\pi \dot{u}_c M_o , \quad P = f \dot{u}_c \pi R^2 \quad \text{and } f' = 8 \, {}^{M_o}\!/\!_{R^2} .$$

b) built-in Johansen (reinforced concrete) plate subjected to a uniformly distributed load f.

If the plate with isotropical reinforcement has a positive plastic moment M_o and a negative plastic moment M'_o , its yield locus is given in fig. 6. For a simply supported plate the stress profile AB' can be used to provide the same solution as for the Tresca plate. On the other hand, with built-in support, the plastic regime is the same

$\left(M_\theta = M_o, \; -M'_o \leq M_r \leq M_o \right)$ but the stress profile must extend from A (r = 0) to B (r = R).

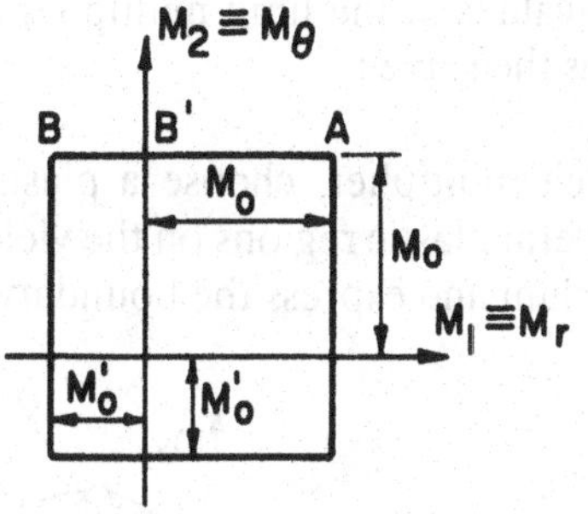

Fig. 6 : Yield locus for isotropic plate

Integration of the equation of equilibrium with this stress profile results in

$$M_\theta = M_o \quad \text{(a)}$$

$$M_r = M_o - \left(M_o + M'_o \right) \frac{r^2}{R^2} \quad \text{(b)} \tag{24}$$

$$f = \frac{6 \left(M_o + M'_o \right)}{R^2} . \tag{25}$$

The corresponding licit mechanism is again given by (19), but it now contains a *negative hinge circle* at the built-in support, with a rotation rate (discontinuity in the slope of $\dot{u}$) $\dot{\theta} = \frac{d\dot{u}}{dx} = -\frac{\dot{u}_c}{R}$. This hinge circle produces a (concentrated) dissipation

$$D_c = \int_0^{2\pi} \left(-M'_o \right) \dot{\theta} \,.\, R \, d\theta = \dot{u}_c \, 2\pi \, M'_o ,$$

to be added to D as given by (22). As expression (23) for P remains valid, the licit kinematic load is also given by (25), proving that this is the exact limit load.

2.3.2. Other Plate and Shell Problems.

If the yield condition of von Mises is regarded as a better approximation it can be used for circular plates ; however, this introduces a complicating nonlinearity to the problem. For a Tresca material with piecewise linear yield condition in principal stresses

$\sigma_1, \sigma_2, \sigma_3$, the nonlimearity of the yield condition appears in non-axisymmetrical plate problems both for metals and reinforced concrete, as well as in shells of revolution, and a fortiori in shells of general shape. Typical yield loci are shown in fig. 7 to 10. For mathematical simplicity, it is often necessary to linearize these yield loci, and satisfy oneself with approximate values of the limit multiplier as given by (12). However, the general procedure remains the same :

- to obtain a static licit multiplier, choose a plastically admissible stress profile (involving one or several plastic regions on the yield locus) ; use it to integrate the equations of equilibrium and express the boundary conditions ;

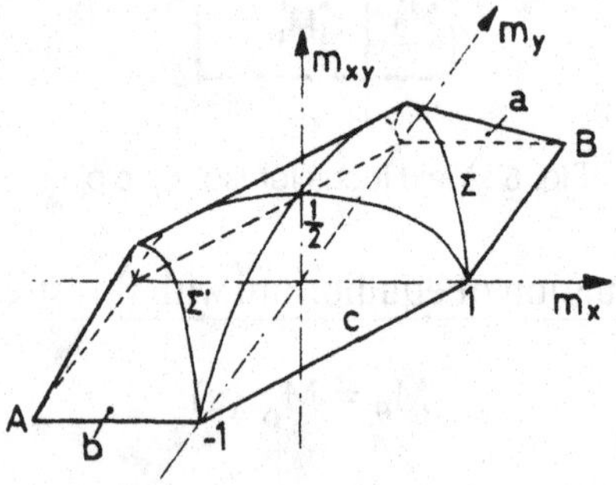

Fig. 7 : Yield locus : Tresca plate (rectangular coordinates)

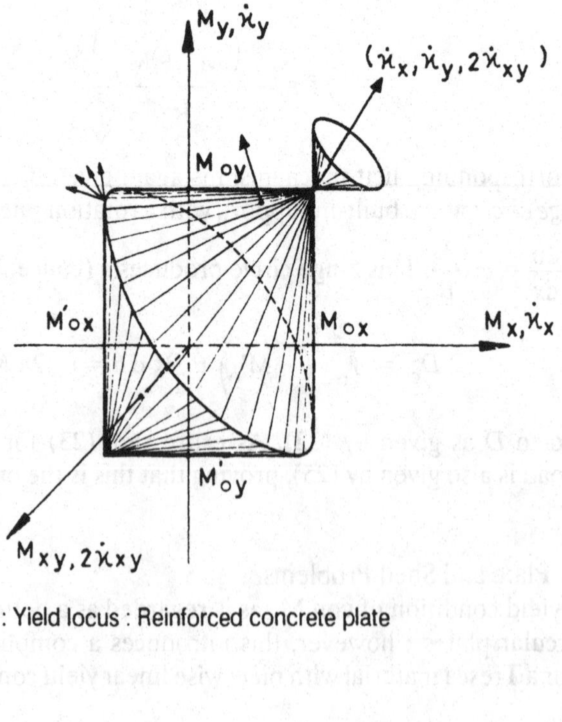

Fig. 8 : Yield locus : Reinforced concrete plate

- to obtain a kinematic licit multiplier, construct a licit mechanism, and express its power equation $P = D$;
- to obtain a complete solution (exact value of s_L) use either approach above in such a way that the existence of a corresponding kinematic or static field can be proved.

Note that, with increasing complexity of the problems, it is often impossible to obtain a complete solution, even when a linearized yield locus is adopted. Only bounds can be obtained, which sometimes have very close values [1]. When difficulties of analytical solutions appear too large, numerical procedures must be adopted.

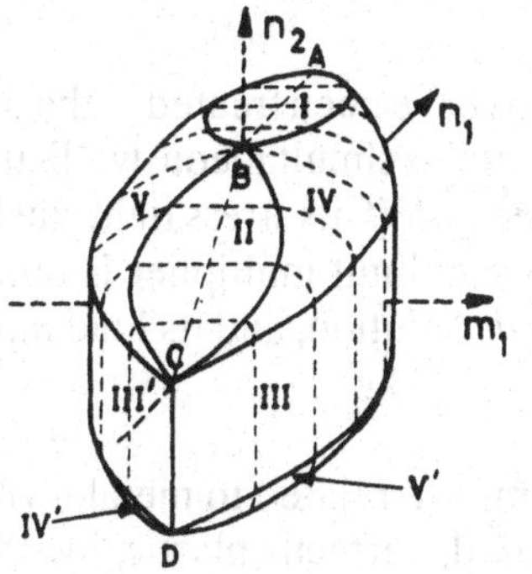

Fig. 9 : Tresca cylindrical shell

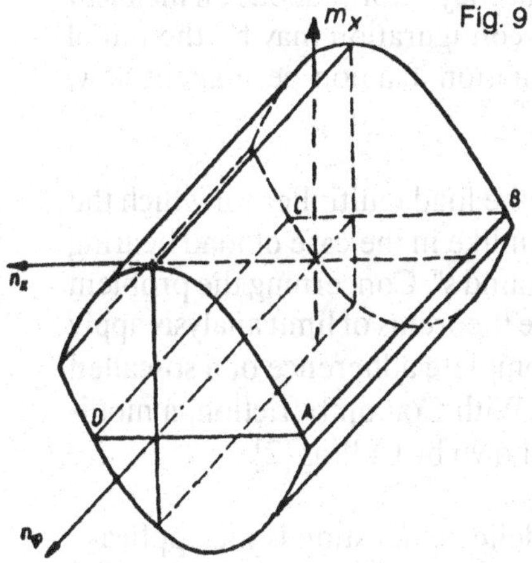

Fig. 10 : Yield surface for reinforced-concrete cylindrical shells

2.3.3. Disks, Metal Forming Processes and Limit States of Soils.

In disks problems, that is, in states of plane stress or plane strain, the variables are the components of the plane stress and strain rate tensors, together with the velocities in the plane of action of the forces. The stress normal to this plane is principal and either vanishes (plane stress) or is a reaction (plane strain). Whereas the yield conditions of Tresca and von Mises differ in plane stress, in plane strain they have the same simple linear form $\sigma_1 - \sigma_2 = 2k$, where σ_1 and σ_2 are the principal stresses acting in the plane of the disk and 2k stands for σ_o or $\frac{2\sigma_o}{\sqrt{3}}$, for the Tresca and von Mises conditions, respectively.

The procedures are as above. In the statical approach, stress fields with statically admissible discontinuities are often used, resulting in good lower bounds of the limit multiplier despite their rather crude representation of the real limit state of stress. Similarly, mechanisms concentrating the plastic strain rates in surfaces of discontinuity of the tangential velocity component often give very good upper bounds. Some remarks are worth making here :

- many plane strain problems have been treated in the literature by the slip-line fields theory. A slip-line field contains simultaneously a licit stress field and its associated licit strain-rate field. If, *and only if*, its stress field can be extended in a licit manner to the rigid regions, the exact limit multiplier is obtained. Otherwise, it must be regarded as a licit *kinematic* solution, and its load multiplier s" as an upper bound of s_L ;

- many plane strain problems are related to metal forming processes. The material being considered as standard, perfectly plastic, the theorems of rigid-plastic limit analysis apply to these processes to the extent that they inform about an *incipient flow from a given configuration* of the body. This configuration may be the initial undeformed shape, some known intermediate situation in a non-permanent flow, or the permanent flow configuration ;

- in metal forming process, it is desired to evaluate the load multiplier for which the forming process will certainly take place. Hence, unlike in the case of load bearing structures, a safe estimate of s_L here is an upper bound s". Concerning the problem of the friction at the contact of the forming tool, the theorems of limit analysis apply directly when there is either no friction at all or complete adherence or a so-called layer friction with fixed friction tangential stress. With Coulomb friction, a modification of the kinematic approach is needed, as shown by Collins [2].

Limit state of soils often are plane strain problems. A delicate question is the applicability of the normality law to soils. Note that, even if this is supposed to apply, with the generally used yield conditions for soils, it will result in a non-vanishing plastic volume change.

2.4. MULTIPLE LOADING

When several loading cases may occur independently or be combined in various ways, *the domain of rigidity* of the structure should be constructed in the loading space.

Consider, for example, a circular simply supported plate subjected aither to a uniformly distributed load f_1 over its entire area, or f_2 over a central circle with radius a. In the latter case, the limit load is $f_2 = 6\ Mp / (3 - 2a/R)$, with the same stress profile AB as under f_1. When both loadings act together, using again profile AB we find :

$$M_r = M_o - \left(f_1 + f_2\right) \frac{r^2}{6}, \quad M_\theta = M_o, \quad \text{for } 0 <= r \leq a \ \text{(a)}$$

$$M_r = M_o - f_1 \frac{r^2}{6} - f_2 \frac{a^2}{2} + f_2 \frac{a^3}{3r}, \quad M_\theta = M_o, \quad \text{for } a \leq r \leq R \ \text{(b)} \tag{26}$$

The boundary condition $M_r = 0$ in $r = R$ yields the relation

$$\frac{f_1}{6\frac{M_o}{R^2}} + \frac{f_2}{\frac{6\,M_o}{\left(3 - \frac{2a}{R}\right)}} = 1 \tag{27}$$

represented in fig. 11.

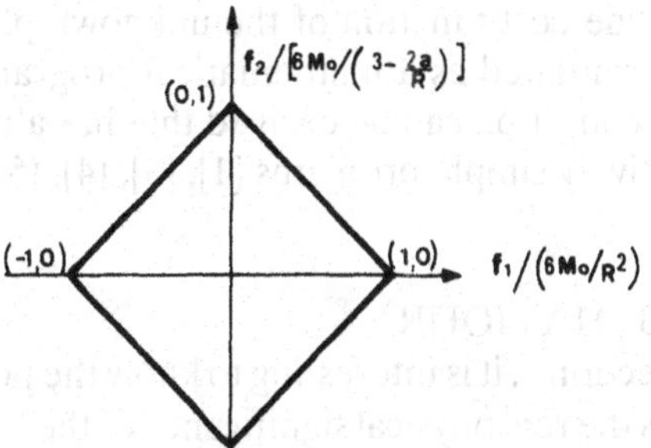

Fig. 11 : Domain of rigidity

This type of domain of rigidity, which is the smallest possible convex domain, occurs when the collapse mechanism is the same for the various separate loading cases. A linear combination of the limit loads f_1 and f_2 will not exceed the carrying capacity of the plate as long as its non-negative coefficients a_1 and a_2 are such that $a_1 + a_2 \leq 1$. When the mechanisms for the various loading cases are different, a linear combination of the limit loads with non-negative coefficients a_i such that $\sum_i a_i \leq 1$ gives a lower bound to the carrying capacity of the structure.

2.5. DESIGN VERSUS ANALYSIS, OPTIMAL DESIGN

The problem most commonly encountered is that of design, in which the collapse loads $s_L \mathbf{F}_\alpha$ are assigned and the strength of the structure must be determined to secure this value s_L or a higher value of the limit multiplier.

In a prismatic beam, if we denote by M'_0 and M''_0 the plastic moments from a licit stress field and a licit mechanism, respectively, both for loads $s_L \mathbf{F}_\alpha$, it is readily seen that the exact plastic moment M_0 for these loads is bounded as follows :

$$M''_0 \leq M_0 \leq M'_0 \tag{28}$$

Indeed, it suffices to express M'_0 and M''_0 by the theorem of virtual powers and by the power equation, respectively, and use inequalities (1). In a framed structure made of prismatic beams, the design consists of determining the various plastic moments of the beams that are allowed to exhibit different cross sections. If all these plastic moments are set to be proportional to one of them, the design problem has only one variable and remains similar to the problem of analysis. When several plastic moments are left to the designer, he is in a position not only to secure the desired carrying capacity but also to get an additional benefit. Often as a measure of economic quality of the structure, the minimization of some cost function of the unknown plastic moments is required. This problem can be formulated as a mathematical programming problem.Alternatively, a general optimality condition can be derived that has already been successfully appplied to a variety of relatively simple problems [1], [3], [4], [5].

3. POST-YIELD BEHAVIOUR

As noted in section 1, it is interesting to know the post-yield behaviour of a structure because it indicates the real physical significance of the limit multiplier. In the very simple case of the frame in fig. 12, whatever the deformation of the frame, the collapse mechanism obtained with the initial geometry (hinges at B and C) is the only one possible. Hence, it is the exact mechanism in any deformed situation. Application of the power equation for increments of rotation rates in a generic deformed shape results in the load-deflection curves of fig. 13, where the rising curve corresponds to the vertical load acting upward. A similar situation occurs in a simply supported circular Tresca plate that collapses into a cone that collapses itself into a cone of higher rise. However, these cases are exceptional. In general, "mode changes" occur with increasing deflections. A step-by-step procedure can then be applied : a limit analysis of the undeformed structure gives the first mechanism that is used to generate a neighbouring deformed situation, etc. This sequence of limit analyses gives a corresponding number of points of the load-deflection

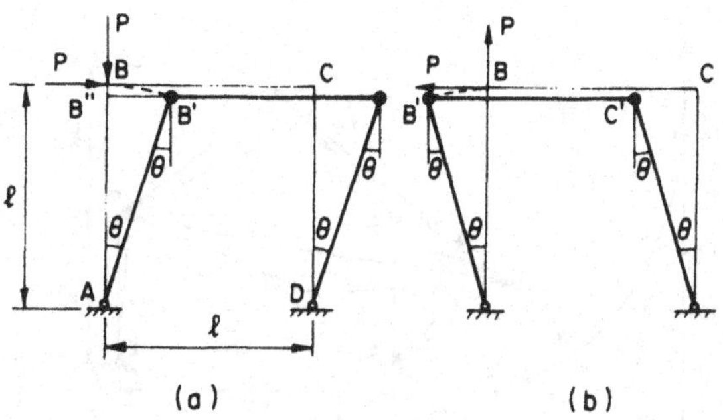

Fig. 12 : Frame collapse mechanisms : (a) downward vertical load (b) upward vertical load

Application of the power equation for increments of rotation rates in a generic deformed shape results in the load-deflection curves of fig. 13, where the rising curve corresponds to the vertical load acting upward. A similar situation occurs in a simply supported circular Tresca plate that collapses into a cone that collapses itself into a cone of higher rise. However, these cases are exceptional. In general, "mode changes" occur

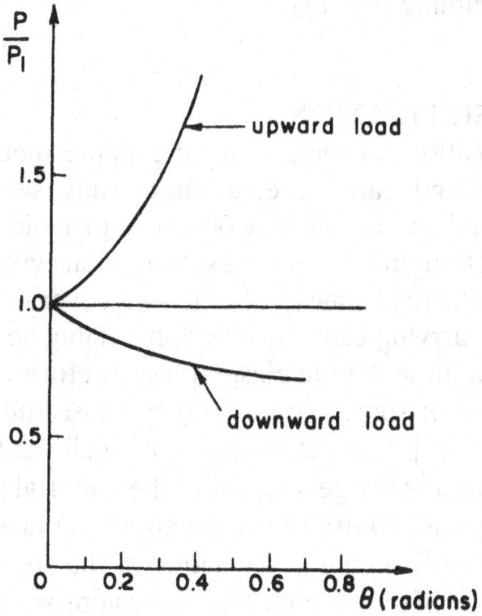

Fig. 13 : load-deflection curves for frame in fig. 12

with increasing deflections. A step-by-step procedure can then be applied : a limit analysis of the undeformed structure gives the first mechanism that is used to generate a neighbouring deformed situation, etc. This sequence of limit analyses gives a corresponding number of points of the load-deflection curve. This procedure was used by

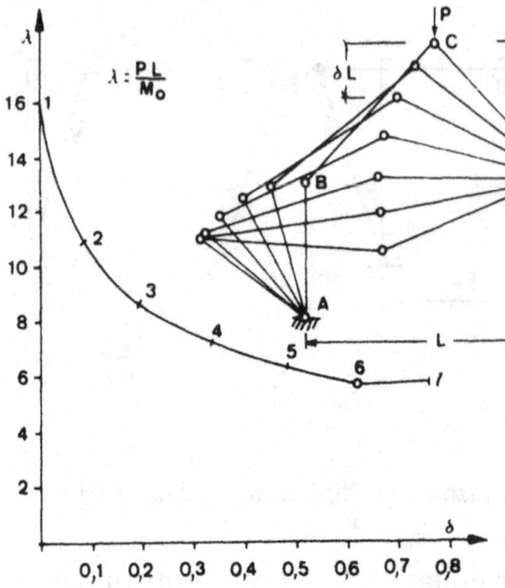

Fig. 14 : Load-deflection curve for a frame via linear programming [6]

Gavarini [6] who performed the successive limit analyses by linear programming (fig. 14).
curve. This procedure was used by Gavarini [6] who performed the successive limit analyses by linear programming (fig. 14).

4. EXPERIMENTAL VERIFICATION

Experimental verifications of the predicted collapse mechanism and of the limit load estimation for beams and frames are very numerous (see for example [3] or [7]). They show that, provided adequate rules are observed to avoid premature local instability, the predictions are well supported by the experimental evidence. On the other hand, it also appears that, for multistory frames with side-sway, overall instability is the relevant phenomenon : the actual carrying capacity may appreciably be reduced with respect to the limit load, due to the non-negligible elastic-plastic deformations. Hence, a step-by-step elastic-plastic analysis, or some semi-empirical substitute like the Rankine-Merchant formula are required [7]. For the theory to be well verified, some fundamental assumptions must be satisfied : 1) large ductility of the material ; 2) moderate strain-hardening (a minimum amount is required to compensate the fact that the ductility is not infinite [7], [8], ; 3) sufficiently high elastic rigidity of the structure, (its elastic-plastic deformed shape on the verge of plastic collapse can be approximated by the undeformed shape). The widely discussed Stussi-Kollbrunner experiments [3], [9], which were used for some time to criticize limit analysis, only demonstrate this fact. A similar situation arises in metal plates. The experiments of Cooper and Shifrin [10] in 1954 were misused by their authors to cast doubt on the practical applicability of the theoretical analysis of Hopkins and Prager [11] published one year before. Indeed, the behaviour of the plate of diameter D and thickness t is strongly influenced by its slenderness ratio $\mu = D/t$.

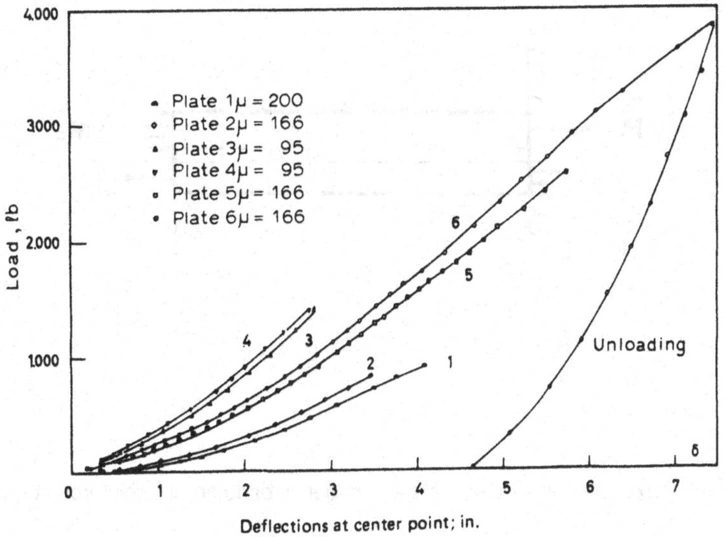

Fig. 15 : Behaviour of a circular plate of diameter D and thickness t for various slenderness ratios $\mu = \frac{D}{t}$, [10]

For μ varying from 95 to 200 as in the Cooper and Shifrin experiments (fig. 15), membrane effects are prominent from the very beginning of loading, and bending analysis is meaningless even in the elastic range (except for extremely small loads). Subsequent experiments on circular steel plates [12], [13] have shown that bending limit analysis has a real physical significance (with an accuracy of about 10%), in the approximate range $10 \leq \frac{D}{t} \leq 40$. A similar order of magnitude of the range of the parameter L/t was obtained for square or rectangular steel plate with largest side length L [14]. Important reserve of strength beyond the limit load must be noted, especially for built-in plates. Experiments by Save [1], [13] on cylindrical and torispherical pressure vessels also confirm the real physical significance of the limit load.

Applicability of limit analysis to reinforced concrete plates is now well accepted. Enhancement of the limit flexure load in continuous or built-in plates, as discussed by Wood [15], was shown by Janas [16], [1] to come from the fact that the mechanism involving only bending was not licit. Indeed, it is seen in fig. 16 that the neutral axes in positive and negative bending are not at the same level. Hence, the mechanism cannot form, except if suitable normal (membrane) forces appear that, at zero deflections, enable the neutral axes to become coplanar. With this correct application of the theory, the limit load is seen to come much closer to the experimental values.

Experiments on reinforced concrete shells [17], [1] have shown that, as in the case of plates, the collapse mechanisms consist of yield lines along which the plastic deformation is concentrated. The yield lines in shells exhibit not only relative rotation rates but also discontinuities in tangential velocities.

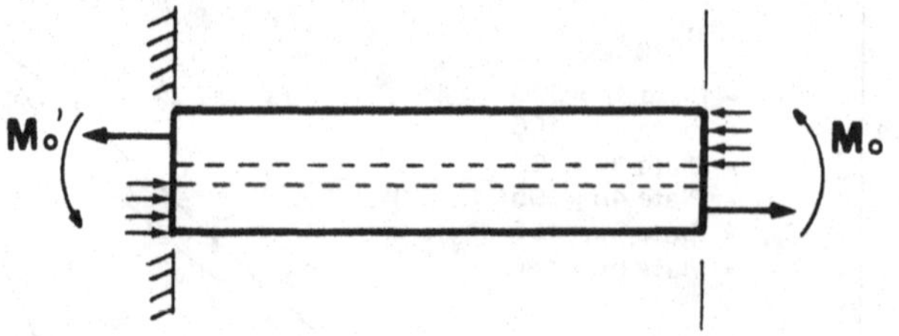

Fig. 16 : Neutral axis positions for positive and negative bending of a reinforced concrete plate.

In reinforced concrete beams and frames, the key question is the rotation capacity of the hinges, that strongly depends not only on the reinforcement ratio but also on the type of reinforcement. Although the practical application of limit analysis and design to these structures is still an open question, it appears to be only a matter of time for a clear definition of the range of applicability of the theory to be achieved.

5 : FINAL REMARKS

It is out of the scope of this introductory text to discribe the developments of limit analysis and design of structure up to the present time, especially if we include optimal design for assigned limit load, a subject which has grown enormously in the last two decades. For this last subject, the books under references [18] to [23] will certainly be of help, together with the vast lists of original papers they contain. Both in optimal limit design and in limit analysis, numerical approaches are increasingly important. Here again we must restrict ourselves to a few titles ([24] to [26]). We wish also to point out that an up-dated and improved version of the book under ref. [1] will be published in the near future by Elsevier Pub. Co, as well as a collection of solutions for metal structures [27].

REFERENCES

[1]: SAVE, M.A. and MASSONNET, C.E., Plastic Analysis and Design of Plates, Shells and Disks, North-Holland Pub., 1972.

[2]: COLLINS, I.F., "The Upper Bound Theorem for Rigid-Plastic Solids Genereli-zed to Include Coulomb Friction", Jour. Mech. Phys. Solids, Vol. 17, 1969, p. 323.

[3]: MASSONNET, C.E. and SAVE, M.A., Plastic Analysis and Design of Beams and Frames, Blaisdell Pub., Waltham, Mass., U.S.A., 1965.

[4]: SAVE, M.A., A Unified Fromulation of the Theory of Optimal Plastic Design with Convex Cost Function, Journ. Struct. Mech., Vol. 1, 1972, p. 267-276.

[5]: ROZVANY, G.I.N., Optimal Design of Flexural Systems, Pergamon Press, Oxford, England, 1976.

[6]: GAVARINI, C., "Une Méthode Générale pour l'Etude du Comportement des Structures après la Ruine Rigide-Plastique", Séminaire de Mécanique des Solides et des Structures, Faculté Polytechnique, Département d'Architecture, Mons, 1969.

[7]: MASSONNET, C.A. and SAVE, M.A., Calcul Plastique des Constructions, Vol. 1, 3ème édition, Nelissen Pub., Liège, 1977.

[8]: LAY, M.G. and SMITH, P.D., "The Role of Strain-Hardening in Plastic Design", Journ. Struct. Div., A.S.C.E., Vol. 91, 1965, p. 25.

[9]: STUSSI, F. and KOLLBRUNNER, C.F., "Beitrag zum Traglastverfahren", Bautechnik, Vol. 13, 1935, p. 264.

[10]: COOPER, R.M. and SHIFRIN, G.A., "An Experiment on Circular Plates in the Plastic Range", Proc. 2nd U.S. Nat. Cong., Appl. Mech. A.S.M.E., Ann Arbor, Michigan, 1954.

[11]: HOPKINS, H.G. and PRAGER, W., "The Load-Carrying Capacity of Circular Plates", J. Mech. Phys. Solids, Vol. 2, 1953, p. 1.

[12]: LANCE, R.H. and ONAT, E.T., "A Comparison of Experiments and Theory in the Plastic Bending of Plates", J. Mech. Phys. Solids, Vol. 10, 1962, p. 301.

[13]: SAVE, M.A., "Vérification Expérimentale de l'Analyse des Plaques et des Coques en Acier Doux" (Experimental Verification of Plastic Limit of Mild

Steel Plates and Shells), C.R.I.F. Report, M.T. 21, February, Fabrimetal, Brussels, 1966.

[14]: DEL RIO, L., "Analyse Limite des Plaques Rectangulaires" These de Maitrise en Sciences Appliquées, Faculté Polytechnique, Mons, Belgique, 1970.

[15]: WOOD, R.H., Plastic and Elastic Analysis and Design of Slabs and Plates, Thames an Hudson, London, 1961.

[16]: JANAS, M., "Large Plastic Deflections of Reinforced Concrete Slabs", Int. J. Solids and Struct., Vol. 3, November, 1967, p. 4.

[17]: BOUMA, A.L., RIEL, A.C., VAN KOTEN, H. and BERANEK, W.J., Investigations on Models of Eleven Cylindrical Shells made of Reinforced and Prestressed Concrete, Shells Research, North-Holland Publ., Amsterdam, 1961, p. 79-101.

[18] : SAVE, M. and PRAGER, W., Editors : Structural Optimization, Vol. 1 Optinalily criteria, by SAVE, PRAGER and SACCHI, Plenum Press, New-york, 1985.

[19] : SAVE, M. and PRAGER, W., Editors : Structural Optimization, Vol. 2 Mathematical Programming, by BORKOWSKI, JENDO, and REITMAN, Plenum Press, New-york, 1990.

[20] : SAVE, M. and PRAGER, W., Editors : Structural Optimization, Vol. 3 Applications to metal and concrete structures, by COHN, FRANGOLOL and ESCHENAUER, to appear at Plenum Press, New-york.

[21] : ROZVANY, G.I.N., Optimal Design of Flexural Systems, Pergamon Press, Oxford, 1976.

[22] : ROZVANY, G.I.N., Structural design via optimal criteria, Kluwer, Doordrecht, 1989.

[23] : BRANDT, A., Editor : Criteria and Methods of structural Optimization, PWN - Polish Scientific Publishers, Warszawa, and Martinus Nijhoff Pub, The Hague - Boston - Lancaster, 1984.

[24] : COHN, M.Z. and MAIER, Editors : G., Engineering Plasticity by Mathematical Programming, Pergamon Press, 1979.

[25] : CHEN, W.F. and LIU, X.L., Limit Analysis in Soil Sechanics, Developments in Geotechnical Engineering, Vol. 52, Elsevier, 1990.

[26] : NGUYEN DANG, H., The Plasticity and the Limit State Analysis and design by finite element method, Birhauser-verlag, 1990.

[27] : SAVE, M. Editor : Atlas of Limit Loads of Metal Plates, Shells and Disks, to be published by Elsevier Pub. Co.

APPENDIX 1

If the yield condition is $\Phi(\mathbf{Q}) \leq 1$, where $\mathbf{Q}$ is the stress vector, it defines q yield locus with equation $\Phi(\mathbf{Q}) = 1$ in the stress space (see fig. 1 for a two-dimensional case).

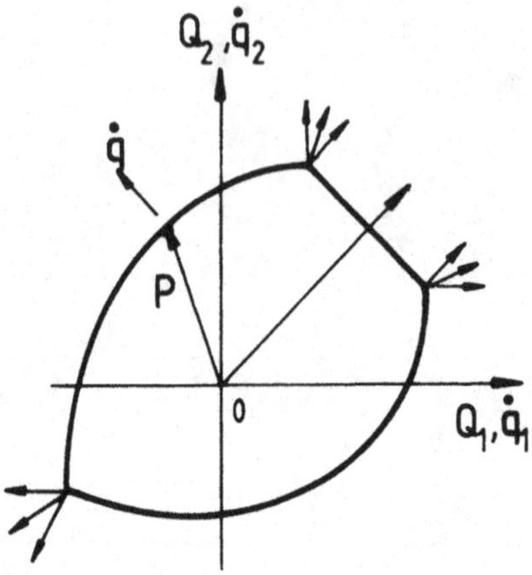

fig. 1

If the extremity P of vector $\mathbf{Q}$ is inside the yield locus, only rigid behaviour is possible. If it is on the yield locus, plastic flow can occur. If the strain rate space $\dot{\mathbf{q}}$ is superimposed on the stress space, the *flow rule* states that $\dot{\mathbf{q}}$ has the direction of the *outward pointing normal* to the yield locus at point P but is otherwise undetermined in magnitude. The plastic dissipation connected to $\dot{\mathbf{q}}$ is $D = \mathbf{Q} \cdot \dot{\mathbf{q}}$, where the point denotes scalar product. Assuming that the yield locus is fixed in the stress-space and convex, we readily see that :

a) D is a single-valued function of $\dot{\mathbf{q}}$, whatever the shape of the yield locus

b) from the convexity of the yield locus and the normality law defining the direction of $\dot{\mathbf{q}}$, we obtain immediately the *theorem of maximum dissipation* expressed by

$$D(\dot{\mathbf{q}}) \geq \dot{\mathbf{q}} \cdot \mathbf{Q}_{pa}$$

where $\mathbf{Q}_{pa}$ is any plastically admissible state of stress. The fundamental theorems of limit analysis are then proved as follows :

Satic (lower bound) theorem

At the limit state we have (either from the theorem of virtual powers or by energy balance)

$$s_L \sum_{\alpha} \mathbf{F}_{\alpha} \cdot \dot{\mathbf{u}}_{\alpha} = \int \mathbf{Q} \cdot \dot{\mathbf{q}} \, dx, \tag{1}$$

where the integral extends over the structure.

Consider a static licit state of stress **Q'** and loads s' $\mathbf{F}_\alpha$.

By the theorem of virtual powers we may write

$$s' \sum \mathbf{F}_\alpha \cdot \dot{\mathbf{u}}_\alpha = \int \mathbf{Q}' \cdot \dot{\mathbf{q}}\, dx, \tag{2}$$

But, because **Q'** is plastically admissible but is in general not associated to $\dot{\mathbf{q}}$ by the normality law, the theorem of maximum dissipation tells us that right-hand of (1) is not smaller than that of (2). Hence

$$s' \leq s_L. \tag{3}$$

Kinematic (upper bound) theorem

Considering now a licit collapse mechanism $\dot{\mathbf{u}}''$ and $\dot{\mathbf{q}}''$, its load parameter is given (by definition) by

$$s'' \sum_\alpha \mathbf{F}_\alpha \cdot \dot{\mathbf{u}}''_\alpha = \int D(\dot{\mathbf{q}}'')\, dx \tag{4}$$

Expressing the equilibrium in the limit state by the theorem of virtuel powers with the mechanism $\dot{\mathbf{u}}''$ and $\dot{\mathbf{q}}''$, we have

$$s_L \sum_\alpha \mathbf{F}_\alpha \cdot \dot{\mathbf{u}}''_\alpha = \int \mathbf{Q} \cdot \dot{\mathbf{q}}''\, dx \tag{5}$$

Again, because **Q** is plastically admissible but does not in general correspond to $\dot{\mathbf{q}}''$ by the fow law, the theorem of maximum dissipation tells us that the right-hand side of 4 is not smaller than that of (5). Hence

$$s_L \leq s''. \tag{6}$$

APPENDIX 2 : STATIC AND KINEMATIC VARIABLES

In this appendix some basic concepts of structural theory will be reviewed. Throughout the general discussion the example of a propped cantilever beam will be used to illustrate the concepts. This horizontal beam is built-in at its left end x = 0 and has a fixed simple support at its right end x = 1 (fig. 1).

Whereas the members of a structure are three-dimensional bodies, structural theory, in general, treats them as one- or two-dimensional continua. Rods, beams, and arches belong to the first class ; disks, plates, and shells, to the second. A point of a one or two-dimensional continuum may be specified by a single parameter or by a pair of parameters. For an arch, for instance, x may be the arc length measured from a reference point on the arch, and dx will then be used to denote the line element of the arch. For a plate, x may stand for the pair of rectangular coordinates specifying a point, and dx will denote the area element of the plate. Adopting a unified terminology, we shall call dx the *volume* of the considered element and use the term specific in the sense of *per unit volume*. Let P_k (x) (k = 1,2,...) denote the specific intensities of the *distributed generalized loads* acting at the point x. They are the components of the load vector **F** (x) at the abscissa x. Assuming, for instance, that our example beam is subject to distributed rightward horizontal and downward vertical forces and a distributed counter-clockwise couple, we may denote their specific intensities by $P_1(x)$, $P_2(x)$ and $lP_3(x)$. The generalized loads P_1, P_2 and P_3 then have the same dimension. For the sake of brevity, concentrated forces and couples will not be explicitly introduced. They would add a sum $\sum_\alpha \mathbf{F}_\alpha \cdot \mathbf{u}_\alpha$ to the integration of $w^{(e)}(x)$ as given by (1).

To the generalized loads $P_k(x)$ there corresponds *generalized displacements* $p_k(x)$ in the sense that the work of the former on the latter is given by the integral of

$$w^{(e)}(x) = \sum_k P_k(x)\, p_k(x) \tag{1}$$

over the structure. The scalars $p_k(x)$ are the components of the vector **u**(x). The quantity $w^{(e)}(x)$ will be called the *specific external work* at the point x. For our example beam the generalized displacements are the rightward horizontal and downward vertical displacements an l times the counter-clockwise rotation of the cross section. Note that the definition that led to dimensionally homogeneous generalized loads resulted in dimensionally homogeneous generalized displacements.

The generalized displacements, which are supposed to be small in comparison to the dimensions of the structure, are subject to *kinematic conditions of continuity*. For the example beam we shall require the displacements to be continuous and to have piecewise continuous first derivatives with respect to x.

The generalized displacements are also subject to *kinematic constraints*, which may be *external or internal. For* our beam the external kinematic constraints are

$$p_1(0) = p_2(0) = p_3(0) = 0, \quad p_1(l) = p_2(l) = 0. \tag{2}$$

To each constraint corresponds a reaction. For the example beam, the reaction corresponding to, say, the constraint $p_2(l) = 0$ is a vertical force at x = l. Because the work of the reactions on the displacements $p_k(x)$ vanishes, the constraints are called *workless*.

External loads and displacements produce internal generalized stresses Q_j (x) and generalized strains q_j (x), which must be such that the *specific virtual internal work* be given by

$$w^{(i)}(x) = \sum_j Q_j(x)\, q_j(x) \tag{3}$$

For the static variables $P_k(x)$, $Q_j(x)$ and the kinematic variables $p_k(x)$ and $q_j(x)$ to describe properly the behaviour of the structure, *we stipulate the applicability of the principle of virtual work*

$$\int \left[\left(w^{(i)}(x) - w^{(e)}(x) \right) dx \right] = 0 \qquad (4)$$

where the integral is extended over the structure.

Application of (4) will enable, if necessary, to relate generalized stresses to generalized loads, or generalized strains to generalized displacements. Indeed, *if we start from the kinematic description of the structure*, we shall, in our beam example in which geometry and loading are supposed to be symmetrical with respect to the xy plane (fig. 1), make the basic assumption that *plane sections remain plane* (but possibly not normal to the deformed axis). Hence, the deformation of an element of beam with length dx between two cross sections is made of (see fig.2) :

- a relative displacement q_1 of the end points along ox
- a variation q_2 of the initially right angle between the beam axis and the trace of the section of abscissa x in the xy plane
- a relative rotation q_3 around oz.

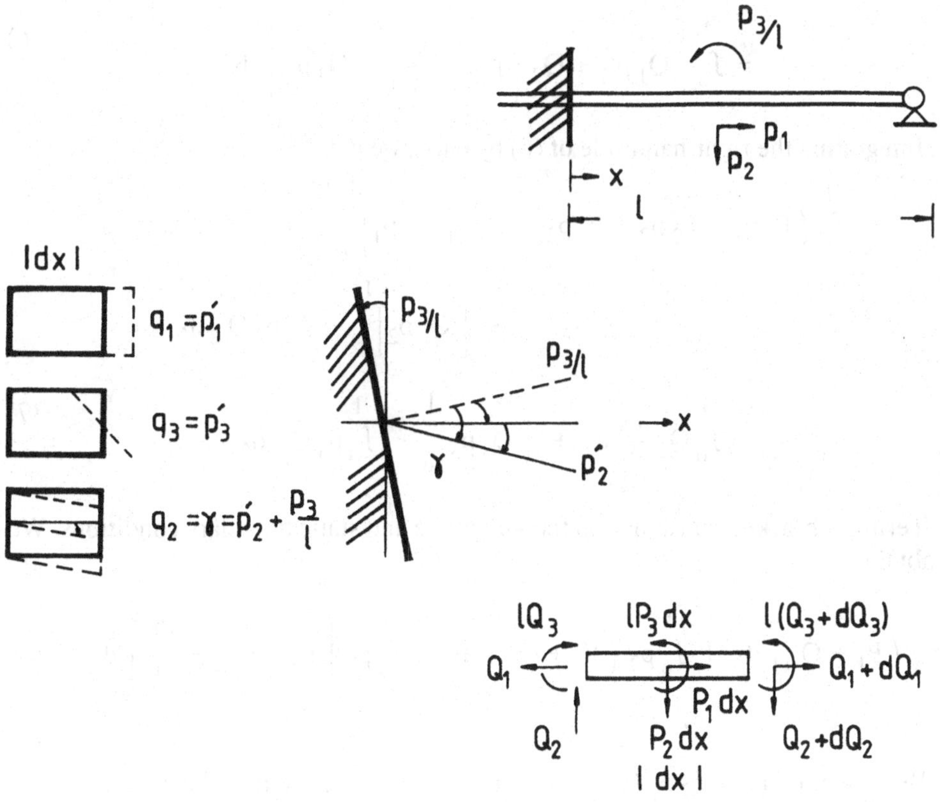

Fig. 1. 2. 3.

We thus obtain

$$q_1 = p'_1 \quad (a)$$

$$q_2 = p'_2 + \frac{p_3}{l} \quad (b) \qquad (5)$$

$$q_3 = p'_3 \quad (c)$$

Relations (5)a and c are obvious. Relation (5)b takes into account the fact that γ and p'_2 are positive clockwise whereas $p_3$is positive counter-clockwise.

The generalized loads and stresses consistent with this kinematic description are seen in fig. 3. Now, if we wish to derive the equilibrium equations relating the generalized stresses Q_j to the loads P_k, we apply equation (4) and use relations (5) to obtain

$$\int_o^l \left(P_1 p_1 + P_2 p_2 + P_3 p_3 \right) dx = \int_o^l \left(Q_1 q_1 + Q_2 q_2 + Q_3 q_3 \right) dx$$

$$= \int_o^l \left[Q_1 p'_1 + Q_2 \left(p'_2 + \frac{p_3}{l} \right) + Q_3 p'_3 \right] dx \qquad (6)$$

Integrating the right-hand side of (6) by parts, we get

$$\int_o^l \left(P_1 p_1 + P_2 p_2 + P_3 p_3 \right) dx = \left[Q_1 p_1 \right]_o^l - \int_o^l p_1 Q'_1 \, dx$$

$$+ \left[Q_2 p_2 \right]_o^l - \int_o^l p_2 Q'_2 \, dx$$

$$+ \int_o^l Q_2 \frac{p_3}{l} dx + \left[Q_3 p_3 \right]_o^l - \int_o^l p_3 Q'_3 \, dx \qquad (7)$$

Terms in bracket vanish due to the kinematic and static boundary conditions. We thus obtain

$$\int_o^l p_1 \left(P_1 + Q'_1 \right) dx + \int_o^l p_2 \left(P_2 + Q'_2 \right) dx + \int_o^l p_3 \left(P_3 + Q'_3 - \frac{Q_2}{l} \right) dx = 0 \qquad (8)$$

Because the functions $p_1(x)$, $p_2(x)$, $p_3(x)$ are arbitrary, we must have

$$\begin{cases} Q'_1 = -P_1 & \text{(a)} \\ Q'_2 = -P_2 & \text{(b)} \\ \text{and } Q'_3 = \frac{Q_2}{l} - P_2 & \text{(c)} \\ \text{or} & \\ Q''_3 = \frac{P_2}{l} - P'_3 & \text{(c')} \end{cases} \qquad (9)$$

We see that (a), (b), (c') are the equations of equilibrium of translation along x, along y and rotation about z, where Q_1 is the axial force, Q_2 the shear force and $l\,Q_3$ the bending moment.

If we now assume, according to BERNOULLI, that the cross sections remain plane and *normal to the deformed axis*, we must impose $p'_2 = -\frac{p_3}{l}$, that is $q_2 = 0$. The shear force Q_2 is then a *workless internal reaction*. In this case, we can obtain *directly* the equilibrium equations (9)a and c', where only Q_1 and Q_3 appear, without recourse to (9)b. Indeed, because p_1 is arbitrary in (8), we have $P_1 + Q'_1 = 0$. On the opposite p_2 and p_3 are not independently arbitrary because $p_3 = -p'_2\,l$. Hence, the other two terms of (8) must be taken together. Using relation $p_3 = -p'_2\,l$ and integrating the second term of (8) by parts, we get

$$\int_0^l p_2 P_2\,dx + \Big[p_2 l\big(P_3 + Q'_3\big)\Big]_0^l - \int_0^l \big(P'_3 + Q''_3\big)\,p_2 l\,dx = 0$$

The bracket vanishes according to the boundary conditions

$$p_2(l) = p_2(o) = 0$$

Hence,

$$\int_0^l p_2 P_2\,dx + \int p_2 l\big(P'_3 + Q''_3\big)\,dx = 0$$

where p_2 is arbitrary. We obtain

$$P_2 + lP'_3 + lQ''_3 = 0$$

or equivalently

$$Q''_3 = - P'_3 - \frac{P_2}{l}$$

which is equation (9)c'.

If we prefer to start from the static description of the structure, we first define our static variables as shown in fig. 3 and derive the equilibrium equations (9). We now apply the principle of virtual work to relate the generalized displacements to the generalized strains. We use in (4) the equations of equilibrium to express the generalized loads in the definition of $w^{(e)}$ by the generalized stresses and their derivatives, remove the latter by integration by parts, and remember that the exterior kinematic constraints are workless. The desired formulas for the generalized strains then follow from the requirement that the coefficient of each generalized stress in the integrand must vanish because the variation of the generalized loads with x, and hence that of the generalized stresses, is arbitrary.

If we first assume the loads on the example beam to be continuous, the generalized stresses and their first derivatives are continuous. We then have

$$\int_0^l w^{(e)}\, dx = - \int_0^l \left[Q'_1 p_1 + Q'_2 p_2 + \left(Q'_3 - \frac{Q_2}{l} \right) p_3 \right] dx$$

$$= - \left(Q_1 p_1 + Q_2 p_2 + Q_3 p_3 \right)_0^l + \int_0^l \left[Q_1 p'_1 + Q_2 \left(p'_2 + \frac{p_3}{l} \right) + Q_3 p'_3 \right] dx \tag{10}$$

On account of the external kinematic constraints and the static constraint, the first term in the second line of eq. (10) vanishes. Substitution of the integral in the second line of eq. (10) into eq. (4) then furnishes

$$\int_0^l \left[Q_1 \left(q_1 - p'_1 \right) + Q_2 \left(q_2 - p'_2 - \frac{p_3}{l} \right) + Q_3 \left(q_3 - p'_3 \right) \right] dx = 0 \tag{11}$$

and, hence, the following definitions of the generalized strains :

$$q_1 = p'_1$$

$$q_2 = p'_2 + \frac{p_3}{l}$$

$$q_3 = p'_3.$$

If, say, the load P_1 and, hence, Q'_1 have discontinuities at $x = a$, separate integrations by parts must be performed over the beam segments to the left and right of $x = a$. In the second line of eq. (10) we must then add the term $\left[Q_1 \, p_1 \right]_{a-o}^{a+o}$, which does, however, vanish because Q_1 and p_1 are continuous at $x = a$.

In treating a specific problem the analyst may decide to neglect certain generalized strains. For example, if the right-hand support of the example beam had horizontal mobility, we might decide to treat the material centre line of the beam as inextensible. This amounts to setting $q_1 = 0$ and thus to introducing the *internal kinematic constraint* $p'_1 = 0$, whose imposition changes the status of the axial force from that of a generalized stress to that of a *reaction* to this constraint. The considered constraint is *workless* because the reaction to it (that is, the axial force) does no work on displacement satisfying the constraint.

If we assume mobility of the right-hand support, the axial force vanishes at this end. The variation of the axial force along the beam follows from this boundary condition and the first equation of equilibrium. For our example beam, however, with its fixed right-hand support, we can no longer determine the variation of the axial force along the beam if we treat its material centre line as inextensible.

A field $Q_j(x)$ of generalized stresses that satisfies the static conditions of continuity, the static constraints, and the equations of equilibrium for the given loads is said to be *statically admissible* for these loads. A field $p_k(x)$ of generalized displacements is called *kinematically admissible* if it satisfies the kinematic conditions of continuity and the external and internal kinematic constraints. A field $q_j(x)$ of generalized strains is called kinematically admissible if it is derived from a kinematically admissible field of generalized displacements.

Generalized loads $P_k(x)$ and generalized stresses $Q_j(x)$ that are statically admissible for these loads will be said to constitute a set of *associated static variables*. Similarly, kinematically admissible displacements $p_k(x)$ and the generalized strains $q_j(x)$ derived from them will be said to constitute a set of *associated kinematic variables*. In view of the way in which the generalized strains are derived from the generalized displacements, the principle of virtual work is valid for *independent* sets of associated static and kinematic variables ; there is no need for these two sets to obey the relation of cause and effect.

APPENDIX 3 : LIMIT LOAD EVALUATION BY COMBINATION OF BASIC MECHANISMS

The method will be explained on the example of fig. 1. In this portal frame, plastic hinges can occur in the six critical cross-sections 1,2,...6, where the location of section 3 is not known but will tentatively be set at mid-span. The degree of redundancy r of the structure being 3, a one-degree of freedom mechanism will necessitate 4 hinges. Combinations of the six possible locations of hinges 4 by 4 results in 15 mechanisms, each of which will give a value s'' by its power equation. Because screaning the whole set of mechanisms to choose that with lowest s" is a fastidious task, we wish to find a shortcut to this value.

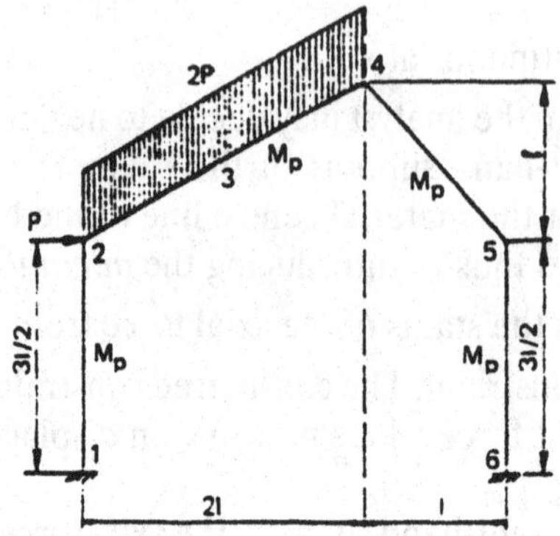

Fig. 1

The number of linearly independent equations of equilibrium relating the moments M_i in the m (here 6) critical sections is m-r, which is also the number of *linearly independent mechanisms* with which these equilibrium equations can be obtained using the theorem of virtual powers. The mechanisms of such a set will be called *basic*. All licit mechanisms can be constructed by linear combination of the basic mechanisms. In our example, we may choose the 6-3 = 3 mechanisms of fig. 2, of which we verify that no one can be obtained by combination of the others.

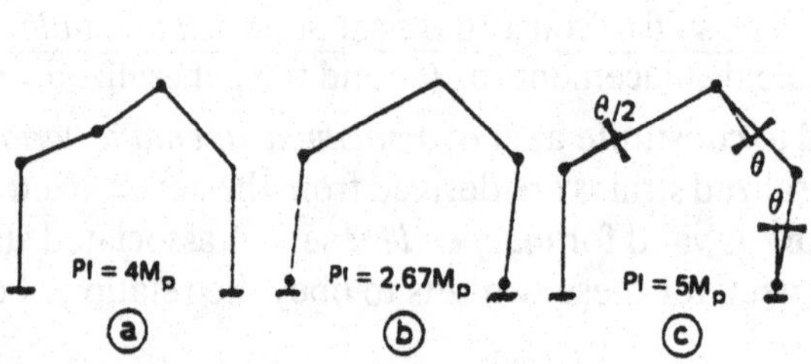

Fig. 2

For the *beam mechanism* a), obtaining the rotation rates of two rigid parts 2-3 and 3-4 is straightforward. For mechanisms b) and c) the rotation rates of the various rigid parts are obtained locating the instantaneous centre of absolute rotation I (see fig. 3 for mechanism c)).

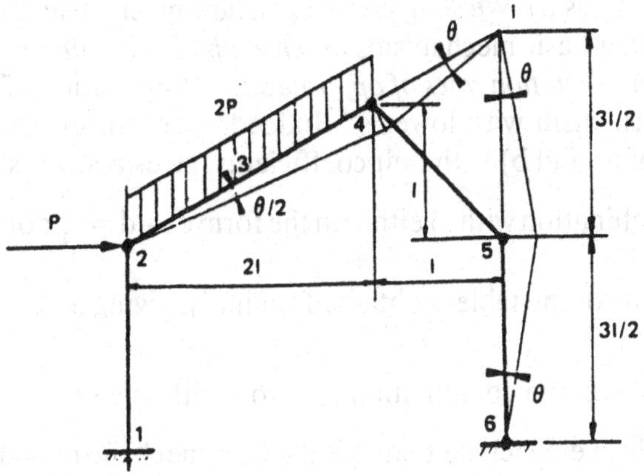

Fig. 3

The absolute value of a hinge rotation rate is the sum of the absolute rotation rates of the adjacent parts.

	Rotations						P_e	D	s''
Sections / Mechanisms	1	2	3	4	5	6			
a	0	-1	2	-1	0	0	Pl	$4M_p$	$\frac{4M_p}{l}$
b	-1	+1	0	0	-1	+1	$\frac{3}{2}$ Pl	$4M_p$	$\frac{2{,}67M_p}{l}$
c	0	$-\frac{1}{2}$	0	$\frac{3}{2}$	-2	1	Pl	$5M_p$	$\frac{5M_p}{l}$
d=a+b	-1	0	2	-1	-1	+1	$\frac{5}{2}$ Pl	$6M_p$	$\frac{2{,}4M_p}{l}$
$e=d+\frac{2}{3}c$	-1	$-\frac{1}{3}$	2	0	$-\frac{7}{3}$	$+\frac{5}{3}$	$\frac{19}{6}$ Pl	$\frac{22}{3}M_p$	$\frac{2{,}31M_p}{l}$
$f=e+\frac{1}{3}b$	$-\frac{4}{3}$	0	2	0	$-\frac{8}{3}$	$\frac{6}{3}$	$\frac{22}{6}$ Pl	$8M_p$	$\frac{2{,}18M_p}{l}$

Table 1

The proper sign being assigned to the rotation rates of the hinges (in relation to the sign convention on bending moments), and if we give arbitrarily the unit value to one of them in each basic mechanism, we obtain the three first lines of table 1. The "best" mechanism, that is with lowest s'', is b). We shall construct a new mechanism d) by combination of b)

The proper sign being assigned to the rotation rates of the hinges (in relation to the sign convention on bending moments), and if we give arbitrarily the unit value to one of them in each basic mechanism, we obtain the three first lines of table 1. The "best" mechanism, that is with lowest s", is b). We shall construct a new mechanism d) by combination of b) with one of the other basic mechanisms *which enables to eliminate the rotation rate in one section by addition of rotation rates of opposite signs.* Application of this rule will result in an improved mechanism with lower s". Indeed, mechanism d) of line 4 obtained by simple addition of a) and b) with unit coefficients, gives a lower s". Improvement of d) is obtained by combination with c) either in the form $e = d = \frac{2}{3}c$ of line 5, or equivalently by $\frac{3}{2}d + c$. Last line of the table is obtained similarly, giving mechanism f with $s" = 2.18 \frac{M_o}{l}$. Because applying the combination rule to f with any other preceding mechanism appears impossible, we conclude that f) is the best mechanism under the assumption of hinge 3 being at mid-span of bar 2-4. To ascertain this conclusion, we compute the bending moments in the non-plastified critical sections, under the loading with $s" = 2.18 \frac{M_o}{l}$. We know, from mechanism f, that $M_1 = -M_o$, $M_3 = M_o$, $M_5 = -M_o$ and $M_6 = M_o$.

To obtain M_2 we apply the theorem of virtual powers to mechanism b), which reads

$$-M_1 + M_2 - M_5 + M_6 = \frac{3}{2} s"l.$$

Introducing the known values of M_1, M_5, M_6 and s" we obtain

$$M_2 = 0.27M_o < M_o$$

Similarly, with mechanism a) we get

$$M_4 = 0.45M_o > -M_o \quad .$$

Hence, we conclude that there exists a licit bending moment diagram corresponding to mechanism f, except for the fact that we have arbitrarily placed the hinge 3 at mid-span. Indeed, if we trace the bending moment diagram in the beam 2-4, we find that the maximum value of M does not occur at mid-span and is larger than M_o. If we now locate hinge 3 at the section of maximum bending moment under the load factor $s" = 2.18 \frac{M_o}{l}$ and keep the other hinges, the mechanism modified in this manner gives s" =

$2.14\,\frac{M_o}{l}$. Repeating the same "adjustment" of the hinge location for section 3 will not result in an appreciable modification of s". Indeed, the value obtained with section 3 at mid-span is already an acceptable approximation of s_L because the error is only about 2%.

APPENDIX 4 : LIMIT LOAD EVALUATION BY LINEAR PROGRAMMING FOR SYSTEMS OF BEAMS

Let us first consider the static approach to limit load evaluation. For more generality assume that the limit moments in positive and negative bending may be different. Denote them by M_{oj} and M'_{oj} for section j.

The unknowns are the bending moments M_j at the critical cross-section (j = 1, ..., m) and the load factor s_L at collapse.

The data of the problem are the values of M_{oj} and M'_{oj} for j = 1, ..., m, together with the coefficients θ_{kj} of the unknowns M_j and independent terms e_k (k = 1, 2, ...(m-r)) of the linearly independent equations of equilibrium obtained by application of the theorem of virtual powers with m-r basic mechanisms.

The static theorem tells us that, provided we satisfy the constraints

$$\sum_{j=1}^{m} \theta_{kj} M_j - s' e_k = 0 \qquad k = 1, ...(m\text{-}r)$$

of static admissibility and

$$\begin{cases} M_j \le M_{oj} \\ -M_j \le M'_{oj} \end{cases}$$

of plastic admissibility, the load factor s_L is obtained by the maximization

$$\max s' = 1 \,.\, s' + 0 \,.\, M_1 + \ldots + 0 \,.\, M_m.$$

Such a maximization problem of a *linear* function of the variables (called objective function), subject to *linear equality and inequality conditions*, is called a linear programming problem and can be solved by existing ad hoc algorithms.

Now, to each such linear programming problem (L.P), called *primal*, there exists a corresponding L.P problem called *dual* such that

- the number of variables of the dual is equal to the number of constraints of the primal (and conversely)

- variables of one problem are free in sign if they correspond to equality conditions of the other, or are non-negative if they correspond to non – strict inequality constraints of the other problem. In the latter case the variables are non-vanishing only when the corresponding inequality condition is satisfied as equality

- right-hand side constants of the constraints of one problem are the coefficients of the objective function of the other (and conversely)

- the matrix of the coefficients of the constraints of one problem is the transpose of that of the other problem

- if one problem is a maximization, the other is a minimization, and both extrema coincide.

If we take the static formulation as written above and formally dualize it, we can discover that the variables are the rotation rates of the collapse mechanism, each being obtained by linear combination with unknown coefficient, which are also variables of the dual, of the corresponding rotation rates of the basic mechanisms.

The objective function is the total dissipation in the collapse mechanism, which is minimized subject to normalization of the power of the applied forces. Hence the dual problem is the kinematic approach of limit analysis.

For more information of the application of mathematical programming to limit analysis and design we refer the reader to the literature ([] to []).

LIMIT DESIGN: FORMULATIONS AND PROPERTIES

C. Cinquini
University of Pavia, Pavia, Italy

ABSTRACT

In the present paper a general formulation of optimal design problems is firstly proposed, and the variational method founded on Lagrangian multiplier technique is shown. As an application a simple example for optimal plastic design of beams is solved.
The variational formulation is discussed in detail for the case of plastic design of circular plates: optimality criterion is shown and special features of optimal solutions are discussed.
Then the same problem is proposed in a discretized form, which makes use of a finite different technique and leads to a linear programming formulation. The features of such an approach are discussed and numerical solutions are shown as well.

1. INTRODUCTION

Analytical approaches to optimum structural design problems, enormously increased over the past years, were developed, in the first pioneering papers, on the basis of two different fundamental methods.
In the first one, some of the properties of the optimized structures are defined on the basis of the relevant structural mechanics theorems. The optimal design of a structure generally needs some physical predictions on the expected solution. Thus, the solution technique, strictly dependent on the particular problem, sometimes becomes laborious for complex structures. Nevertheless elastic optimization (for prescribed compliance, for given deflection, etc.) optimum design for prescribed collapse load, optimization for given buckling load were studied (see e.g.[1][6]) and unified formulations proposed [7][10]).
The second way rests on the variational formulation of the problem. Plastic ([11][12]) and elastic ([13][14]) optimization problems have been studied. As known, this approach allows for the implementation of numerical methods; thus, theoretically, any physical prediction on the expected solution can be neglected. A large qualified literature can be found concerning numerical approaches (see e.g. [19][21]).
In the present paper a general formulation of optimal design problems is firstly proposed, and the variational method founded on Lagrangian multiplier technique is shown. As an application a simple example for optimal plastic design of beams is solved.
The variational formulation is discussed in detail for the case of plastic design of circular plates: optimality criterion is shown and special features of optimal solutions are discussed.
Then the same problem is proposed in a discretized form, which makes use of a finite different technique and leads to a linear programming formulation. The features of such an approach are discussed and numerical solutions are shown as well.

2. GENERAL REMARKS ON OPTIMAL DESIGN

2.1. Formulation of the problem

A structural optimization problem can be seen, from a mathematical point of view, as a constrained minimum problem.
The design variable will be denoted by h; the other functions (state functions) by q, the physical meaning of which depends on the features of the specific problem. The domain defined by structure geometry will be denoted by Ω ($\Omega \in R^n$, with n = 1,2 for one- ore two-dimensional structures respectively).
Suitable functional spaces must be considered as follows

$$h \in Y; \qquad q \in Z$$

The objective function $\Gamma(h)$, to be minimized, is obtained by integrating on Ω a specific "cost" function $\gamma(h)$

$$\Gamma(h) = \int_{\Omega} \gamma(h)\, d\Omega \qquad (2.1)$$

The mathematical constraints of the minimum problem are defined by physical performances, that is by state equations and behavioral and technological constraints. For the sake of simplicity only local constraints are considered in the present formulation.
Then, the general form of the set of constraints is

$$F_i(h, q) = 0 \text{ in } \Omega \qquad (i = 1...n), \qquad (2.2)$$

$$G_j(h, q) \le 0 \text{ in } \Omega \qquad (j = 1...m) \qquad (2.3)$$

which define in YxZ a set HQ, representing the set of the feasible solutions. In Y the set H of feasible design solutions is defined as

$$H = \{h \in Y \mid \exists q \in Z \text{ for which } (h, q) \in HQ\}$$

The constrained minimum problem may be expressed as

$$\inf_{h \in H} \Gamma(h) \qquad (2.4)$$

2.2. Existence and uniqueness of the solution

In order to discuss the conditions for existence and uniqueness of the solution, the results of convex analysis (see, e.g. [20]) can be used, which are considered in [12][21].
If Y and Z are able to be defined as Hilbert spaces, the above mentioned results may be applied.
In this hypothesis the following conditions are necessary:
(a) $\Gamma(h)$ continuous and convex;
(b) HQ convex, closed and not empty.
Condition (a) is rather restrictive, because convexity of $\Gamma(h)$ is not fulfilled in several problems. As condition (b) is concerned, the set HQ may be considered not empty, according to a correct formulation of the physical problem and in most of cases there is no difficulty in investigating for such formulations where the set HQ is closed.
The discussion on convexity is more complex: frequently this condition may be fulfilled in plastic design problems [2], while it is very hard for elastic ones.
In complying with hypotheses (a) and (b) the existence of an optimal solution for the optimization problem is assumed if one between the following conditions is fulfilled:
(c1) H is limited;
(c2) $\Gamma(h) \to \infty$ if $\|h\| \to \infty$ for $h \in H$
Condition (c1) is certainly verified if an upper bound is prescribed for h.
Condition (c2) may be verified, depending on the particular form of $\Gamma(h)$.
Under the hypotheses (a), (b) and (c1) or (c2), the uniqueness of the

solution (at least for the design variable) can be guaranteed, if function $\Gamma(h)$ is strictly convex.

2.3. Variational formulation

The minimum problem defined in (2.4), may be reformulated by defining $\phi(h,q) = \Gamma(h)$, as

$$\inf_{(h,q) \in HQ} \phi(h,q) \tag{2.5}$$

Indeed q may be seen as a "dead" variable.
Let a function $\psi_{HQ}(h,q)$, be considered defined as follows:

$$\psi_{HQ}(h,q) = \begin{cases} 0 & \text{if } (h,q) \in HQ \\ +\infty & \text{otherwise} \end{cases} \tag{2.6}$$

Then problem (2.5) is equivalent to

$$\inf_{(h,q)\ YxZ} \phi(h,q) + \psi_{HQ}(h,q) \tag{2.7}$$

A general form for $\psi_{QH}(h,q)$ is suggested by the Lagrangian multipliers technique

$$\psi_{HQ}(h,q) = \sup_{\lambda_i \in U_i} \sup_{\mu_j \in V_j} \int_\Omega \left(\sum_{i=1}^{n} \lambda_i F_i + \sum_{j=1}^{m} \mu_j G_j \right) d\Omega \tag{2.8}$$

where the multipliers μ_j are to be assumed non-negative.
Then, the optimization problem becomes

$$\inf_{(h,q)\ YxZ} \sup_{\lambda_i\ U_i} \sup_{\mu_j\ V_j} L = \Gamma(h) + \int_\Omega \left(\sum_{i=1}^{n} {}_i F_i + \sum_{j=1}^{m} \mu_j G_j \right) d\Omega \tag{2.9}$$

The stationarity conditions of the functional L defined in (2.9) with respect to λ_i and μ_j furnish relations (2.1), (2.2), and also

$$\mu_j\ G_j = 0 \quad \text{in} \quad \Omega \quad (j=1...m) \tag{2.10}$$

so- called orthogonality or switching conditions.
The stationarity conditions with respect to q usually provide further behaviour constraints [11], that are, in some cases, the constraints of the dual problem [12]. The stationarity condition of with respect to the design variable h furnishes the optimality criterion.
The solution of the optimization problem may be sought among the solutions of the stationarity condition system (system of partial

derivative equations or inequalities).
In the convexity hypotheses seen in Sect 2.2, the stationarity condition system (system of partial derivative equations or inequalities).
In the convexity hypotheses seen in Sect. 2.2, the stationarity conditions of L provide one and only one solution which is the solution of the optimization problem. If the convexity of the objective function , or of the set HQ, is not fulfilled, the uniqueness of solution for the stationarity problems is no larger guaranteed.

2.4. Physical aspects

The above-exposed considerations imply that, from a mathematical viewpoint, the existence and uniqueness of solution is assured and this solution may be found by the variational approach seen in sect. 2.3 only for a rather limited class of problems.
However, the existence of solution can be frequently assured at least on the basis of physical considerations, so that a larger class of problems may be dealt with and, moreover, a variational approach may be formulated, even if the convexity hypotheses are not fulfilled: the stationarity conditions of the functional L can be considered at least as necessary optimality conditions.
A suitable problem discretization is required for numerical solutions. In order to assure the existence and uniqueness of the solution analogous difficulties are met, although the discretization allow the problem to be studied in a finite dimension subspace of YxZ.
A lot of problems can be numerically investigated, by using conditions which are only necessary , as they were sufficient taking into account the fact that the actual solutions are limited to the physically meaningful ones.
From a mathematical point of view it can be emphasized that prescribed upper and lower bounds on h assure the set H to be limited (see sect. 2.2) and excludes some possible singular points in the solution.
If a shape function is imposed on the variation of h, the optimal criterion is to be modified. In particular, if the design function is prescribed to be piecewise constant, the optimality criterion is integrated on each part having the same design variable.

3. OPTIMAL DESIGN OF BEAMS FOR PRESCRIBED PLASTIC COLLAPSE LOAD

3.1. Static formulation

Numerous papers dealt with the problem of the optimal plastic design, in particular for bending structures. According with the approach seen in section 2 a variational formulation is proposed here for rigid perfectly-plastic bending beams. Equilibrium equation and plastic rule read:

$$p + M'' = 0, \tag{3.1}$$

$$-M_L^- \le M \le M_L^+ \tag{3.2}$$

where p and M ar load an moment function respectively. A prime

represents the derivative with respect to be abscissa x. M_L^+ and M_L^- are limit moments, depending on the design variable h, which can be a geometrical parameter of beam cross section.
Obviously, if l is the length of beam, $\Omega \equiv [0,l]$ and x Ω are to be assumed.
The functional L, defined in section 2 becomes

$$L = \int_0^l \gamma(h)dx + \int_0^l \lambda(p+M'')dx + \int_0^l \mu_1(M-M_L^+)dx + \int_0^l \mu_2(-M-M_L^-)\,dx \qquad (3.3)$$

and the optimization problem may be written as

$$\inf_{\substack{h \geq 0 \\ M}} \; \sup_{\substack{\lambda \\ \mu_1,\mu_2 \geq 0}} \; L \qquad (3.4)$$

The stationarity conditions of L provide Rel.s (3.1)(3.2) and

$$\mu_1(M - M_L^+) = 0 \qquad (3.5)$$

$$\mu_2(-M - M_L^-) = 0 \qquad (3.6)$$

$$\gamma_{,h} - \mu_1 M_{L,h}^+ - \mu_2 M_{L,h}^- = 0 \qquad (3.7)$$

$$\lambda'' + \mu_1 - \mu_2 = 0 \qquad (3.8)$$

Rel.(3.8) is obtained by taking into account Green formula, and suitable boundary conditions.
In particular it has been assumed

$$M\lambda' = 0, \qquad M'\lambda = 0 \qquad \text{in } x = 0,\ x = l \qquad (3.9)$$

With suitable assumptions of functional spaces, the results seen in section 2 can be applied: in plastic optimization problems, a suitable choice of the design variable h allows the set HQ to be convex; in many cases also a convex objective function may be defined as well and conditions (c1), or (c2), seen in section 2, may be fulfilled.
The Lagrangian multipliers λ, μ_1, μ_2 represent the variables of the dual problem. Their physical interpretation is suggested by Rel.s (3.5),(3.6),(3.8), and boundary conditions: if the material shows an associated flow rule, λ is a function proportional to the displacement rate at the collapse, and $(\mu_1 - \mu_2)$ is proportional to the rates of plastic curvatures.
On the basis of this interpretation, Eq.(3.8) assures the compatibility of the collapse mechanism, Eq.s (3.5) and (3.6) represent the complementarity rule. Then, the analytical minimum, reached by means of mathematical theorems, represents a correct solution in order to verify the limit analysis theorems: a static solution, for which a collapse mechanism can be found in complying with compatibility and normality rules is achieved.

Note that, in case of non-convexity hypotheses, all the analytical solutions are correct plastic collapse solutions. The optimal design is found among the designs obtained in this way.
Finally, Rel. (3.7) represents the optimality criterion for optimal plastic design problems: the derivative of the cost function with respect to the design variable is equal to the derivative of the dissipated power at collapse.

3.2. Kinematic formulation

On the basis of the kinematic theorem of limit analysis, a kinematic formulation of the optimal plastic design may also be proposed.
Analytical considerations allow this formulation to be expressed by starting from the functional L defined in (3.3).
The compatibility condition of the collapse mechanism is fulfilled , if the plastic displacement rate function is assumed as follows

$$y = \lambda / c; \qquad y'' = -(\mu_1 - \mu_2)/c \tag{3.10}$$

where c is a positive constant.
Accordingly, the dissipated power (always non-negative) may be defined

$$D = \begin{cases} -y''\, M_L^{+} & \text{if } y'' < 0 \\ 0 & \text{if } y'' = 0 \\ +y''\, M_L^{-} & \text{if } y'' > 0 \end{cases}$$

Then, by using Green formula

$$L = \int_0^1 \gamma(h)\,dx + c \int_0^1 (py - D)\,dx \tag{3.12}$$

If side constraints on design variable are omitted for the sake of simplicity, the optimization problem becomes

$$\inf_{\substack{h \ge 0 \\ y}} \ \sup_{c \ge 0} \ L \tag{3.13}$$

The maximum condition with respect to c furnishes

$$\int_0^1 (py - D)\,dx \le 0 \tag{3.14}$$

while with respect to h and y (taking into account Green formula and suitable boundary conditions) one has

$$\gamma_{,h} - cD_{,h} = 0 \tag{3.15}$$

$$p - D_{,y} = 0 \tag{3.16}$$

The kinematic formulation is pointed out in relation (3.14): among all possible mechanisms, the optimum design is such that the power performed by external loads does not exceed the internal dissipated power.
Applying the variational method to the kinematic formulation the optimality criterion of Rel. (3.15) is obtained, the meaning of which is the same of Rel. (3.7). The equilibrium is assured too: the relationship between external loads and moments is provided by Rel. (3.16).
Therefore, starting from a kinematic formulation all the theorems of limit analysis are fulfilled: for a material with associated flow rule a correct plastic solution is achieved.

3.3. An example

Numerous examples have been proposed for plastic optimization problems of beams; some classical absolute minimum cost solutions are given on the basis of the optimality condition, obtained by using limit analysis Theorems (see e.g.[1]-[6]). Starting from the variational formulation, some examples are shown in [22], with prescribed piecewise constant design function. Generally speaking, analytical solutions may be obtained only for rather simple examples.
No considerable difficulties are found, if numerical methods of solution are investigated, in consequence of the convex set of the feasible solutions. The linearity of relations (3.1),(3.2)(and (3.14) for kinematic methods) allows techniques of convex programming to be used in particular linear programming with linear cost functions (see e.g. [15]-[16]).
An example of absolute minimum cost solution will be shown here, starting from the variational formulation above-exposed.
Let $\gamma(h) = h^2$, and $M_L^+ = M_L^- = kh$ (sandwich beams) be assumed. Thus, the hypotheses required for existence and uniqueness solution are fulfilled.
A fixed-end beam is considered, which is subjected to a uniformly distributed load p (see Fig.1).
Then

$$M = -(px^2/2) + (pl/2)x - X$$

where X is the unknown value of the moment in x = 0 and x = l.

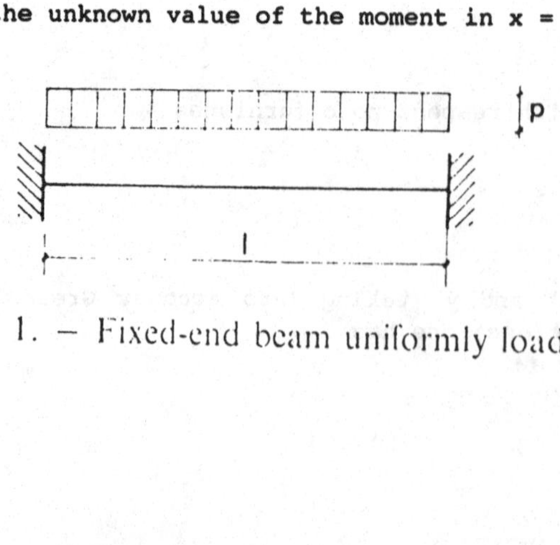

Fig. 1. – Fixed-end beam uniformly loaded.

Equation (3.7) reads

$$2h - k(\mu_1 + \mu_2) = 0$$

In the absence of technological constraints, the design function is to be expected equal to $|M|/k$.
Then, for $M > 0$

$$M = k\,h, \quad \mu_2 = 0, \quad \mu_1 = 2h/k, \quad \lambda'' = -\mu = -\ 2h/k = -\ 2M/k^2$$

For $M < 0$

$$M = -\ kh, \quad \mu_1 = 0, \quad \mu_2 = 2h/k, \quad \lambda'' = \mu_2 = 2h/k = -\ 2M/k^2$$

Finally

$$\lambda'' = 1/k^2\ (px^2 - plx + 2X)$$

may be assumed, almost everywhere. Taking into account boundary conditions, function and unknown parameter X can be calculated

$$\lambda = p/6k^2\ (1/2\ x^4 - lx^3 + 1/2\ l^2x^2) \qquad X = pl^2/12$$

Moment function

$$M = p(-x^2/2 + 1/2x - l^2/12)$$

shows two zero-points at

$$x = 1/2 \pm \sqrt{3}/6\ l$$

The form of $h(x) = |M(x)|/k$ is drawn in Fig.2.

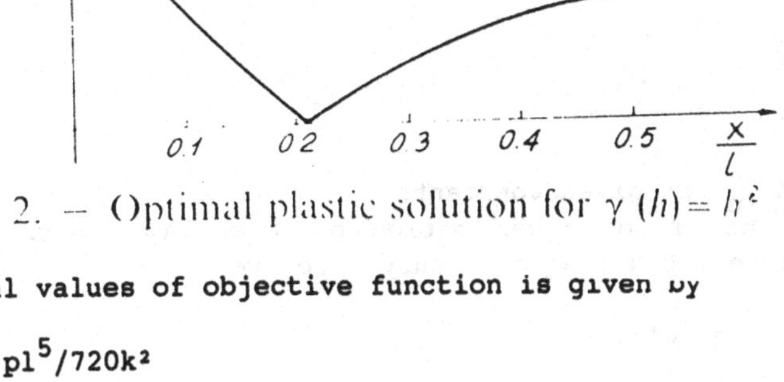

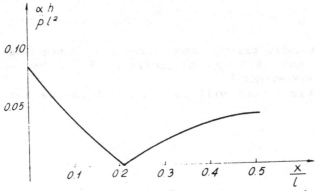

Fig. 2. – Optimal plastic solution for $\gamma\ (h) = h^2$

The optimal values of objective function is given by

$$\Gamma(h) = pl^5/720k^2$$

4. VARIATIONAL FORMULATION OF THE OPTIMAL PLASTIC DESIGN OF CIRCULAR PLATES

4.1. Preliminary remarks

In this section the optimal plastic design of circular (or annular) plates is dealt with, by using continuum formulation and a variational approach. So, by means of Lagrangian multiplier technique the optimality criterion is shown. Axisymetric loads are considered and Mises yield condition is adopted.

The "absolute" minimum cost design is the problem firstly discussed, having only lower and upper bouds on design variable as technological constraint; then the problem is formulated looking for a design composed by rings of constant thickness.

4.2. Formulation of the problem

Circular or annular plates simply supported or built-in at outer edge are considered. For simplicity, the inner edge of annular plates is assumed to be free.

The usual assumptions of thin-plate theory regarding the smallness of the plate thickness and deflection are adopted. Plate material is assumed to obey to a rigid, perfectly plastic flow rule with the same behaviour in both tensile and compressive states.

Rotational symmetry conditions are assumed to be fulfilled and a prime will denote the differentiation with respect to the radius r, which is the only significant geometric variable. For brevity, only distributed loads p(r) are considered. The state of stress for a generic value of r will be specified by the radial and circumferential bending moments M(r) and N(r), the shear force being regarded as a "reaction" in the sense of Prager [9]. The equilibrium equation is

$$M'' + (2M' - N')/r + p = 0 \tag{4.1}$$

and the yield condition

$$f(M,N) - M_p \le 0 \tag{4.2}$$

where M_p is the fully plastic moment, and f is homogeneous of first order in M and N. In particular, for the von Mises condition, $f(M,N)=(M^2+N^2-MN)^{1/2}$.

The fully plastic moment will be subjected to the technological constraints

$$\begin{aligned} &M_p - M_{max} \le 0 \\ &M_{min} - M_p \le 0 \end{aligned} \tag{4.3}$$

where M_{max} and M_{min} are given constants.

If w(r) denotes the rate of deflection of a collapse mechanism, the radial and circumferential rates of curvature are

$$\kappa = -w'' \quad , \qquad \chi = -w'/r \tag{4.4}$$

wherever w, w' and w" are continuous.
Denoting the specific cost (cost per unit area of the middle plane) by $\gamma(M_p)$, a plate design $M_p(r)$ is looked for, which is at the plastic collapse condition under the given load p(r) and has minimum total cost

$$\Gamma(M_p) = 2\pi \int_{R_i}^{R_e} r\gamma(M_p)dr$$

where R_i and R_e are the inner and outer radii of the plate. The relation between the specific cost and the fully plastic moment depends on the type of plate; for a sandwich plate of constant core depth, γ is proportional to M_p, if the cost is assumed to be proportional to the weight of the plate sheets.
To complete the formulation of the problem, static boundary conditions must be added, which depend on the nature of the support at outer edge and on the presence ($R_i > 0$) or absence ($R_i = 0$) of a central hole.

4.3. Optimality conditions

Applying the Lagrangian multiplier $\eta(r)$, which is not sign-restricted, to the equality constraint (4.1) and the nonnegative multipliers $\lambda(r)$, $\alpha(r)$, and $\beta(r)$ to the inequality constraints (4.2) and (4.3), we construct the functional

$$L = \int_{R_i}^{R_e} 2\pi r\{\gamma(M_p) + \eta(M''+(2M'-N')/r+P) + \lambda(f(M,N)-M_p) + \\ + \alpha(M_p-M_{max}) + \beta(M_{min}-M_p)\}\, dr$$

The stationarity conditions for L are necessary conditions for the optimality of the design.
The stationary conditions of L with respect to Lagrangian multipliers provide not only the Rel.s(4.1), (4.2) and (4.3) but also the equations

$$(f(M,N) - M_p) = 0 \tag{4.5}$$

$$\alpha(M_p - M_{max}) = 0 \quad , \quad \beta(M_{min} - M_p) = 0 \tag{4.6}$$

Denoting differentiation with respect to M_p by a superscript dot, we have the following stationarity conditions of L with respect to M_p, M and N

$$\dot{\gamma} - \lambda + \alpha - \beta = 0 \tag{4.7}$$

$$\eta'' + \lambda\partial f/\partial M = 0 \tag{4.8}$$

$$\eta'/r + \lambda\partial f/\partial N = 0 \tag{4.9}$$

for $R_i \le r \le R_e$. The natural boundary conditions are

$$\eta M + \eta r M' - \eta' M r - \eta N = 0 \tag{4.10}$$

for $r = R_i$ and $r = R_e$

It appears worth emphasizing at this point that in the derivation of (4.8) and (4.9) the multiplier η has been assumed to be twice continuously differentiable.

At a free edge $M = 0$ and $rM' - N = 0$, so that (4.10) does not impose any constraints on η and η'. At a simply supported edge $M = 0$ and (4.10) requires that $\eta = 0$. Finally, if $R_i = 0$, we have $M = N$ in the plate center and (4.10) is fulfilled without any conditions on . These facts suggest that the Lagrangian multiplier is proportional to the rate of deflection w

$$\eta = w/c \tag{4.11}$$

where c is a constant. The continuous differentiability of then excludes the presence of yield hinges. The formulation would have to modified for eventual discontinuities of w'.

Substituting (4.4) and (4.11) into (4.8) and (4.9) leads to

$$\kappa = -c\eta'' = c\lambda \partial f/\partial M \tag{4.12}$$

$$\chi = -c\eta'/r = c\lambda \partial f/\partial N$$

Accordingly,

$$\kappa/\chi = (\partial f/\partial M)/(\partial f/\partial N) \tag{4.13}$$

in accordance with the normality rule of plastic flow.

Since f is homogeneous of first order in M and N, it follows from (4.12) that the specific dissipated power is

$$D = \kappa M + \chi N = c\lambda/f \tag{4.14}$$

Solution of (4.14) for λ and substitution into (4.7) yields

$$\acute{\gamma} = (D/cf) - \alpha + \beta \tag{4.15}$$

Since a collapse mechanism is defined to within a constant positive factor, c = 1 in Rel. (4.15) can be assumed. Where $M_{min} < M_p < M_{max}$ it follows $\alpha = \beta = 0$ by $f = M_p$. The modified eq. (4.15) then reads

$$\mathring{\gamma} = D/M_p \tag{4.16}$$

If plastic flow occurs at a circle where $M_p = M_{min}$, one has $\alpha = 0$, $\beta \geq 0$, $f = M_p$, and hence

$$\dot{\gamma} \geq D/M_p \tag{4.17}$$

If $M_p = M_{min}$ and $f < M_p$, again

$$\gamma \geq D/M_p = 0 \qquad (4.18)$$

Finally of plastic flow occurs at a circle where $M_p = M_{max}$, one has $\alpha \geq 0$, $\beta = 0$, $f = M_p$, and hence

$$\gamma \leq D/M_p \qquad (4.19)$$

The necessary optimality conditions (4.16)-(4.19) may then be expressed as follows: the optimal plate admits a collapse mechanism such that, depending on whether the fully plastic moment has a value at the lower bound, between the bounds, or at the upper bound, the specific rate of dissipation per unit fully plastic moment is not greater than, is equal to, or is not smaller than the derivative of the specific cost with respect to the design variable.

4.4. The dual problem

Starting from the Lagrangian L, the dual problem of the original optimization one can be formulated [11][12] and [23].
With the above mentioned hypotheses, the solution can be reached by maximizing L with respect to η, λ, α, β with constraints of Rel.s (4.7), (4.8) and (4.9), and taking (4.10) as boundary condition. These expression can then be used to modify the form of the functional.
From Rel.(4.7)

$$\lambda = \gamma + \alpha - \beta$$

and from (4.8) and (4.9),

$$\lambda f = -\eta'' M - \eta' N/r$$

By substituting these relations in L, and taking (4.10) and Green formula into account, the dual objective function to be maximed for η, λ, α, β can be written in the form

$$\int_{R_i}^{R_e} 2\pi r\{(\gamma-\gamma M_p) + p\eta - \alpha M_{max} + \beta M_{min}\}\, dr \qquad (4.20)$$

It appears worth emaphasizing that the first term of (4.20) vanishes if γ is a linear function; on other hand the solution of (4.7) for M_p allows M_p to be eliminated from Rel.(4.20).
The term

$$\int_{R_i}^{R_e} 2\pi pr\, dr$$

represents (to within a constant positive factor) to power P of the external forces.
The optimal values of the objective functions of the primal and dual problems are equal, i.e.

$$\int_{R_i}^{R_e} 2\pi r\,\gamma\,dr = \int_{R_i}^{R_e} 2\pi r\{(\gamma - \dot{\gamma} M_p) + p\eta - \alpha M_{max} + \beta M_{min}\}\,dr$$

Taking into account Rel.s (4.6) and (4.15), one has

$$\int_{R_i}^{R_e} 2\pi r\,D\,dr = P \tag{4.21}$$

Rel.(4.21) represents the energy balance at the collapse of the structure since the left side is the power dissipated within the structure, while the right side represents the power of the external forces.

4.5. Designing rings of constant thickness

Let the plate be assumed to divide into n rings, in each of which the moment M_p is constant.
Let $x_i\,(i= 1,n+1)$ be the radii of the circles that divide the rings, with $x_1 = R_i$ and $x_{n+1} = R_e$. Moment functions on the i-th ring will be denoted by M_i and N_i; furthermore $f_i = f(M_i, N_i)$.
A superscript i will represent the value assumed by a function at $r = x_i$ (for example $M_{i-1}^{\ i} \equiv M_{i-1}(x_i)$).
In view of the discontinuities that may arise when $r = x_i\,(i = 2,n)$ the optimal design problem can then be formulated as follows

$$\text{Minimize } \Gamma(M_p) = \sum_{i=1}^{n} \pi(x_{i+1}^2 - x_i^2)\quad \gamma(M_{pi})$$

under the conditions

$$f_i - M_{pi} \le 0 \tag{4.22}$$

$$rM_i'' + 2M_i' - N_i' + pr = 0 \qquad x_i \le r \le x_{i+1},\quad i = 1,\ldots n \tag{4.23}$$

$$M_{pi} - M_{max} \le 0,\ M_{min} - M_{pi} \le 0 \tag{4.24}$$

$$M_i^{\ i} - M_{i-1}^{\ i} = 0 \tag{4.25}$$

$$x_i(M_i'^{\,i} - M_{i-1}'^{\,i}) - (N_i^{\ i} - N_{i-1}^{\ i}) = 0 \tag{4.26}$$

$$f_{i-1}^{\ i} - M_{pi-1} \le 0 \qquad i = 2,\ \ldots,\ n \tag{4.27}$$

$$f_i^{\ i} - M_{pi} \le 0 \tag{4.28}$$

Using the Lagrangian multipliers $\eta_i(r), \lambda_i(r) \ge 0,\ \alpha_i \ge 0,\ \beta_i \ge 0$ $(i = 1,\ldots, n)$ and $\xi_i, \zeta_i, \mu_i, \nu_i \ge 0$, $(i = 2,\ldots, n)$ it follows

$$L = \sum_{i=1}^{n} \{\pi(x_{i+1}^2 - x_i^2)\,\gamma(M_{pi}) + \int_{x_i}^{x_{i+1}} 2\pi r\{+\lambda_i(f_i - M_{pi})$$

$$+\eta_i(M_i'' + (2M_i' - N_i')/r + p)\}dr + \alpha_i(M_{pi} - M_{max}) + \beta_i(M_{min} - M_{pi})\}$$
$$+ \sum_{i=2}^{n} (\xi_i(M_i^{\,i} - M_{i-1}^{\,i}) + \zeta_i(x_i(M_i'^{\,i} - M_{i-1}'^{\,i}) - (N_i^{\,i} - N_{i-1}^{\,i}))$$
$$+ \mu_i(f_{i-1}^{\,i} - M_{pi-1}) - \nu_i(f_i^{\,i} - M_{pi}))$$

Since the boundary conditions $x_i(i=2,\dots n)$ have also been collected into the functional L, this should then be completed by the boundary conditions on R_i and R_e. fro the sake of simplicity, these conditions are omitted.

In the same way as in section 3, the next step is to establish the stationary conditions that are necessary for the solution of the minimum problem.

The stationarity conditions of L for η_i, λ_i, α_i, β_i, ξ_i, ζ_i, μ_i, ν_i also include, besides (4.22)-(4.28)

$$\lambda_i(f_i - M_{pi}) = 0 \tag{4.29}$$

$$\alpha_i(M_{pi} - M_{max}) = 0 \qquad i = 1,\dots,n \tag{4.30}$$

$$\beta_i(M_{min} - M_{pi}) = 0$$

$$\mu_i(f_{i-1}^{\,i} - M_{pi-1}) = 0 \qquad i = 2,\dots,n \tag{4.31}$$

$$\nu_i(f_i^{\,i} - M_{pi}) = 0 \tag{4.32}$$

The stationarity conditions of L for M_{pi}, M_i, $N_i(i=1,\dots n)$, as well as $M_{i-1}^{\,i}$, $M_i^{\,i}$, $N_{i-1}^{\,i}$, $M_{i-1}'^{\,i}$, $M_i'^{\,i}(i = 2,\dots n)$, lead to

$$\pi(x_{i+1}^2 - x_i^2)\gamma_i - \int_{x_i}^{x_{i+1}} 2\pi r\lambda_i dr + \alpha_i - \beta_i - \mu_{i+1} + \nu_i = 0 \tag{4.33}$$

$$\eta_i'' + \lambda_i \partial f_i/\partial M_i = 0 \qquad i=1,\dots n \tag{4.34}$$

$$\eta_i'/r + \lambda_i \partial f_i/\partial N_i = 0 \tag{4.35}$$

$$2\pi(\eta_{i-1}^{\,i} - x_i\eta_{i-1}'^{\,i}) - \mathfrak{J}_i + \mu_i \partial f_{i-1}^{\,i}/\partial M_{i-1}^{\,i} = 0 \tag{4.36}$$

$$2\pi(-\eta_i^{\,i} + x_i\eta_i'^{\,i}) - \xi_i + \nu_i \partial f_{i-1}^{\,i}/\partial M_i^{\,i} = 0 \tag{4.37}$$

$$-2\pi\eta_{i-1}^{\,i} + \zeta_i + \mu_i \partial f_{i-1}^{\,i}/\partial N_{i-1}^{\,i} = 0 \qquad i=2,\dots n \tag{4.38}$$

$$2\pi\eta_i^{\,i} - \zeta_i' + \nu_i \partial f_i^{\,i}/\partial N_i^{\,i} = 0 \tag{4.39}$$

$$2\pi\eta_{i-1}^{\,i} x_i - \zeta_i x_i = 0 \tag{4.40}$$

$$-2\pi\eta_i^{\,i} x_i + \zeta_i x_i = 0 \tag{4.41}$$

In (4.33) for i=1 and i=n, respectively, ν_1 and μ_{n+1} should be taken as

zero or else as representing suitable multipliers of the conditions imposed on the boundaries R_i and R_e.
In the same way as before, it can be set

$$\eta_i = w_i/c, \quad (i = 1,\ldots n) \tag{4.42}$$

Rel.s (4.40) and (4.41) prescribe the continuity of the function at the ring edges

$$\zeta_i/2\pi = \eta_{i-1}{}^i = \eta_i{}^i \tag{4.43}$$

where, as will be shown, the continuity of ' is not necessarily required. On the basis of (4.34),(4.35) and (4.43), it follows

$$\kappa_i/\chi_i = (\partial f_i/\partial M_i)/(\partial f_i/\partial N_i) \tag{4.44}$$

in accordance with the normality rule.
With c = 1, and taking (4.29) into account, (4.33) becomes

$$\gamma_i = 2\int_{x_i}^{x_{i+1}} r(\kappa_i M_i + \chi_i N_i)\, dr/(x_{i+1}{}^2 - x_i{}^2)\, M_{pi}$$

$$+(\mu_{i+1} - \nu_i)/\pi(x_{i+1}{}^2 - x_i{}^2) - (\alpha_i - \beta_i/\pi(x_{i+1}{}^2 - x_i{}^2), \quad i=1,\ldots n \tag{4.45}$$

Eq.(4.45) represents the optimality conditions for the optimal design problem with technological constraints, as formulated in this section. The first term on the right side, in fact, represents the average specific dissipated power in the i-th ring divided by the design variable M_{pi}. The second term represents the dissipated power at circles $r=x_{i+1}$ and $r=x_i$, respectively, due to the possible discontinuity of the function w'. The case $\mu_i = \nu_i = 0$ would involve the continuity of w' at $r=x_i$ (see (4.36),(4.37),(4.42) and (4.43)).
With $M_{pi} \neq M_{pi-1}$, from the continuity of M and the discontinuity of N it follows $\mu_i \nu_i = 0$. With $\mu_i = 0$ and $\nu_i \neq 0$, (4.32)(4.36)(4.37)(4.39) and (4.43) lead to the expression

$$\nu_i = -2\pi\, x_i\, M_i(\eta_i{}'^{\,i} - \eta_{i-1}{}'^{\,i})/M_{pi}$$

where the dissipated power at the edges of the ring is shown, due to the plasticity at of the outer fiber of the i-th ring.

5. APPLICATION OF LINEAR PROGRAMMING TO THE OPTIMAL PLASTIC DESIGN OF CIRCULAR PLATES

5.1. Preliminary remarks

A discretized approach to the optimal plastic design of circular or annular plates is the subject of this section (see e.g. [24]).
A finite difference technique is employed and the variational formulation is discussed for the discretized form of the problem. Having

assumed a linear cost function, a linear programming problem is achieved, if Tresca yield condition is adopted.
Special features of the problem are discussed and some numerical examples are shown.

5.2. Formulation of the problem

Circular or annular plates simply supported or built-in at outer edge are considered. For simplicity, the inner edge of annular plates is assumed to be free.
The usual assumptions of thin-plate theory regarding the smallness of plate thickness and deflection are adopted. Plate material is assumed to obey to a rigid perfectly plastic flow rule, and in particular Tresca yield condition is adopted.
Rotational symmetry conditions are assumed to be fulfilled and, for brevity, only distributed loads p(r) are considered.
Let the plate be divided into n rings, in each of which the yield moment M_p is constant. Let r_0 and r_n be the radii of the concentric boundaries of the rings.
Each plastic moment $M_{pi}(i=1,\dots n)$ will be subjected to the technological constraints

$$M_{min} \le M_{pi} \le M_{max} \qquad i = 1,\dots n \tag{5.1}$$

where M_{min} and M_{max} are given constants.
A set of design variables M_{pi} is looked for, in such a way the plate is at the plastic collapse condition under the given load and has the minimum total cost.

$$C = \sum_{i=1}^{n} \pi(r_i^2 - r_{i-1}^2)\, M_{pi} = \sum_{i=1}^{n} A_i\, M_{pi} \tag{5.2}$$

For a sandwich plate of constant core depth the expression (5.2) is seen to be proportional to the volume of the sheets
The equilibrium equation for the axially symmetric problem is

$$M + r\, dM/dr - N = rT \tag{5.3}$$

where T, M and N, respectively, are the shear force and the radial and circumferential bending moments. The differential equation (5.3) will be replaced by an equivalent finite difference equation. On purpose the i-th ring is divided into s rings of identical width δ_i (Fig.3) and the following assumptions are made

$$x_i^j = r_{i-1} + j\delta_i, \quad T_i^j = T(x_i^j), \quad M_i^j = M(x_i^j), \quad N_i^j = N(x_i^j) \qquad j = 0,\dots s \qquad i = 1,\dots n \tag{5.4}$$

As the finite difference analog of (5.3) the following equations are adopted

$$M_i^j + x_i^j \frac{M_i^{j+1} - M_i^{j-1}}{2\delta_i} - N_i^j = x_i^j T_i^j, \quad j=1,\dots s-1 \tag{5.5}$$

$$(1-r_{i-1}/\delta_i)M_i^j + r_{i-1}/\delta_i \, M_i^{j+1} - N_i^j = r_{i-1}T_i^j, \quad j=0 \quad i=1,\dots n \tag{5.6}$$

$$-r_i/\delta_i \, M_i^{j-1} + (1+r_i/\delta_i)M_i^j - N_i^j = r_i T_i^j, \quad j=s \tag{5.7}$$

Since the plate is supposed to be supported along only one edge, T_i^j can be calculated from given loads. By assuming, without loss of generality, $N_i^j \geq 0$ everywhere, the yield condition (Fig.4) is defined as follows

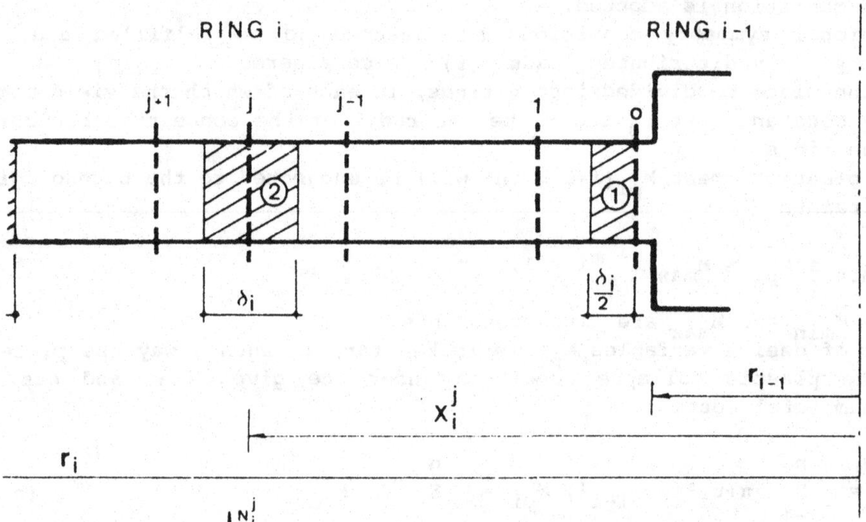

Fig. 3. Part of a circular or annular plate divided into rings.

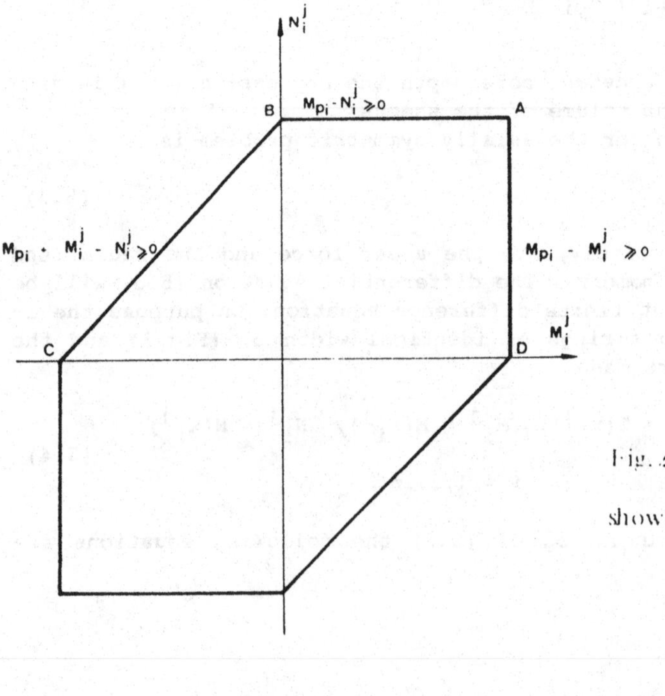

Fig. 4. Tresca hexagon

showing yield conditions.

$$
\begin{aligned}
&-M_i^j - M_{pi} \le 0 \\
&-N_i^j - M_{pi} \le 0 \qquad j = 0,\dots s, \qquad i = 1,\dots n \qquad (5.8)\\
&N_i^j - M_i^j - M_{pi} \le 0
\end{aligned}
$$

The technological constraints of Rel.s(1), static boundary conditions and continuity conditions for the radial moment at r_i are to be added. The latter are

$$M_i^s = M_{i+1}^0 \qquad \text{for} \qquad i = 1,\dots n-1 \qquad (5.9)$$

In matrix notation the optimal plastic design problem can be formulated as follows:

$$\min \sum_{i=1}^{n} A_i M_{pi}$$

subject to

$$c_i^t M_i - N_i = T_i \qquad (5.10)$$

$$M_i - M_{pi} e \le 0 \qquad (5.11)$$

$$N_i - M_{pi} e \le 0 \qquad (5.12)$$

$$N_i - M_i - M_{pi} e \le 0 \qquad i = 1,\dots n \qquad (5.13)$$

$$M_{pi} - M_{max} \le 0 \qquad (5.14)$$

$$M_{min} - M_{pi} \le 0$$

$$D_i^t M_i - B^t M_{i+1} = 0, \qquad i = 1,\dots n-1 \qquad (5.15)$$

to which boundary conditions for the specific problem are to be added. The (s+1)th-order vector **e** has all components equal to one, the vectors **B,D,M,N** and **T** are defined as

$B^t = \{1, o_s^t\}$, $D^t = \{o_s^t, 1\}$ (o_s is the sth-order zero vector),
$M_i^t = \{M_i^0,\dots M_i^s\}$; $N_i^t = \{N_i^0,\dots N_i^s\}$, $T_i^t = \{r_{i-1}T_i^0,\dots r_i T_i^s\}$

and the (s+1)th-order matrix c_i^t is (e.g. if s = 3)

$$
c_i^t = \begin{bmatrix}
1-r_{i-1}/\delta_i & r_{i-1}/\delta_i & 0 & 0 \\
-x_i^1/(2\delta_i) & 1 & x_i^1/(2\delta_i) & 0 \\
0 & -x_i^2/(2\delta_i) & 1 & x_i^2/(2\delta_i) \\
0 & 0 & -r_i/\delta_i & 1+r_i/\delta_i
\end{bmatrix}
$$

5.3. Optimality criterion and dual problem

Applying the Lagrangian multiplier η_i and Θ_i, which are not sign-restricted, to the equality constraint (5.10) and (5.15) and the nonnegative multipliers λ_i, μ_i, ν_i, α_i and β_i to the inequality constraints (5.11) to (5.14), one has

$$L = \sum_{i=1}^{n} \{A_i M_{pi} + \eta_i^T(T_i - C_i^T M_i + N_i) + \lambda_i^T(M_i - M_{pi}e) - \mu_i^T(N_i - M_{pi}e)$$

$$+ \nu_i^T(N_i - M_i - M_{pi}e) + \alpha_i(M_{pi} - M_{max}) + \beta_i(M_{min} - M_{pi})\} + \sum_{n+1}^{n-1} \Theta_i^T(D^T M_i - B^T M_{i+1} \tag{5.16}$$

The stationarity conditions for L are necessary conditions for the optimality of the design. The stationarity conditions of L with respect to Lagrangian multipliers provide not only Rel.s (5.10) to (5.15) but also the equations

$$\lambda_i^d(M_i - M_{pi}e) = 0 \tag{5.17}$$

$$\mu_i^d(N_i - M_{pi}e) = 0 \tag{5.18}$$

$$\nu_i^d(N_i - M_i M_{pi}e) = 0 \tag{5.19}$$

$$\alpha_i(M_{pi} - M_{max}) = 0 \tag{5.20}$$

$$\beta_i(M_{min} - M_{pi}) = 0 \tag{5.21}$$

when a superscript d denotes suitable diagonal matrix, such that, e.g., $\lambda_i = \text{diag}(\lambda_i^d)$.

Stationarity conditions of L with respect to M_{pi}, **M** and **N** are

$$A_i - e^t(\eta_i + \mu_i + \nu_i) + \alpha_i - \beta_i = 0 \tag{5.22}$$

$$C_i\eta_i + \lambda_i - \nu_i + D_i\Theta_i - B\Theta_{i-1} = o \qquad i = 1,\dots n \tag{5.23}$$

$$\eta_i + \mu_i + \nu_i \geq o \tag{5.24}$$

In Rel. (5.23) the terms $-B\Theta_{i-1}$ and $D_i\Theta_i$ need not be considered for i=1 and i=n respectively. By means of a suitable substitution of the stationarity conditions here obtained into the functional of Rel.(5.16) the dual form of the problem can be obtained, which reads

$$\max Z = \sum_{i=1}^{n} (T_i^t\eta_i - M_{max}\alpha_i + M_{min}\beta_i) \tag{5.25}$$

Under the constraints of Rel.s (5.22) to (5.24). Let the transverse deflection rate of section j in ring i be denoted by w_i^j. η_i^j can be verified to be proportional to increments of rates. Thus to within a common positive factor

$$\eta_i^j = \pi(w_i^{j+1} - w_i^{j-1}), \qquad j=2,\dots s-1 \tag{5.26}$$

$$\eta_i^0 = \pi(w_i^1 - w_i^0) \qquad i=1,\dots n \tag{5.27}$$

$$\eta_i^s = \pi(w_i^s - w_i^{s-1}) \tag{5.28}$$

or, in a more concise form

$$\eta_i^j = \pi \Delta w_i^j$$

From Rel.s (5.26)-(5.28) it is easy to see that the first term of the objective function of Rel.(5.25) is a discretized form of

$$P = 2\pi \int_{r_0}^{r_n} r\, T(dw/dr)\, dr$$

With the assumed hypotheses (only one supported edge and distributed load) P is the dissipated power of load. The objective function Z of the dual problem represents the external dissipated power to within a factor, that is only a function of α_i, β_i, M_{max} and M_{min}.
When N_i^j is different from zero, constraint (5.24) is fulfilled as equality and becomes

$$(\mu_i^j + \nu_i^j) = -\eta_i^j$$

or, using the appropriate relation (5.26)(5.27) or (5.28)

$$(\mu_i^j + \nu_i^j) = -2\pi\delta_i x_i^j((1/x_i^j)(\Delta w_i^j/2\delta_i)), \qquad j=2,\dots s-1 \tag{5.29}$$

$$(\mu_i^0 + \nu_i^0) = -2\pi(\delta_i/2)r_{i-1}((1/r_{i-1})(w_i^0/2\delta_i) \tag{5.30}$$

$$(\mu_i^s + \nu_i^s) = -2\pi(\delta_i/2)r_i((1/r_i)(w_i^s/\delta_i) \tag{5.31}$$

In the RHS of Rel.s (5.29)-(5.31) the dissipated powers per unit yield moment can be recognized for the dashed zones noted 1, 2 and 3, respectively, in Fig.3 when the plastic profile (Fig.4) is AB($\mu_i^j \geq 0$, $\nu_i^j = 0$), BC($\mu_i^j = 0$, $\nu_i^j \geq 0$) or is characterized by point N($\nu_i^j + \mu_i^j > 0$). Then

$$(\mu_i^j + \nu_i^j) = A_i^j \chi_i^j \tag{5.32}$$

with A_i^j equal to the mean surface of the plate element of section j in ring i, and χ_i^j equal to the discretized curvature in the same element. The general form of an equation of Rel.(5.23) for $j \neq 0$ and $j \neq s_i$ is

$$\lambda_i^j - \nu_i^j = (x_i^{j-1}/2\delta_i)\, \eta_i^{j-1} + \eta_i^j - (x_i^{j+1}/2\delta_i)\, \eta_i^{j+1} \tag{5.33}$$

or

$$\lambda_i^j - \nu_i^j = 2\pi\delta_i x_i^j\{\Delta^2 w^i/\Delta r_j^2\} \tag{5.34}$$

According to Rel (5.26) it is readily seen that the term in brackets is the discretized form of the second derivative w that is of the radial curvature κ_i^j to within a sign.

$$\lambda_i^j - \nu_i^j = A_i^j \kappa_i^j \tag{5.35}$$

The RHS represents the dissipated power per unit yield moment for the element of the plate surrounding section j of ring i when the plastic profile is the side DA($\lambda_i^j \geq 0$, $\nu_i^j = 0$) or BC($\lambda_i^j=0$, $\nu_i^j \geq 0$) of the Tresca hexagon. Let the equalities of Rel.(5.23) obtained for j=0 in ring i+1 for j=s in ring i be considered. By eliminating $_i$ one obtains

$$(\lambda_{i+1}^0 - \nu_{i+1}^0) + (\lambda_i^s - \nu_i^s) = 2\pi r_i \left[\left(\frac{\Delta w}{\Delta r}\right)_i^s - \left(\frac{\Delta w}{\Delta r}\right)_{i+1}^0\right]$$

$$- 2\pi(\delta_i/2)r_i \left(\frac{\Delta^2 w}{\Delta r^2}\right)_i^s - 2\pi(\delta_{i+1}/2)r_i \left(\frac{\Delta^2 w}{\Delta r^2}\right)_i^0 \tag{5.36}$$

where the quantities

$$\left(\frac{\Delta w}{\Delta r}\right)_i^j \quad \text{and} \quad \left(\frac{\Delta^2 w}{\Delta r^2}\right)_i^j$$

are the discretized expression of the first and second derivatives of w respectively, in section j of ring i. In the second and third terms of the RHS one finds expressions similar to (5.35), which represent the dissipated power per unit yield moment in elements 0 and 1, respectively of rings i+1 and i for a plastic profile identical to sides DA or BC of the Tresca hexagon.
Denote by H_i^{i+1} the first term of the second member of Rel (5.36) (H_i^{i+1} may be different from zero only if there is a relative rotation between rings i and i+1) and suppose

$$M_i^{si} = M_{i+1}^0 > 0 \tag{5.37}$$

if follows

$$\nu_{i+1}^0 = \nu_i^s = 0 \tag{5.38}$$

If $H_i^{i+1} > 0$, the last relation imply that at least one of the λ_{i+1}^0, λ_i^s must be positive. The Kuhn-Tucker conditions imply that one must be equal to zero if the plastic Moments of rings i and i+1 are different. Therefore H_{i+1}^i represents the dissipated power per unit yield moment in the hinge circle formed in the ring with the smaller plastic moment. When the radial moment is negative at the boundary of the two rings (λ=0, $\nu \neq 0$), a similar discussion shows that

$$-M_i^s = -M_{i+1}^0 = \min\{M_{pi}, M_{pi+1}\} \tag{5.39}$$

and that H_i^{i+1} is again a dissipated power in a hinge circle.

According to the previously given interpretation of the dual variables λ_i^j, μ_i^j, ν_i^j, the product

$$e^t(\lambda_i + \mu_i + \nu_i) = z_i/M_{pi} \tag{5.40}$$

in Rel. (5.22) represents the total dissipated power per unit yield moment in the i-th ring.
If $M_{pi} = M_{min}$, one has $\alpha_i>0$ and $\beta_i=0$, and Rel.(5.22) becomes

$$z_i/M_{min} \le A_i \tag{5.41}$$

and similarly, if $M_{pi} = M_{max}$ then $\beta_i \ge 0$, $\alpha_i=0$, and Rel (5.22) becomes

$$z_i/M_{max} \ge A_i \tag{5.42}$$

When $M_{min}<M_{pi}<M_{max}$, then $\alpha_i=\beta_i=0$, and Rel (5.22) yields

$$z_i/M_{pi} \le A_i \tag{5.43}$$

The condition (5.41)(5.43) are the optimality criteria [24], as derived in different ways in [25].

6. COMPUTATIONAL REMARKS

It is will-known that the computer time required for the solution of a linear programming problem is roughly proportional to m^3n, where m is the number of constraints and n the number of variables.
For ring i, subdivided into s elements, one has for the primal problem 4s+7 constraints and 2s+3 variables, and accordingly $m \underset{\sim}{-} 2n$. Since the number of variables of the primal becomes the number of constraints of the dual, it has to be expected that the computer time required for solving the dual is only about a quarter of the time required for the primal. Furthermore, to start the solution of a linear programming problem, a basic feasible solution is recognized. For the dual an obvious feasible solution is obtained by setting all variables equal to zero: this corresponds to the undeformed plate(all η, λ, μ, ν, equal to zero). While on the one hand the dual is computationally advantageous, on the other hand the formulation and data input for the primal are easier. The above remarks suggest that it is convenient to use linear programming codes capable of forming and processing the problem dual of the problem defined in the input and capable of supplying in output both primal and dual solutions.
In the computer program used here variables N_i^j were extracted from equations (5.10) and substituted into constraints (5.12)-(5.13). Furthermore advantages are taken of equalities (5.15) to reduce the number of unknowns in the vector $\mathbf{M}_i$ and hence the number of constraints in the dual problem. For sandwich plates the structural weight can readily be considered by properly evaluating $\mathbf{T}_i$, which will contain the plastic moment M_{pk}(k=1,...i) as unknowns.
A formulation can be used for limit analysis problems, where the M_{pi} are

given, and the scalar multiplier of the loads has to be maximized [30].

7. NUMERICAL EXAMPLES

7.1. Simply supported uniformy loaded circular plate

Let the plate represented in Fig.5 be considered for which no technological constraints are assumed. The first and second rings are respectively divided into 6 and 4 pieces (s=6 and s=4). The analytical solution for this problem has been given in [25] and [27]. The computed results are compared in tables 1 and 2 with the exact analytical values. The relative errors are very small and less than 1 percent for the yield moments and the cost and less than 3.4 percent for the radial bending moments with a mean error of 1 percent. The accuracy can be improved by increasing the values of s. With s=8 for both rings the relative errors are reduced to less than 0.4 percent.

In Fig.6 we have plotted the design obtained for a simply supported plate divided into 16 rings of equal width taking s=2. Here again the computed design is close to the analytical design [26]. Furthermore, Fig.6 shows that the computed design well approximates the absolute minimum-volume design [28], that is the design without any technological constraints and with continuously variable yield moment.

7.2. Built-in, uniformly loaded circular plate

For a plate divided into 2 rings limited by the circles of radii 0, R/2 and R, Table 3 collects the results obtained by using s=5 for both rings. As for the previous example the values furnished by linear programming agree very well with the exact results given in [29].

The linear programming approach above described can be used to design a plate of constant thickness (one ring for given load or, equivalently, to obtain the limit load of a plate with given constant thickness. For the built-in, uniformly loaded plate $pR^2/M_p=11.19$ for s=10 and n=1 is obtained. The exact value is $pR^2/M_p=11.26$.

7.3. Simply supported, uniformly loaded annular plate

The plate sketched in Fig.7 is considered. It is well known [25][31] that for a plate of this kind the optimal design is characterized by an "edge effect", that is by M_{p2} much larger than M_{p1}. To obtain a realistic design, technological constraints must be introduced. Taking s=10 and s=5 for rings 1 and 2, respectively, and setting the constraints

$$M_{pi} \le 0.25\ pR^2, \qquad i = 1,2$$

the results given in Table 4 have been obtained. The analytical solution of this problem is found in [30].

REFERENCES

1. Drucker, D.C. and R.T.Shield, Design for minimum weight, Proc. 9th Int. Congr. of Applied Mechanics, Brussels, 5, 1956, 212-222
2. Gross, O. and W.Prager, Minimum-Weight Design for Moving Loads, Proc.4th U.S. Nat. Congr. Appl.Mech.ASME, New York, 2, 1962, 1047
3. Prager, W. and R.T.Shield, A General Tehory of Optimal Plastic Design, J.Appl. Mech. Trans. A.S.M.E., 34, 1, 1967
4. Chern, J.M. and W.Prager, Optimum Design for Prescribed Compliance Under Alternative Loads, J. Opt.Th.Appl., 5 (1970), 424-431
5. Save, M., A Unified Formulation of the Theory of Optimal plastic Design with Convex Cost Function, J.Struct. Mech., 1(1972), 267-276
6. Rozvany, G.I.N., Optimal Design of Flexural Systems, Pergamon Press, Sydney, 1976
7. Prager, W. and J.E.Taylor, Problems of Optimal Structural Design, J.Appl. Mech., 35, (1968), 102-106
8. Prager, W. Conditions for Structural Optimality, Computers and Structures, 2, (1972), 833-840
9. Save, M., A general criterion for Optimal Structural Design, J.Opt. Tehory Appl., 15, (1975), 119-129
10. Rozvany, G.I.N., Optimal design of Flexural Systems, Oxford, Pergamon Press (1976)
11. Hemp, W.S., Optimum Structures, Claredon Press, Oxford, 1973
12. Cinquini, C. and B.Mercier, Minimal Cost in Elastoplastic Structures, Meccanica, 11, 4, 1976, 219-226
13. Huang, N.C., Optimal Design of Elastic Beams for Minimum-Maximum Deflection, J. Appl. Mech.Trans. A.S.M.E., Dec., 1971, 1078-1081
14. Cinquini. C. Optimal Elastic Design for Prescribed Maxium Deflection, J. Struct. Mech., Vol. 7, 1, 1979, 21-34
15. Cohn, M.Z., Analysis and Design of Inelastic Structures, Univ. of Waterloo Press, Waterloo, 1972
16. Sawczuk A. and Z.Mroz, Optimization in Structural Design, Proc. IUTAM Symp., Warsaw 1973, Springer Verlag, Berlin, 1975
17. Haug, E.J. and J. Cea, Optimization of Distributed Parameter Structures, Proc. NATO ASI, Iowa City, Iowa, 1980, Noordhoff, The Netherlands, 1981
18. Gallager, R.H., Proceedings International Symposium on Optimum Structural Design, Univ. of Arizona, Tucson, Arizona, 1981
19. Morris, A.J., Foundations of Structural Optimization: A Unified Approach, Proc. NATO ASI, Liege, Belgium, 1980, Chichester, 1982
20. Ekeland, I. and R.Temam, Analyse convexe et problemes variationnels, Dunod, Paris, 1973
21. Cinquini, C. and G.Sacchi, Problems of optimal design for elastic and plastic structures, J. de Mecanique Appl., 4, (1980), 31-59
22. Guerlement, G., Lamblin, D. and C. Cinquini, Dimensionnement plastique de cout minimal avec contraintes technologiques de poutres soumines a plusieurs ensembles de charges, J. de Mec. Appl., 1, 1, 1977, 1-25

23. Guerlement, G., Lamblin, D. and C. Cinquini, Variational formulation of the optimal plastic design of circular plates, Comput. Meth. Appl. Mech. Eng. 11, 1977, 19-30
24. Guerlement, G., Lamblin, D. and C. Cinquini, Application of linear programming to the optimal plastic design of circular plates subject to technological constrains, Coput. Meth. Appl. Mech. Eng., 13, 2, 1978, 233-243
25. Sheu,C.Y. and W. Prager, Optimal Plastic Design of circular and annular sandwich plates with piecewise constant cross section, J. Mech.Phys. Solids, 17. 1969,, 11-16
26. Hadley, G., Nonlinear and dynamic programming, Addinson Wesley, Chicago, 1965
27. Hopkins, H. and W.Prager, Limits of economy of material in plates, J. Appl. Mech., 22, 1955, 372-374
28. Hopkins, H. and W.Prager, The load carrying capacity of circular plates, J. Mech. Phys. Solids, 2, 1953, 372-374
29. Guerlement, G. and D. Lamblin, Dimensionnement plastique de volume minimal sous contraintes de plaques sandwhich circulaires soumises a des charges fixes ou mobiles, J.Mec. , 15,1 1976, 55-84
30. Lamblin, D. Analyse et dimensionnement plastique de cout minimum de plaques circulaires, These de Doctorat en Sciences Appl., Faculte Polytechnique de Mons, 1975
31. Mgarefs, G.J., Method for minimal design of axisymmetric plates, Asce J. Eng.Mech. Div., 92, 1966, 79-99

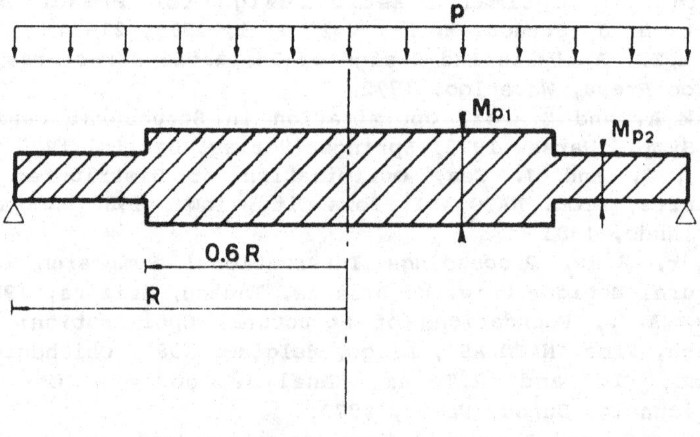

Fig. 5. Circular plate (sec. 7.1).

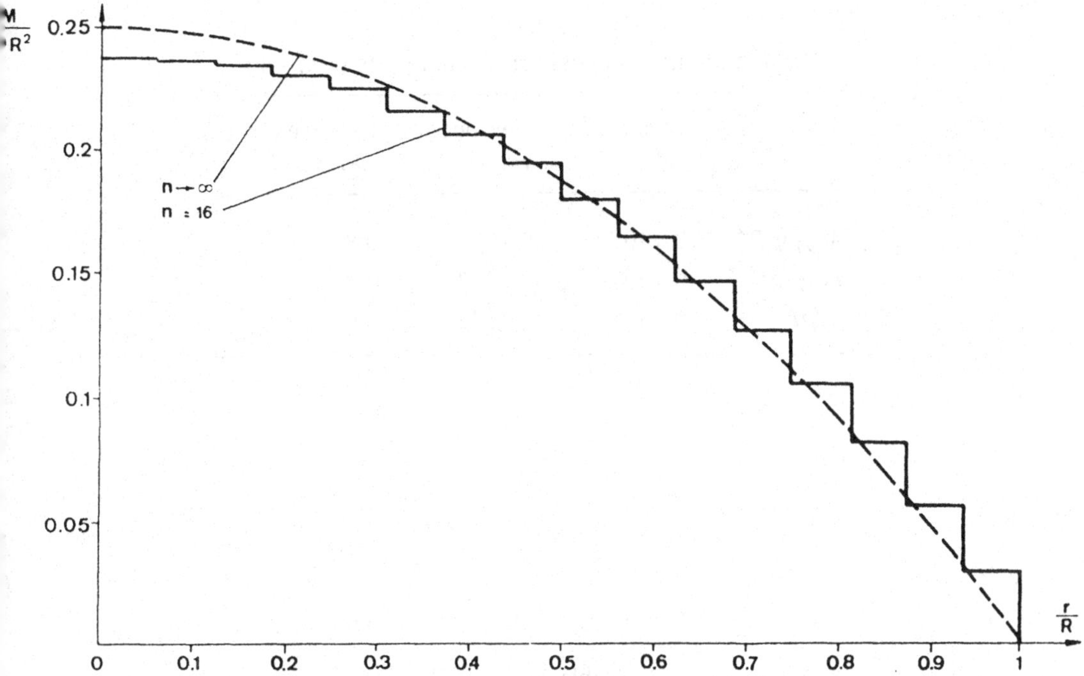

Fig. 6. Radial bending moment for circular plate divided into 16 rings (sec. 7.1).

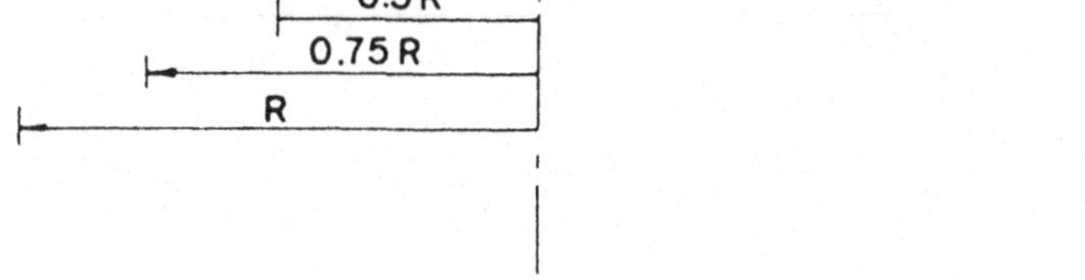

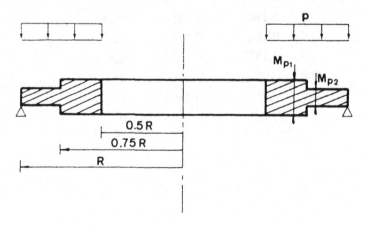

Fig. 7. Annular plate (sec. 7.3).

Table 1
Yield moments and cost for circular plate (sec. 7.1)

	Exact values	Computed values	Relative error (%)
M_{p1}/pR^2	0.1907	0.1921	0.8
M_{p2}/pR^2	0.1307	0.1313	0.5
C/pR^4	0.1523	0.1532	0.6

Table 2
Radial bending moment for circular plate (sec. 7.1)

r/R	M_i^j/pR^2 exact	M_i^j/pR^2 computed	Relative error (%)
0	0.1907	0.1921	0.8
0.2	0.184	0.185	0.65
0.4	0.164	0.1649	0.55
0.6	0.1307	0.1313	0.5
0.8	0.069	0.069	0
0.9	0.0357	0.0369	3.4
1	0	0	0

Table 3
Yield moments and cost for circular plate (sec. 7.2)

	Exact values	Computed values	Relative error (%)
M_{p1}/pR^2	0.1184	0.1197	1.1
M_{p2}/pR^2	0.0768	0.0772	0.5
C/pR^4	0.08718	0.08777	0.7

Table 4
Yield moments and cost for annular plate (sec. 7.3)

	Exact values	Computed values	Relative error (%)
M_{p1}/pR^2	0.25	0.25	0
M_{p2}/pR^2	0.0833	0.0845	1.4
C/pR^4	0.1146	0.1151	0.4

OPTIMAL DESIGN
OF FIBER-REINFORCED ORTHOTROPIC BODIES

G. Sacchi Landriani
Polytechnic of Milan, Milan, Italy
M. Rovati
University of Trento, Trento, Italy

ABSTRACT
These lecture notes concern optimal plastic design of plane structures orthotropically reinforced by long fiber systems. These structures are considered as submitted to a plane state of stress or in bending. The problems are seen as distributed parameter optimization problems and are dealt with by means of a variational technique. In such a way, necessary conditions for optimal solutions are found and their mechanical interpretation discussed as well. In conclusion, an approach for approximate solutions fulfilling the optimality criteria is presented and the analogy with elastic solutions pointed out.

1. INTRODUCTION

Design of composite materials, because of their superior mechanical properties, represents a promising field of engineering, especially in the design of lightweight structures having particular stiffness and strength requirements. Consequently, designing with composites can become a challenge for the designers because of a wide range of parameters that can be varied, and because of the complex behaviour of these structures, that requires sophisticated analysis techniques. An efficient composite structural design that meets some given requirements can be achieved not only by sizing the cross-sectional areas and member thicknesses, but also by global or local tayloring of the material properties through selective use of orientation, number and stacking sequence of laminae in the laminate. All these aspects, in last years, have attracted the interest of many researchers, so that optimal design of composite materials and structures can be considered now as a new field in structural mechanics. In these notes only those aspects related to optimization will be taken into account, referring to the literature for the strictly mechanical behaviour of the material considered.

One of the earliest results on optimization of composites can be found in a work of Banichuk [1], where necessary conditions for an optimal orientation of mechanical properties in orthotropic bodies are given. Such results, which are still interesting and topical, in very recent times have found new impulses, and have been reproposed and applied by Pedersen (see [2] to [5]), Bendsøe ([6] and [7]), Olhoff and Thomsen [8]. In all of these papers, the problems are essentially dealt with by means of numerical techniques and new tools of optimal design and several solutions for different structural problems are proposed. It must be pointed out that such examples indicate how the optimal solutions are characterized by suitable concentration of fibers in special regions of the body, with a very coarse distribution of material in the remaining parts. As a consequence it seems reasonable to deal with the problem of optimal orientation and distribution of reinforcing fibers as a topological problem, where the structural material inside the body plays the role of a structure itself, but with unknown layout. In these notes, which go in the same direction, some results formerly obtained by Sacchi and Rovati (see [9] and [10]) are presented, and some ideas on a possible solution procedure are discussed. Such a quite general approach is in agreement with the indications given by the above mentioned numerical results and also shows several analogies with the classical solutions obtained for analogous problems in the isotropic case. Although the fabrication processes and therefore the application fields of composite materials, up to now, privilege considerably the use of laminae with straight fibers, it is desirable that in the future the performances of such materials can be improved, by suitable local orientations of the reinforcements embedded into the matrix.

Section 2, in the field of limit analysis of structures, deals with the problem of finding the optimal area distribution of reinforcements in a two dimensional fiber-reinforced body, subject to a plane state of stress. In the optimal design problem, for a prescribed load distribution, the total volume of the fibers is minimized, according to equilibrium conditions and to the strength of the material.

A variational formulation is applied in order to solve the problem, in which both fiber orientation and fiber density are adopted as design variables, and the set of necessary conditions for optimality are presented. By virtue of such conditions, the property of local collinearity between stresses, strain rates and strengths is pointed out, and the fully stressed behaviour of the optimal solution is shown as well, while some kinematical aspects of the optimal solution complete the discussion of the problem.

The same problem is then studied for the case of plates in bending (Sec. 4), showing how analogous results and related mechanical interpretations can be found.

On the other hand, because of a well established theory of optimization problems for flexural systems (see e.g. Rozvany [11]), in these notes particular attention is paid only to plane problems, which, up to now, do not find in literature comprehensive contributions.

In Sections 3 and 5 some remarks on the optimal orientation of fibers in a elastic bodies are presented as distributed parameter optimization problems, with the aim to highlight some particular mechanical features of the optimal solution, which easily suggest possible practical applications. The problem of minimizing the elastic compliance of a structure is dealt with, and, through the application of a variational technique, meaningful optimality conditions are obtained. In particular, collinearity between principal directions of stress, strain and orthotropy is shown to be conditions for an optimal design.

In such a way, it is outlined that the elastic and plastic problems show optimal solutions which are characterized by similar properties, and then in Section 6 the unified approach for approximate solutions of the problem is presented. In particular it is shown how the necessary conditions for both the problems agree with the conditions fulfilled by a Michell structure or by a truss-like continuum as defined by Prager [12] for the isotropic case. Therefore, this remark justifies the choice, on the other hand already empirically pointed out by numerical results (see, e.g., Bendsøe [7]), of dealing with the problem, seeking the solution in the class of truss-like structures. The procedure is then shown through sample examples concerning cantilever trusses, for which some optimal solutions are easily found. In particular, it is pointed out how in this way the optimization problem becomes a topological problem, where the solution is characterized by an optimal layout of the structure. The influence of the structural topology on the solution itself is then discussed.

2. OPTIMAL LIMIT DESIGN OF PLANE ORTHOTROPIC BODIES

Consider an orthogonal reference frame $x_1 - x_2$, where a solid body of domain $\Omega \in \mathcal{R}^2$ is defined, with boundary $\partial\Omega \equiv \partial\Omega_1 \cup \partial\Omega_2$ on which tractions and displacements are prescribed, respectively. The body is made by a non structural matrix component and reinforced by a system of fibers orthogonal each others. In such a way the solid is regarded as locally orthotropic, and the local orthogonal frame $z_1 - z_2$ denotes the principal directions of orthotropy, (refer to Fig. 1). Moreover, the stresses in the material are supposed to be entirely carried by the fibers, both in traction and in compression.

For a prescribed load distribution $\underline{p}$ on $\partial\Omega_1$, the stress field in the body, expressed in terms of characteristics, is specified through the tensor $\underline{\underline{N}}$, which satisfies the equilibrium equations

$$div\,\underline{\underline{N}} = \underline{0} \quad \text{in } \Omega$$

$$\underline{\underline{N}}\,\underline{n} = \underline{p} \quad \text{on } \partial\Omega_1$$

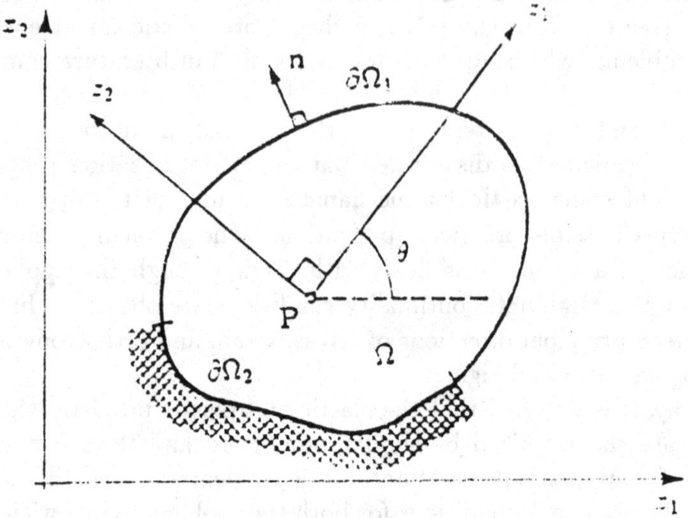

Figure 1

with $\underline{\underline{N}} = \underline{\underline{N}}^T$, under the hypotesis of vanishing body forces, and where $\underline{n}$ denotes the outward unit normal on $\partial\Omega_1$.

The fibers are supposed to have cross sectional area variable along the fiber length. In principal directions of orthotropy, at each point of the body, the strength of the fibers – and therefore of the material – is defined as

$$R^+{}_{I,II} = A_{I,II}\,\sigma^+ \tag{1}$$

$$R^-{}_{I,II} = A_{I,II}\,\sigma^- \tag{2}$$

where A_I and A_{II} represent the cross sectional areas of the fibers in direction z_1 and z_2 respectively, while σ^+ and σ^- are the ultimate stresses in tension and compression. A number of experimental results suggests to assume (Johansen) the strength of the material to be defined through the tensors $\underline{\underline{R}}^+$ and $\underline{\underline{R}}^-$ (see the symbolic representations of Fig.s 2a and 2b), whose scalar components in the $x_1 - x_2$ frame are $R^\pm_{11}$, $R^\pm_{22}$, $R^\pm_{12}$.

Finally, the strength criterion, in orthogonal coordinates, can be stated as follows (see Fig. 3)

$$R^+{}_{nn} \geq N_{nn} \geq -R^-{}_{nn} \quad \forall n \tag{3}$$

under the conditions of non–negativeness

$$R^+{}_{nn} \geq 0, \qquad R^-{}_{nn} \geq 0 \quad \forall n$$

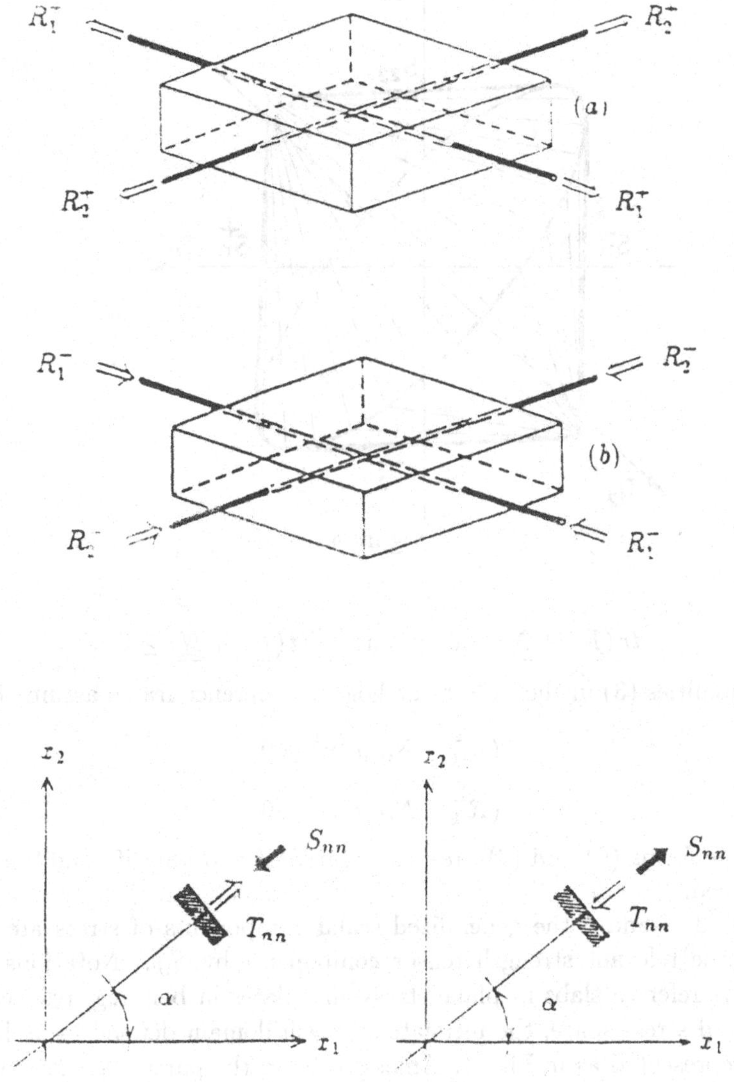

Figure 3

Instead of a representation given in terms of components, the strength criterion can be written in tensor form

$$tr\,(\underline{\underline{R}}^+ - \underline{\underline{N}}) \geq 0 \quad \text{and} \quad det\,(\underline{\underline{R}}^+ - \underline{\underline{N}}) \geq 0 \tag{4}$$

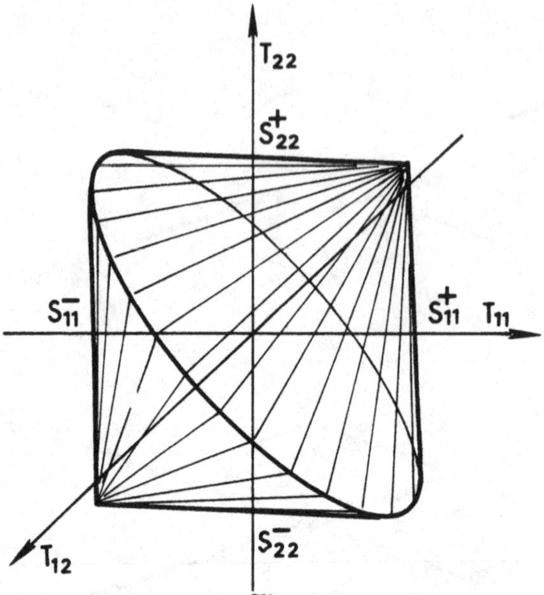

Figure 4

$$tr\,(\underline{\underline{R}}^- + \underline{\underline{N}}) \geq 0 \qquad \text{and} \quad det\,(\underline{\underline{R}}^- + \underline{\underline{N}}) \geq 0 \tag{5}$$

In fact, inequalities (3) in the $x_1 - x_2$ orthogonal reference frame assume the form

$$(R^+_{ik} - N_{ik})n^i n^k \geq 0 \tag{6}$$

$$(R^-_{ik} + N_{ik})n^i n^k \geq 0 \tag{7}$$

The quadratic forms (6) and (7) are non-negative if and only if conditions (4) and (5) are satisfied.

In Figs. 3, 4 and 5 the generalized scalar components of stress are indicated by T_{ik} and the relevant strength tensor components by S_{ik}. Notations N_{ik}, R_{ik} and M_{ik}, K_{ik} refer to slabs in plane stress and plates in bending, respectively. In the generalized stress space, the ultimate strength domain defined by Rel.s (4) and (5) can be represented as in Fig. 4. Analogously, in the particular case of $\underline{\underline{T}}$ and $\underline{\underline{S}}$ collinearity, in the space of the principal stresses T_I, T_{II} the ultimate strength domain can be represented as in Fig. 5.

From Fig. 5 it clearly appears that the ultimate strength domain, for given values of the limit stresses σ^+ and σ^-, depends on the cross sectional areas of the fibers.

By virtue of such a remark, in the present Section the total amount of structural material (fibers), i.e., the size of the strength domain, will be minimized, for prescribed values of the external loads, according to equilibrium and strength criterion.

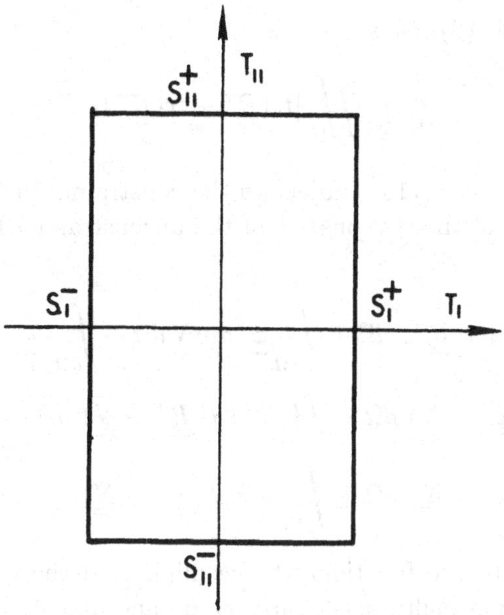

Figure 5

The limit mechanical behaviour of the reinforced material is assumed to be perfectly plastic and obeying to the associated flow rule, so that the strain rate components can be conceived as proportional to the components of the gradient of the plasticity condition.

In such a way, assuming as design variables the local values of the cross sectional areas of the fibers, the optimal design problem can be stated as follows:
PROBLEM 1 *Find:*

$$\min_{A_{I,II}\geq 0} \iint_{\Omega} (A_I + A_{II}) d\Omega \tag{8}$$

such that

$$div\, \underline{\underline{N}} = \underline{0} \qquad \text{in } \Omega \tag{9}$$

$$\underline{\underline{N}} \cdot \underline{n} = \underline{p} \qquad \text{on } \partial\Omega_1 \tag{10}$$

$$tr\,(\underline{\underline{R}}^+ - \underline{\underline{N}}) \geq 0 \quad tr\,(\underline{\underline{R}}^- + \underline{\underline{N}}) \geq 0 \tag{11}$$

$$det\,(\underline{\underline{R}}^+ - \underline{\underline{N}}) \geq 0 \quad det\,(\underline{\underline{R}}^- + \underline{\underline{N}}) \geq 0 \tag{12}$$

By virtue of Rel.s (1) and (2) the design variables A_I and A_{II} are replaced by R_I^+ and R_{II}^+, taking into account that

$$R_I^+ R_I^- = R_{II}^+ R_{II}^- = 0$$

and consequently Rel. (8) reads

$$\min_{\underline{\underline{R}}^{+,-}\geq 0} \iint_\Omega \left[tr(\underline{\underline{R}}^+ + tr(\underline{\underline{R}}^-)\right] \, d\Omega \tag{13}$$

The minimization problem (13) subject to the constraints (9) to (12) can be transformed in the search of the stationarity of the unconstrained Lagrangian functional $\mathcal{L}$, defined as

$$\begin{aligned}\mathcal{L} = & \iint_\Omega \left(tr\underline{\underline{R}}^+ + tr\underline{\underline{R}}^-\right) d\Omega + \iint_\Omega \underline{\phi}^T div \underline{\underline{N}} \, d\Omega + \int_{\partial\Omega_1} \underline{\phi}^T \left(\underline{\underline{N}} \cdot \underline{n} - \underline{p}\right) dS + \\ & + \iint_\Omega \mu_1 tr\left(\underline{\underline{R}}^+ - \underline{\underline{N}}\right) d\Omega + \iint_\Omega \alpha_1 det\left(\underline{\underline{R}}^+ - \underline{\underline{N}}\right) d\Omega + \\ & + \iint_\Omega \mu_2 tr\left(\underline{\underline{R}}^- + \underline{\underline{N}}\right) d\Omega + \iint_\Omega \alpha_2 det\left(\underline{\underline{R}}^- + \underline{\underline{N}}\right) d\Omega\end{aligned}$$

obtained appending to the functional to be minimized the constraints (9) to (12) through the Lagrangian multipliers $\underline{\phi}$, μ_1, μ_2, α_1 and α_2.

In such a way, the optimization problem can be rewritten in the following way: PROBLEM 2 *Find:*

$$\mathcal{V} = inf \; sup \; \mathcal{L}$$

where the *infimum* must be computed with respect to the state variables $\underline{\underline{R}}^+ \geq 0$, $\underline{\underline{R}}^- \geq 0$, $\underline{\underline{N}}$, and the *supremum* with respect to Lagrangian multipliers $\underline{\phi}$, $\mu_1 \leq 0$, $\mu_2 \leq 0$, $\alpha_1 \leq 0$ and $\alpha_2 \leq 0$.

The following orthogonality constraints hold

$$\mu_1 \, tr\left(\underline{\underline{R}}^+ - \underline{\underline{N}}\right) = 0 \tag{14a}$$

$$\mu_2 \, tr\left(\underline{\underline{R}}^- + \underline{\underline{N}}\right) = 0 \tag{14b}$$

$$\alpha_1 \, det\left(\underline{\underline{R}}^+ - \underline{\underline{N}}\right) = 0 \tag{15a}$$

$$\alpha_2 \, det\left(\underline{\underline{R}}^- + \underline{\underline{N}}\right) = 0 \tag{15b}$$

$$\mu_1 \mu_2 = 0 \tag{16}$$

$$\alpha_1 \alpha_2 = 0 \tag{17}$$

The necessary stationarity conditions with respect to a variation of the state variables N_{11}, N_{22} and N_{12} are respectively

$$-\phi_{1,1} - \mu_1 - \alpha_1\left(R_{22}^+ - N_{22}\right) + \mu_2 + \alpha_2\left(R_{22}^- + N_{22}\right) = 0 \tag{18a}$$

$$-\phi_{2,2} - \mu_1 - \alpha_1\left(R_{11}^+ - N_{11}\right) + \mu_2 + \alpha_2\left(R_{11}^- + N_{11}\right) = 0 \tag{18b}$$

$$-\phi_{1,2} + \alpha_1\left(R_{12}^+ - N_{12}\right) - \alpha_2\left(R_{12}^- + N_{12}\right) = 0 \tag{18c}$$

Then, the stationarity conditions with respect to R^+_{11}, R^+_{22} and R^+_{12} read respectively

$$1+\mu_1+\alpha_1\left(R^+_{22}-N_{22}\right)\geq 0 \tag{19a}$$
$$1+\mu_1+\alpha_1\left(R^+_{11}-N_{11}\right)\geq 0 \tag{19b}$$
$$\alpha_1\left(R^+_{12}-N_{12}\right)=0 \tag{19c}$$

and, analogously, the stationarity with respect to R^-_{11}, R^-_{22} and R^-_{12} implies

$$1+\mu_2+\alpha_2\left(R^-_{22}+N_{22}\right)\geq 0 \tag{20a}$$
$$1+\mu_2+\alpha_2\left(R^-_{11}+N_{11}\right)\geq 0 \tag{20b}$$
$$\alpha_2\left(R^-_{12}+N_{12}\right)=0 \tag{20c}$$

Such necessary conditions suggest some phisical considerations, concerning the optimal solution of the problem.

REMARK 1. When the cross sectional areas of the reinforcing fibers are not vanishing (i.e., when $R^+_{11}>0$ and $R^+_{22}>0$), then the body, at the optimum, shows a fully stress and a corner stress behaviour, simultaneously.

In fact, if $R^+_{11}>0$ and $R^+_{22}>0$, Rel.s (19a,b) read

$$1+\mu_1+\alpha_1\left(R^+_{22}-N_{22}\right)=0 \tag{21a}$$
$$1+\mu_1+\alpha_1\left(R^+_{11}-N_{11}\right)=0 \tag{21b}$$

from which one has

$$\alpha_1\left(R^+_{22}-N_{22}\right)=\alpha_1\left(R^+_{11}-N_{11}\right)$$

and multiplying both the sides by $\left(R^+_{11}-N_{11}\right)$

$$\alpha_1\left(R^+_{11}-N_{11}\right)\left(R^+_{22}-N_{22}\right)=\alpha_1\left(R^+_{11}-N_{11}\right)^2 \tag{22}$$

Now, it is worth noting that the left hand side of Rel. (22), by virtue of the orthogonality condition (15a), vanishes, for any value of $\alpha_1\leq 0$. This means that

$$\alpha_1\left(R^+_{11}-N_{11}\right)^2=0$$

or

$$\alpha_1\left(R^+_{11}-N_{11}\right)=0 \tag{23}$$

Eq. (21b) can be written $1+\mu_1=0$ and, multiplying by $tr(\underline{\underline{R}}^+-N)$:

$$tr(\underline{\underline{R}}^+-\underline{\underline{N}})+\mu_1 tr(\underline{\underline{R}}^+-\underline{\underline{N}})=0$$

Now, by virtue of the orthogonality condition (14a) and taking inequalities (3) into account, Rel. (23), which holds again for any $\alpha_1 \leq 0$, shows that also the following conditions hold

$$R_{11}^{+} = N_{11} \qquad \text{and} \quad R_{22}^{+} = N_{22} \tag{24}$$

In such a way, Rel.s (24) are simultaneously satisfied and it is possible to prove that both the fully stress and the corner stress conditions hold at the optimum.

REMARK 2. The optimal solution is characterized by local collinearity of stress, strength and strain rate.

In fact, in principal directions of stress, i.e., when $N_{12} = 0$, and for $\alpha_1 \leq 0$, Rel. (19c) implies also $R_{12} = 0$. Moreover, such a result, substituted in Rel. (18c) and taking into account the orthogonality condition (17), returns $\phi_{1,2} = 0$.

By virtue of its own nature, the Lagrangian multiplier $\underline{\phi} = \{\phi_1, \phi_2\}^T$ can be seen as a displacement rate vector, and consequently its derivatives as strain rates. Hence, the principal directions of stress, strength and strain rate must be collinear in the optimal solution.

REMARK 3. The stationarity conditions allow also for some kinematical considerations on the mechanism of the optimal solution. In fact, Rel. (19a) can be rewritten as

$$\mu_1 \geq -1 - \alpha_1 \left(R_{22}^{+} - N_{22}\right)$$

The last inequality, after substitution in Rel. (18a), gives

$$\begin{aligned} &- \phi_{1,1} - \alpha_1 \left(R_{22}^{+} - N_{22}\right) + \mu_2 + \\ &+ \alpha_2 \left(R_{22}^{-} + N_{22}\right) \geq -1 - \alpha_1 \left(R_{22}^{+} - N_{22}\right) \end{aligned}$$

which, by virtue of the fully stress conditions (24) and of the orthogonalities (16) and (17), furnishes

$$1 - \phi_{1,1} \geq 0. \tag{25a}$$

Following the same considerations also for the other analogous equations, one obtains the following set of inequalities

$$1 + \phi_{1,1} \geq 0 \tag{25b}$$

$$1 - \phi_{2,2} \geq 0 \tag{25c}$$

$$1 + \phi_{2,2} \geq 0 \tag{25d}$$

By virtue of the fact that in Eq.s (25a) and (25c) the equality holds if $R_{11}^{+} > 0$ and $R_{22}^{+} > 0$ (or, analogously, if in Eq.s (25b) and (25d) $R_{11}^{-} > 0$ and $R_{22}^{-} > 0$)

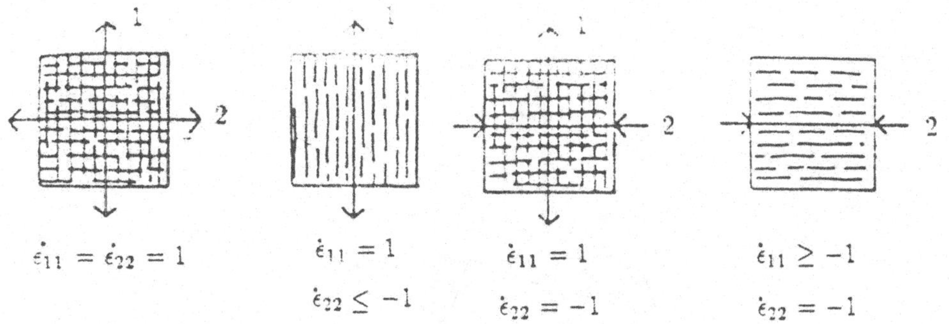

Figure 6

and, conversely, the inequality is true if the fiber vanish, the kinematical conditions schematically shown in Fig. 6 are satisfied at the optimum.

This means that the optimal solution shows a mechanism characterized by constant strain rate in the fiber direction and by a sign–restricted strain rate if the fibers vanish.

Such a result, with the equilibrium conditions, allows one to find a possible mechanisms and a possible paths for the reinforcing fibers in some very simple cases, as illustrated in Fig. 7.

3. REMARKS ON THE OPTIMAL DESIGN OF FIBER ORIENTATION AND FIBER DENSITY FOR GIVEN ELASTIC COMPLIANCE

Let us consider now an elastic optimization problem in order to compare the elastic optimality conditions with the results obtained in the limit plastic design problem. In an orthogonal reference frame x_1–x_2–x_3, we consider a solid body defined on the open domain $\Omega \in \mathcal{R}^2$ (in the plane x_1–x_2) with boundary $\partial\Omega \equiv \partial\Omega_1 \cup \partial\Omega_2$, sufficiently smooth. Moreover, let the body be subjected to a plane state of stress, with tractions and displacements prescribed on $\partial\Omega_1$ and $\partial\Omega_2$ respectively. Moreover, the body is regarded as locally orthotropic in the plane x_1–x_2 and the principal directions of orthotropy z_1, z_2 are specified through the angle $\theta(x_1, x_2)$, made by the z_1 and x_1 axes (see Fig. 8).

This elastic optimization problem can be solved minimizing the work done by the external loads, with an upper bound on the total amount of the structural cost or, in other words, with a bound on the density of reinforcement, which can be regarded as responsible for the cost of the structure. If we assume as meaningful parameters

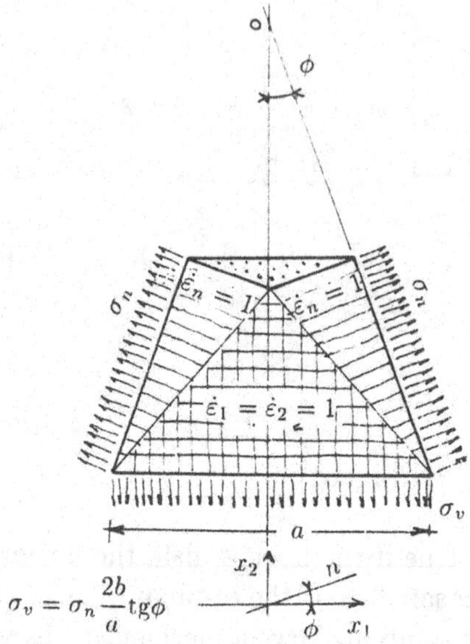

Figure 7

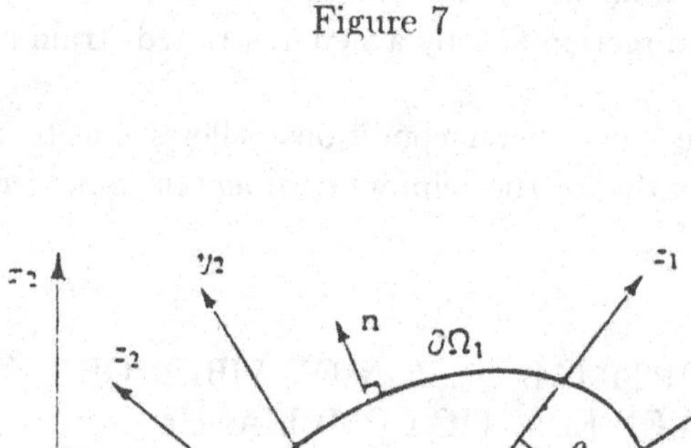

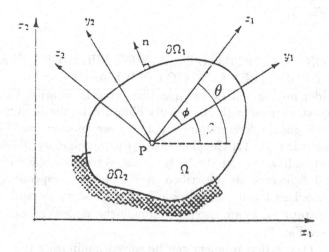

Figure 8

the stiffnesses Q^0_{11} and Q^0_{22}, the problem, taking as design variables both the angle

$\theta(x_1, x_2)$ and the functions $Q_{11}^0(x_1, x_2)$ and $Q_{22}^0(x_1, x_2)$, can be stated in the following way:

PROBLEM 3 *Find:*

$$J^0(u_i, \theta, Q_I^0, Q_{II}^0) = \min_{\theta, Q_{I,II}^0 \geq Q^0} \left(\frac{1}{2} \iint_\Omega p_i u_i \, ds \right)$$

subject to the equilibrium and compatibility conditions and to the global constraint

$$\iint_\Omega (Q_{11}^0 + Q_{22}^0) \, d\Omega \leq C^0$$

with Q^0 and C^0 prescribed values. Here Q^0 can be seen as the contribution to the stiffness due to the matrix in a composite.

Making use of the Lagrange multiplier method (see Save, Prager [17]) Problem 6 can be stated in the form:

PROBLEM 4 *Find:*

$$J^0(u_i, \theta, Q_I^0, Q_{II}^0, \alpha) = \min_{\theta, Q_{I,II}^0 \geq Q^0} \; \max_{u_i, \alpha \geq 0} \left(\int_{\partial\Omega_1} p_i u_i \, ds - \iint_\Omega \mathcal{E} \, d\Omega \right.$$

$$\left. + \alpha \iint_\Omega (Q_{11}^0 + Q_{22}^0) \, d\Omega - \alpha C^0 \right)$$

where $\mathcal{E}$ represents the elastic strain energy density (see [9]) and α is a Lagrange multiplier, constant with respect to x_1 and x_2.

In order to compute the optimality condition with respect to the design variable θ it appears convenient to introduce a new local reference frame y_1–y_2 coinciding, at each point of the body, with the principal directions of strain. Let us define such reference system through the angle $\phi(x_1, x_2)$ made by the z_1 and y_1 axes (Fig. 8).

From Fig. 8 it appears that the design variable is given by $\theta = \beta + \phi$; nevertheless the angle $\beta(x_1, x_2)$ is univocally determinated when the strain state is known. Thus it is possible to assume as new design variable the angle $\phi(x_1, x_2)$.

Now, the stationarity conditions with respect to displacements u_i return the equilibrium equations, in Ω and on the boundary. The stationarity with respect to the orientation angle of orthotropy axes furnishes

$$\sin\phi \cos\phi \, (2\mathcal{A} \cos^2\phi + \mathcal{B}) = 0 \tag{26}$$

where $\mathcal{A}$, $\mathcal{B}$ are suitable functions of stiffness parameters and principal components of strain tensor (see [9]). Finally, as a consequence of a variation of the stiffness coefficients Q_I^0 and Q_{II}^0, we obtain the following inequalities that must be fulfilled by the strain field at the optimum

$$(\epsilon_I - \epsilon_{II})^2 \cos^4\phi + 2(\epsilon_I \epsilon_{II} - \epsilon_{II}^2) cos^2\phi + \epsilon_{II} \leq \frac{2}{\alpha} \tag{27a}$$

$$(\epsilon_I - \epsilon_{II})^2 \cos^4\phi + 2(\epsilon_I\epsilon_{II} - \epsilon_I^2)\cos^2\phi + \epsilon_I^2 \leq \frac{2}{\alpha}. \tag{27b}$$

For the two cases $\sin\phi = 0$ and $\cos\phi = 0$, corresponding to local minima, both Rels. (27a) and (27b) give

$$\epsilon_{II}^2 \leq \frac{2}{\alpha} \quad \text{and} \quad \epsilon_I^2 \leq \frac{2}{\alpha}. \tag{28}$$

If the stiffnesses Q_I^0, Q_{II}^0 do not attain their lowest value Q^0, then inequalities (28), at the optimum, become equalities and the following relation holds

$$|\epsilon_I| = |\epsilon_{II}| = \text{const.} \qquad \text{in } \Omega. \tag{29}$$

This strong constraint on the strain field can be, on the other hand, avoided if Q_{11}^0 (or Q_{22}^0) is equal to Q^0 (i.e. if the reinforcement of the composite in the z_1 (resp. z_2) direction vanishes), and one of the inequalities (28) can be recovered. A related problem has been dealt with by Olhoff and Thomsen [8], where some numerical solution are also given.

So, it has been shown that in the minimum compliance problem, with a constraint on the total structural cost, the condition of collinearity between principal directions of stress, strain and orthotropy still holds, but an optimal design can be obtained only for particular values of the components of strain.

4. OPTIMAL LIMIT DESIGN OF REINFORCED PLATES IN BENDING WITH ASSIGNED LIMIT LOAD

Consider an orthogonal reference frame $x_1 - x_2$, where a solid body of domain $\Omega \in \mathcal{R}^2$ is defined. We adopt the usual Kirchhoff's kinematic assumption of the structural theory of thin plates in bending and we use the reference frame indicated above.

We consider a solid plate with constant thickness, with boundary $\partial\Omega \equiv \partial\Omega_1 \cup \partial\Omega_2$ on which no tractions and generalized displacements are prescribed, respectively. On $\partial\Omega_2$ displacements are prescribed in such a way that rigid motions of all the body are prevented.

The given load is supposed orthogonal to $x_1 - x_2$ plane, so that in Ω a scalar p can be defined, as a function of the coordinates $x_1 - x_2$, on the midsurface of the structure.

The generalized stresses are the tensor moments $\underline{M}$, the scalar components of which are: M_{ik} $(i = 1,2)$ bending moments corresponding to $i = k$ and twisting moments to $i \neq k$ (with $M_{ik} = M_{ki}$). In the way of Kirchhoff's assumption, we assume that the collapse mechanism of the body can be represented exhaustively by the deflection rate $\dot{w}$ of the points of the midplane of the plate (Fig. 9). The deflection

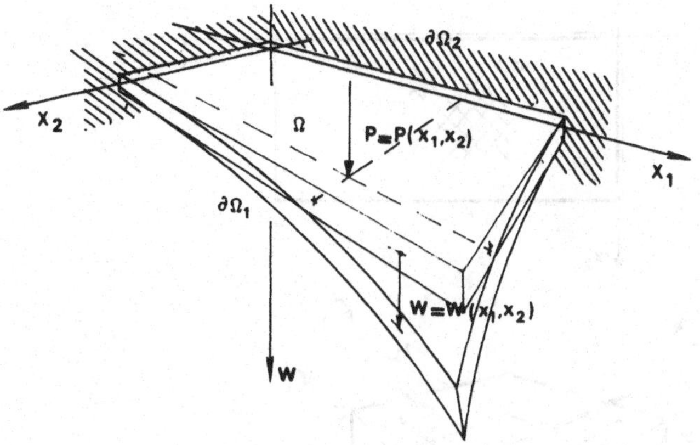

Figure 9

rate field being $\dot{w} = \dot{w}(x_1, x_2)$, the generalized strain rates are $\kappa_{ik} = -\dot{w}_{,ik}$. Bending curvature rates are obtained with $i = k$; twisting curvature rates, with $i \neq k$.

The body is made by a non structural matrix component and reinforced by a system of fibers orthogonal each others. In such a way the solid is regarded as locally orthotropic, and the local orthogonal frame $z_1 - z_2$ denotes the principal directions of orthotropy, (refer to Fig. 1). Moreover, the stresses in the material are supposed to be entirely carried by the fibers, both in traction and in compression.

For a prescribed load distribution p on Ω, the stress field in the body, expressed in terms of characteristics, is specified through the tensor $\underline{\underline{M}}$, which satisfies the equilibrium equations:

$$div(div\underline{\underline{M}}) = p \text{ in } \Omega \tag{30a}$$

$$\underline{\underline{M}} \cdot \underline{n} = \underline{0} \text{ on } \partial\Omega_1 \tag{30b}$$

$$(div\underline{\underline{M}}) \cdot \underline{n} = 0 \text{ on } \partial\Omega_1 \tag{30c}$$

with $\underline{\underline{M}} = \underline{\underline{M}}^T$, under the hypotesis of vanishing body forces, and where $\underline{n}$ denotes the outward unit normal on $\partial\Omega_1$.

The fibers are supposed to have cross sectional area variable along the fiber length. In principal directions of orthotropy, at each point of the body, the strength of the fibers – and therefore of the material – is defined as

$$K^+{}_{I,II} = A_{I,II}\, h\, \sigma^+ \tag{31a}$$

$$K^-{}_{I,II} = A_{I,II}\, h\, \sigma^- \tag{31b}$$

where h has been assumed as a known constant value for the moment arm of the internal forces in the collapse state.

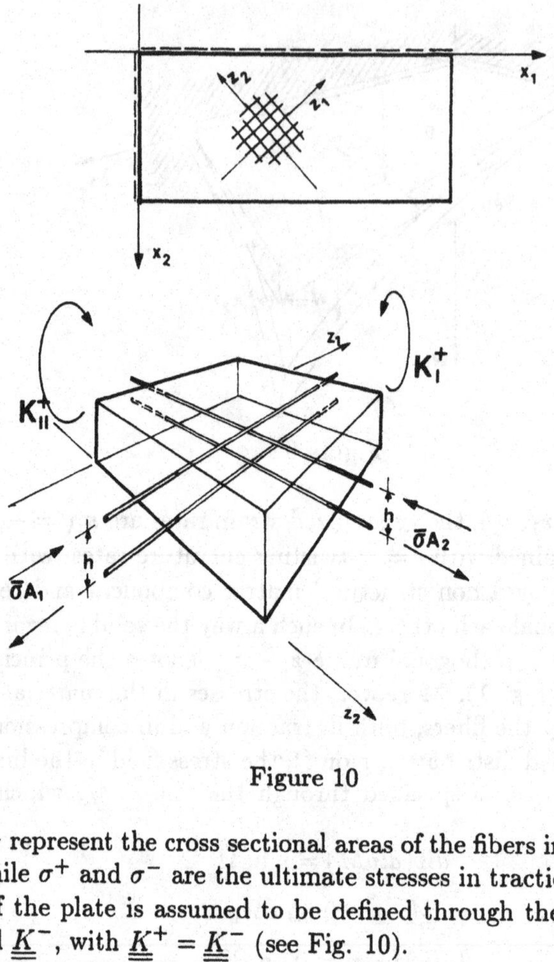

Figure 10

A_I and A_{II} represent the cross sectional areas of the fibers in direction z_1 and z_2 respectively, while σ^+ and σ^- are the ultimate stresses in traction and compression. The strength of the plate is assumed to be defined through the components of the tensors $\underline{\underline{K}}^+$ and $\underline{\underline{K}}^-$, with $\underline{\underline{K}}^+ = \underline{\underline{K}}^-$ (see Fig. 10).

Finally, the strength criterion, in orthogonal coordinates, can be stated as follows (see Fig. 3):

$$K^+{}_{nn} \geq M_{nn} \geq -K^-{}_{nn} \qquad \forall n \tag{32}$$

under the conditions of non-negativeness

$$K^+{}_{nn} \geq 0 \quad \forall n$$

$$K^-{}_{nn} \geq 0 \quad \forall n$$

Instead of a representation given in terms of components, the strength criterion can be written in tensor form

$$tr\,(\underline{\underline{K}}^+ - \underline{\underline{M}}) \geq 0 \qquad \text{and} \qquad det\,(\underline{\underline{K}}^+ - \underline{\underline{M}}) \geq 0 \tag{33a}$$

$$tr\,(\underline{\underline{K}}^- + \underline{\underline{M}}) \geq 0 \qquad \text{and} \quad det\,(\underline{\underline{K}}^- + \underline{\underline{M}}) \geq 0 \tag{33b}$$

In the generalized stress space, the ultimate strength domain defined by Rel.s (33a,b) can be represented as in Fig. 4. Or, analogously, in the space M_I, M_{II} of the principal moments, as in Fig. 5.

From Fig. 5 it clearly appears that the ultimate strength domain, for given values of the limit stresses σ^+ and σ^-, depends on the cross sectional areas of the fibers.

By virtue of such a remark, in the present Section the total amount of structural material (fibers), i.e., the size of the strength domain, will be minimized, for prescribed values of the external loads, according to equilibrium and strength criterion.

In such a way, assuming as design variables the local values of the cross sectional areas of the fibers, the optimal design problem can be stated as follows:
PROBLEM 5 *Find:*

$$\min_{A_{I,II} \geq 0} \iint_\Omega (A_I + A_{II}) d\Omega \tag{34}$$

such that

$$div\,(div\underline{\underline{M}}) = p \quad \text{in } \Omega \qquad + \text{b.c.} \tag{35a}$$

$$tr\,(\underline{\underline{K}}^+ - \underline{\underline{M}}) \geq 0 \quad tr\,(\underline{\underline{K}}^- + \underline{\underline{M}}) \geq 0 \tag{35b}$$

$$det\,(\underline{\underline{K}}^+ - \underline{\underline{M}}) \geq 0 \quad det\,(\underline{\underline{K}}^- + \underline{\underline{M}}) \geq 0 \tag{35c}$$

By virtue of Rel.s (31a,b) the design variables A_I and A_{II} are replaced by K_I^+ and K_{II}^+, taking into account that

$$K_I^+ K_I^- = K_{II}^+ K_{II}^- = 0$$

and consequently Rel. (34) reads

$$\min_{\underline{\underline{K}}^{+,-} \geq 0} \iint_\Omega \left[tr(\underline{\underline{K}}^+) + tr(\underline{\underline{K}}^-) \right] d\Omega \tag{36}$$

The minimization problem (36) subject to the constraints (35a) to (35c) and taking the boundary conditions into account can be transformed in the search for the stationarity of the unconstrained Lagrangian functional $\mathcal{L}$, defined as

$$\begin{aligned} \mathcal{L} = & \iint_\Omega (tr\underline{\underline{K}}^+ + tr\underline{\underline{K}}^-)\, d\Omega + \iint_\Omega \phi(div\,(div\underline{\underline{M}}) - p)\, d\Omega + \\ & + \iint_\Omega \mu_1 tr\,(\underline{\underline{K}}^+ - \underline{\underline{M}})\, d\Omega + \iint_\Omega \alpha_1 det\,(\underline{\underline{K}}^+ - \underline{\underline{M}})\, d\Omega + \\ & + \iint_\Omega \mu_2 tr\,(\underline{\underline{K}}^- + \underline{\underline{M}})\, d\Omega + \iint_\Omega \alpha_2 det\,(\underline{\underline{K}}^- + \underline{\underline{M}})\, d\Omega \end{aligned}$$

obtained appending to the functional to be minimized the constraints (35a) to (35c) through the Lagrangian multipliers ϕ, μ_1, μ_2, α_1 and α_2.

In such a way, the optimization problem can be rewritten in the following way: PROBLEM 6 *Find:*

$$\mathcal{V} = inf\ sup\ \mathcal{L}$$

where the *infimum* must be computed with respect to the state variables $\underline{\underline{K}}^+ \geq 0$, $\underline{\underline{K}}^- \geq 0$, $\underline{\underline{M}}$, and the *supremum* with respect to Lagrangian multipliers ϕ, $\mu_1 \leq 0$, $\mu_2 \leq 0$, $\alpha_1 \leq 0$ and $\alpha_2 \leq 0$.

The following orthogonality constraints hold

$$\mu_1\, tr\, (\underline{\underline{K}}^+ - \underline{\underline{M}}) = 0 \tag{37a}$$

$$\mu_2\, tr\, (\underline{\underline{K}}^- + \underline{\underline{M}}) = 0 \tag{37b}$$

$$\alpha_1\, det\, (\underline{\underline{K}}^+ - \underline{\underline{M}}) = 0 \tag{37c}$$

$$\alpha_2\, det\, (\underline{\underline{K}}^- + \underline{\underline{M}}) = 0 \tag{37d}$$

$$\mu_1 \mu_2 = 0 \tag{37e}$$

$$\alpha_1 \alpha_2 = 0 \tag{37f}$$

The necessary stationarity conditions with respect to a variation of the state variables M_{11}, M_{22} and M_{12} are respectively

$$-\phi_{,11} - \mu_1 - \alpha_1 (K_{22}^+ - M_{22}) + \mu_2 + \alpha_2 (K_{22}^- + M_{22}) = 0 \tag{38a}$$

$$-\phi_{,22} - \mu_1 - \alpha_1 (K_{11}^+ - M_{11}) + \mu_2 + \alpha_2 (K_{11}^- + M_{11}) = 0 \tag{38b}$$

$$-\phi_{,12} + \alpha_1 (K_{12}^+ - M_{12}) - \alpha_2 (K_{12}^- + M_{12}) = 0 \tag{38c}$$

Then, the stationarity conditions with respect to K_{11}^+, K_{22}^+ and K_{12}^+ read respectively

$$1 + \mu_1 + \alpha_1 (K_{22}^+ - M_{22}) \geq 0 \tag{39a}$$

$$1 + \mu_1 + \alpha_1 (K_{11}^+ - M_{11}) \geq 0 \tag{39b}$$

$$\alpha_1 (K_{12}^+ - M_{12}) = 0 \tag{39c}$$

and, analogously, the stationarity with respect to K_{11}^-, K_{22}^- and K_{12}^- implies

$$1 + \mu_2 + \alpha_2 (K_{22}^- + M_{22}) \geq 0 \tag{40a}$$

$$1 + \mu_2 + \alpha_2 (K_{11}^- + M_{11}) \geq 0 \tag{40b}$$

$$\alpha_2 (K_{12}^- + M_{12}) = 0. \tag{40c}$$

Such necessary conditions suggest some physical considerations, concerning the optimal solution of the problem.

REMARK 4. When the cross sectional areas of the reinforcing fibers are not vanishing (i.e., when $K_{11}^{+} > 0$ and $K_{22}^{+} > 0$), then the body, at the optimum, shows a fully stress and a corner stress behaviour, simultaneously.

In fact, if $K_{11}^{+} > 0$ and $K_{22}^{+} > 0$, Rel.s (39a,b) read

$$1 + \mu_1 + \alpha_1 \left(K_{22}^{+} - M_{22}\right) = 0 \tag{41a}$$

$$1 + \mu_1 + \alpha_1 \left(K_{11}^{+} - M_{11}\right) = 0 \tag{41b}$$

from which one has

$$\alpha_1 \left(K_{22}^{+} - M_{22}\right) = \alpha_1 \left(K_{11}^{+} - M_{11}\right)$$

and multiplying both sides by $(K_{11}^{+} - M_{11})$

$$\alpha_1 \left(K_{11}^{+} - M_{11}\right)\left(K_{22}^{+} - M_{22}\right) = \alpha_1 \left(K_{11}^{+} - M_{11}\right)^2 \tag{42}$$

Now, it is worth noting that the left hand side of Rel. (42), by virtue of the orthogonality condition (37c), vanishes, for any value of $\alpha_1 \leq 0$. This means that

$$\alpha_1 \left(K_{11}^{+} - M_{11}\right)^2 = 0$$

or

$$\alpha_1 \left(K_{11}^{+} - M_{11}\right) = 0$$

Eq. (41b) can be written $1 + \mu_1 = 0$ and, multiplying by $tr(\underline{\underline{K}}^{+} - \underline{\underline{M}})$:

$$tr(\underline{\underline{K}}^{+} - \underline{\underline{M}}) = \mu_1 tr(\underline{\underline{K}}^{+} - \underline{\underline{M}}) \tag{43}$$

Now, by virtue of orthogonality condition (37a) and taking into account inequalities (32), Rel. (43), which holds again for any $\alpha_1 \leq 0$, shows that the following condition holds

$$K_{11}^{+} = M_{11} \tag{44a}$$

Following the same way, we can prove that

$$K_{22}^{+} = M_{22} \tag{44b}$$

In such a way, Rel.s (44a,b) are simultaneously verified which means, in other words, that both the fully stress and the corner stress conditions hold at the optimum.

REMARK 5. The optimal solution is characterized by local collinearity of moments, strength and curvature rates.

In fact, in principal directions of moments, i.e., when $M_{12} = 0$, and for $\alpha_1 \leq 0$, Rel. (39c) implies also $K_{12} = 0$. Moreover, such a result, substituted in Rel. (38c) and taking into account the orthogonality condition (37f), returns $\phi_{,12} = 0$.

By virtue of its own nature, the Lagrangian multiplier ϕ can be seen as deflection rate, and consequently its derivatives as curvature rates. Hence, the principal directions of moments, strength and curvature rates must be collinear in the optimal solution.

REMARK 6. The stationarity conditions allow also for some kinematical considerations on the mechanism of the optimal solution. In fact, Rel. (39a) can be rewritten as

$$\mu_1 \geq -1 - \alpha_1 \left(K_{22}^{+} - M_{22}\right)$$

The last inequality, after substitution in Rel. (38a), gives

$$\begin{aligned} &- \phi_{,11} - \alpha_1 \left(K_{22}^{+} - M_{22}\right) + \mu_2 + \\ &+ \alpha_2 \left(K_{22}^{-} + M_{22}\right) \geq -1 - \alpha_1 \left(K_{22}^{+} - M_{22}\right) \end{aligned}$$

which, by virtue of the fully stress condition and of the orthogonalities (37e,f), furnishes

$$1 - \phi_{,11} \geq 0. \tag{45}$$

Following the same considerations also for the other analogous equations, one obtains the following set of inequalities

$$1 + \phi_{,11} \geq 0 \tag{46a}$$
$$1 - \phi_{,22} \geq 0 \tag{46b}$$
$$1 + \phi_{,22} \geq 0 \tag{46c}$$

By virtue of the fact that in Eq.s (45) and (46b) the equality holds if $K_{11}^{+} > 0$ and $K_{22}^{+} > 0$ (or, analogously, if in Eq.s (46a) and (46c) $K_{11}^{-} > 0$ and $K_{22}^{-} > 0$) and, conversely, the inequality is true if the fiber vanish, the kinematical conditions schematically shown in Fig. 11 are satisfied at the optimum.

This means that the optimal solution shows a mechanism characterized by a constant curvature in the fiber direction and by a sign–restricted curvature if the fibers vanish.

Such a result, with the equilibrium conditions, allows one to find a possible mechanisms and a possible paths for the reinforcing fibers in many cases, as illustrated in Figs. 12 and 13.

REMARK 7: Influence of the orthotropy ratio.
The volume minimization problem stated by Eq. (34) can be rewritten as follows:

$$V = \iint_{\Omega} \left[A_1(1+\gamma)\right] d\Omega \longrightarrow \min \tag{47}$$

where $\gamma = A_2/A_1$ is the orthotropy ratio. Thus, the functions A_1 and γ minimizing the functional $V = V[A_1, \gamma]$ have to be determined.

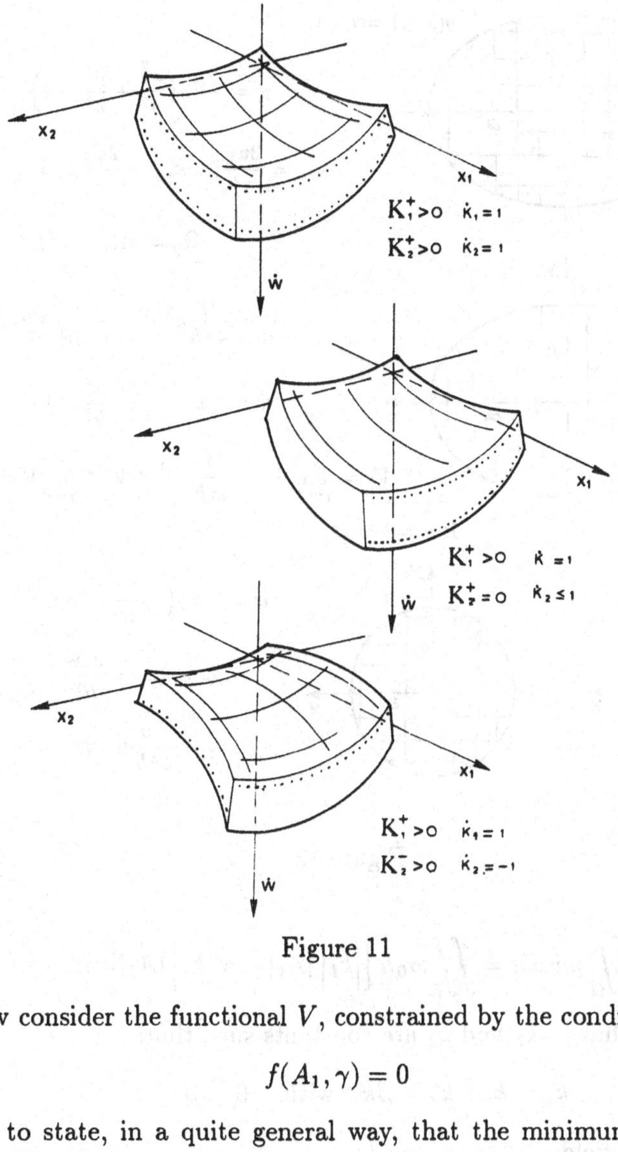

Figure 11

If we now consider the functional V, constrained by the condition

$$f(A_1, \gamma) = 0 \tag{48}$$

it is possible to state, in a quite general way, that the minimum of functional V constrained by Eq. (48) is greater than, or equal to, the unconstrained minimum of the functional.

Thus, in order to attain the absolute minimum volume, it is necessary that the orthotropy ratio is not subjected to any prescribed condition. This remark is useful since in many cases the value of the orthotropy ratio is *a-priori* prescribed by technological requirements.

By making use of the orthotropy ratio γ, the equation of the dissipated power can be written as:

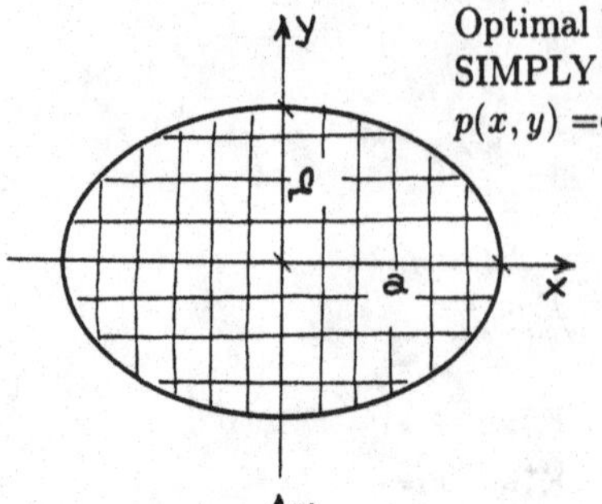

Optimal layouts
SIMPLY SUPPORTED PLATES
$p(x, y)$ =const.

$$\dot{w} = -\dot{w}_0\left(\frac{x^2}{a^2} + \frac{y^2}{b^2} - 1\right)$$

$$\dot{\kappa}_{xx} = \frac{2\dot{w}_0}{a^2}; \quad \dot{\kappa}_{yy} = \frac{2\dot{w}_0}{b^2}; \quad \dot{\kappa}_{xy} = 0$$

$$\Omega_y = \gamma\Omega_x$$

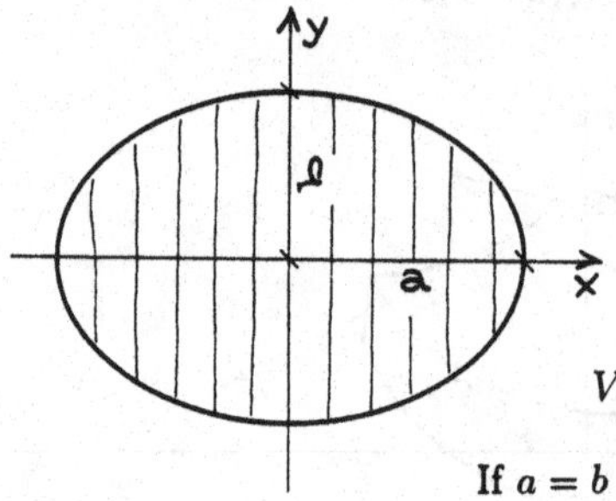

$$V = \frac{\pi p}{4\bar{\sigma}h} a^3 b^3 \frac{1+\gamma}{b^2 + \gamma a^2} \ \forall\gamma$$

$$\dot{w} = -\dot{w}_0\left(\frac{x^2}{a^2} + \frac{y^2}{b^2} - 1\right)$$

$$V_1 = \lim_{\gamma\to\infty} V = \frac{\pi p}{4\bar{\sigma}h} ab^3 < V < \lim_{\gamma\to 0} V = \frac{\pi p}{4\bar{\sigma}h} a^3 b$$

If $a = b = R$

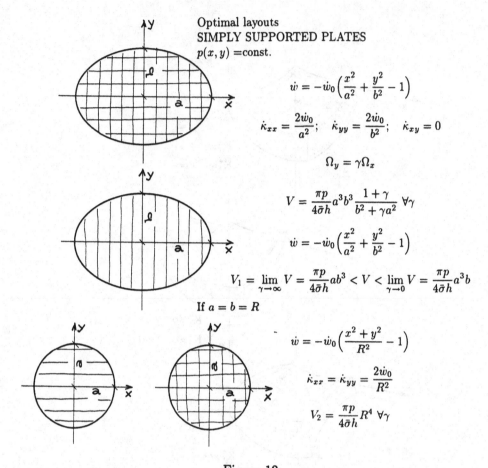

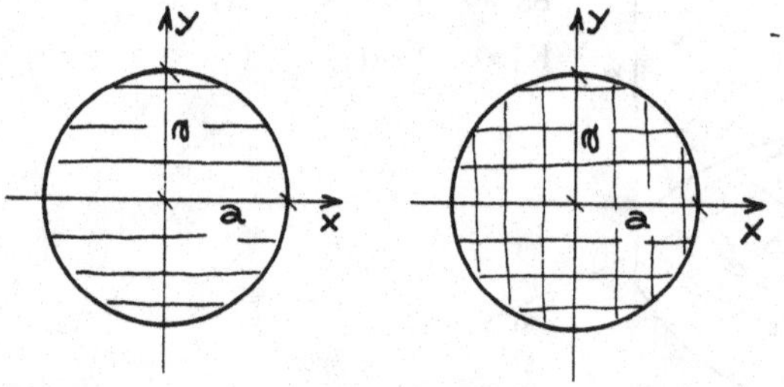

$$\dot{w} = -\dot{w}_0\left(\frac{x^2 + y^2}{R^2} - 1\right)$$

$$\dot{\kappa}_{xx} = \dot{\kappa}_{yy} = \frac{2\dot{w}_0}{R^2}$$

$$V_2 = \frac{\pi p}{4\bar{\sigma}h} R^4 \ \forall\gamma$$

Figure 12

$$\iint_\Omega p\dot{w}\, d\Omega = \iint_\Omega \sigma_0 h \left[\left|\dot{k}_1\right| |A_1| + \gamma \left|\dot{k}_2\right| |A_1|\right] d\Omega \tag{49}$$

Suppose now that γ, $\dot{k}_1$ and $\dot{k}_2$ are constants such that:

$$\dot{k}_1 = k, \quad \dot{k}_2 = \beta k \quad \text{with} \quad 0 \leq \beta \leq 1 \tag{50}$$

Eqs. (49) and (50) yield:

$$\iint_\Omega p\dot{w}\, d\Omega = \sigma_0 hk \frac{(1+\beta\gamma)}{(1+\gamma)} \iint_\Omega A_1(1+\gamma)\, ds \tag{51}$$

By considering the mechanism featured by $\dot{w}$ and k, if the equilibrium conditions in Ω and on the boundary are compatible with $M_{22} = 0$, so that $\gamma = 0$, Eq. (51) can be cast in the following form:

$$\iint_\Omega p\dot{w}\, d\Omega = \sigma_0 hk \iint_\Omega A_1^* d\Omega \tag{52}$$

Comparing Eqs. (51) and (52) one gets:

$$\frac{1+\beta\gamma}{1+\gamma} \iint_\Omega A_1(1+\gamma)\, d\Omega = \iint_\Omega A_1^* d\Omega$$

and, since

$$\frac{1+\beta\gamma}{1+\gamma} \leq 1, \tag{53}$$

$$\iint_\Omega A_1(1+\gamma)\, d\Omega \geq \iint_\Omega A_1^* d\Omega.$$

On the contrary, if the equilibrium conditions are such that it is possible to assume $M_{11} = 0$, so that $A_1 = 0$, by Eq. (49) one gets:

$$\iint_\Omega p\dot{w}\, d\Omega = \sigma_0 hk\beta \int_\Omega A_2^* d\Omega$$

Comparing this equation with Eq. (51), one obtains:

$$\frac{1+\beta\gamma}{1+\gamma} \iint_\Omega A_1(1+\gamma)\, d\Omega = \beta \iint_\Omega A_2^* d\Omega \tag{54}$$

Since

$$\frac{1+\beta\gamma}{1+\gamma} \geq \beta$$

from Eq. (54) one gets:

$$\iint_\Omega A_1(1+\gamma) d\Omega \leq \iint_\Omega A_2^* d\Omega \tag{55}$$

Finally, by virtue of inequalities (53) and (55) one obtains:

$$\iint_\Omega A_1^* d\Omega \leq \iint_\Omega A_1(1+\gamma)\, d\Omega \leq \iint_\Omega A_2^* d\Omega$$

which amounts at saying that, *within the framework of the assumptions made here, the minimum reinforcement volume is obtained if fibers are placed only along the direction of principal curvature with the maximum absolute value. On the contrary, if fibers are placed only along the direction of the principal curvature with minimum absolute value, the volume is greater than the volume obtained by placing fibers along the two principal curvature directions.*

It is worth noting that if $\beta = 1$, the orthotropy ratio has no influence on the value of minimum volume (Fig. 12).

This procedure does not give any guarantee that absolute minima were obtained. However, the computed volumes are the minimum that can be obtained by placing fibers along two orthogonal cartesian axes.

It is easy to verify that, if the specific dissipated power given by Eq. (49) is prescribed to be constant, conditions are obtained for a minimum volume of reinforcement within a class of solutions with varying A_1 and γ.

5. REMARKS ON OPTIMAL ELASTIC DESIGN OF PLATES IN BENDING

Now, it is possible to show the analogy between the results obtained for optimal plastic design and the optimality conditions for the minimum compliance problem of a solid elastic plate in bending. In such a case, and for the same body formerly defined, the constitutive law, in terms of moments M_{ik} and curvatures κ_{ik}, is given by

$$\begin{Bmatrix} M_{11} \\ M_{22} \\ M_{12} \end{Bmatrix} = \begin{pmatrix} D_{11} & D_{12} & D_{16} \\ D_{12} & D_{22} & D_{26} \\ D_{16} & D_{26} & D_{66} \end{pmatrix} \begin{Bmatrix} \kappa_{11} \\ \kappa_{22} \\ 2\kappa_{12} \end{Bmatrix}.$$

It is necessary to point out that if, for the sake of simplicity, the plate is assumed to be orthotropic but not a laminate, the flexural stiffness coefficients D_{ij} are defined as

$$D_{ij} = \frac{h^3}{12} Q_{ij}.$$

In this case, the minimum compliance problem can be stated as follows:

PROBLEM 7 *Find, in agreement with equilibrium and compatibility:*

$$J^0(w,\theta) = \min_\theta \left(\frac{1}{2} \iint_\Omega p\, w \, d\Omega \right)$$

where $p(x_1, x_2)$ and $w(x_1, x_2)$ are the load distribution and the displacement function respectively, both in direction of x_3.

Once again, The *min* Problem 7 can be transformed in the following *min max* problem:

PROBLEM 8 *Find:*

$$J^0(w,\theta) = \min_\theta (-U^0) = \min_\theta \max_w \left(\frac{1}{2} \iint_\Omega p\, w \, d\Omega - \iint_\Omega \mathcal{E} \, d\Omega \right)$$

SQUARE PLATE SIMPLY SUPPORTED ALONG TWO SIDES

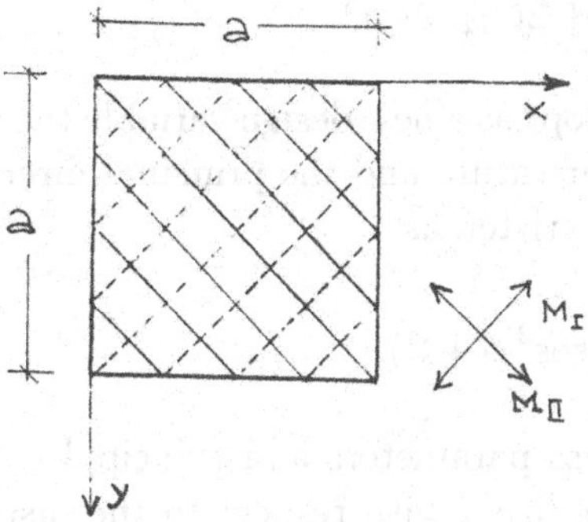

$$\dot{w} = C\frac{xy}{a^2} \quad \text{deflection rate}$$

$$\dot{\kappa}_{xx} = \dot{\kappa}_{yy} = 0: \quad \dot{\kappa}_{xy} = \frac{c}{a^2}$$

$$M_{xx} = M_{yy} = 0$$

$$M_{xy} = \frac{pa^2}{2}\left(1 - \frac{x}{a}\right)\left(1 - \frac{y}{a}\right)$$

$$M_I = -M_{II} = \frac{pa^2}{2}\left(1 - \frac{x}{a}\right)\left(1 - \frac{y}{a}\right)$$

SIMPLY SUPPORTED RECTANGULAR PLATE

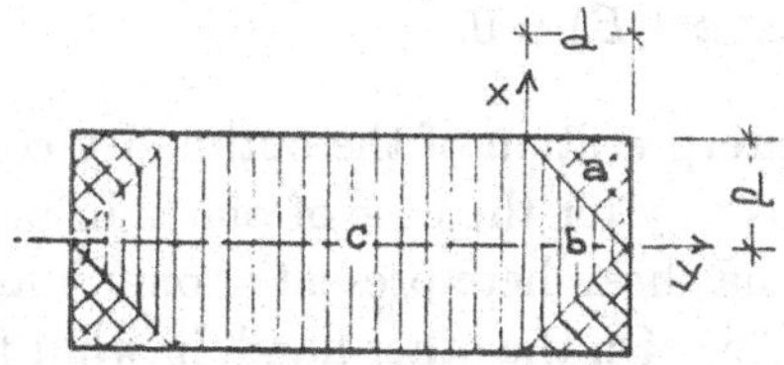

$$\dot{w}_a = \alpha\left(1 - \frac{x}{d}\right)\left(1 - \frac{y}{d}\right)$$

$$\dot{w}_b = \beta\left(1 - \frac{x^2}{d^2} - \frac{y^2}{d^2}\right)$$

$$\dot{w}_c = \gamma\left(1 - \frac{x^2}{d^2}\right)$$

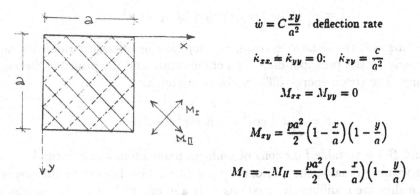

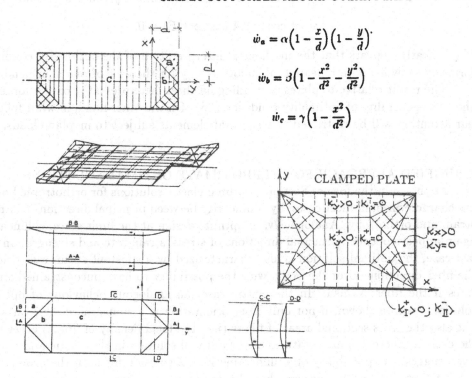

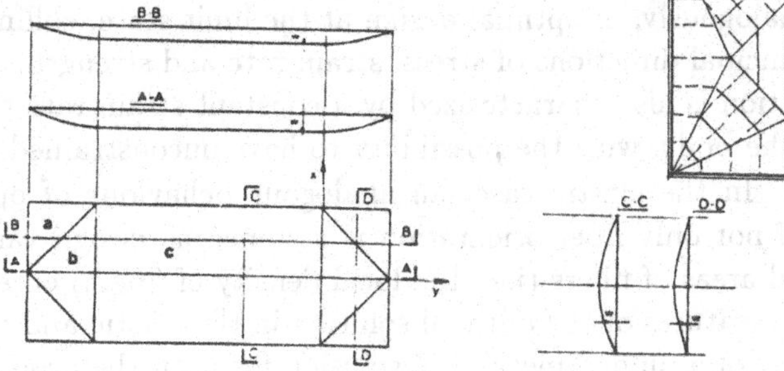

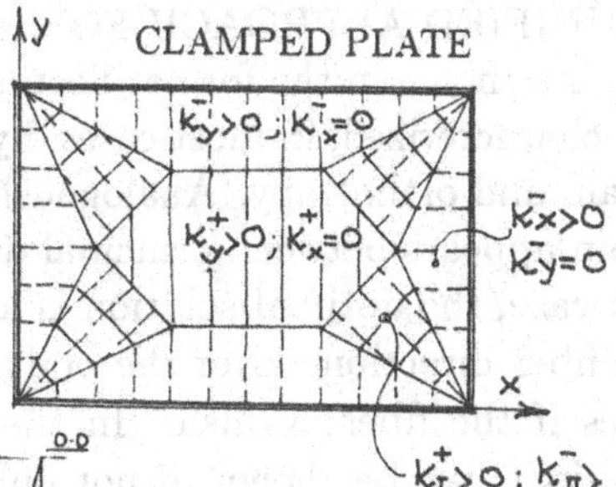

Figure 13

where U is the total potential energy, and $\mathcal{E}$ represents the specific strain energy, given, in principal directions of curvature, by

$$\mathcal{E} = \frac{h^3}{24}\left(Q_{11}\kappa_I^2 + Q_{22}\kappa_{II}^2 + 2Q_{12}\kappa_I\kappa_{II}\right). \tag{56}$$

Proceeding as in the plane stress case, we adopt as a new design variable the angle $\phi(x_1, x_2)$, made by the principal directions of curvature and the principal directions of orthotropy. The strain energy (56) can be rewritten as

$$\mathcal{E} = \frac{h^3}{24}(\mathcal{A}\cos^4\phi + \mathcal{B}\cos^2\phi + \mathcal{C}),$$

where $\mathcal{A}$ and $\mathcal{B}$ are suitable functions of stiffness parameters and principal components of curvature. Finally, the stationarity condition with respect to the displacement w furnishes the equilibrium equations in Ω and on the boundary, whereas the stationarity condition with respect to the angle ϕ (optimality condition) reads again

$$\sin\phi\,\cos\phi\,(2\mathcal{A}\cos^2\phi + \mathcal{B}) = 0. \tag{26}$$

So, it clearly appears that the mechanical interpretation of the optimality condition formerly shown can be repeated in analogous way for the case of plates in bending.

The result relative to plates in bending has been here presented only in order to show the generality of optimality condition (26). On the other hand, in what follows, our attention will be restricted to structural elements subject to in–plane loads.

6. UNIFIED APPROACH FOR APPROXIMATE OPTIMAL SOLUTIONS

As shown in the former Sections, optimal elastic solutions for orthotropic bodies are characterized, in most cases, by collinearity between principal directions of stress, strain and orthotropy. Analogously, in optimal design at the limit state, collinearity again appears between principal directions of stress, strain rate and strength. In the last case, the optimal solution is also characterized by a constant strain rate field in the fiber directions, over the body, with the possibility to have unconstrained strain rates if the fibers vanish. In the elastic case, an analogous behaviour of optimal solutions can be shown, if not only fiber orientation is assumed as design variable, but also the cross sectional areas of fibers (i.e., the local density of fibers) enter into the design. These common features of the optimal solution in the elastic and plastic cases suggest the possibility of a unified method of solution for both the problems.

At first, it must be noticed that the conditions, in terms of strains, or strain rates, ϵ (see Rel.s (27), (28) and (24a) to (24d))

$$|\epsilon| = \text{const.} \qquad \text{along fiber directions} \tag{57a}$$

$$|\epsilon| \leq \text{const.} \qquad \text{if the fibers vanish} \tag{57b}$$

represent a common feature for the two problems.

On the other hand, the possibility of generating solutions with constant strain (or strain rate) fields seems to be possible only for few trivial cases. In such a way, a first possibility to avoid these difficulties can be seen in a re-formulation of the problem, allowing for some lines of discontinuity inside the body, i.e., taking into account the possibility to find an optimal solution characterized by constant strain (or strain rate) over subsets of the domain Ω of the body. Of course, such an hypotesis leads to possible discontinuities of the state variables and of the fiber paths, along the boundaries of discontinuity. Moreover the problem becomes in this way much more difficult, because the number and the shapes of the internal interfaces are both unknowns, and the optimal design problem becomes a free boundary problem.

Nevertheless, a simpler way to proceed to optimal solutions can be found if one observes that the conditions (57) are the same conditions fulfilled by a Michell truss, or, more precisely, by a truss-like continuum, defined by Prager [12], i.e., by a double infinity of bars of infinitesimal length that follow the lines of a dense net formed by two orthogonal families of curves. It is worth noting that this classical result was derived for a problem concerning minimum weight design of isotropic structural elements under plane states of stress, at the limit state. On the other hand, the results obtained show how the same optimality conditions are necessary conditions also for minimum weight plastic design and minimum elastic compliance problems, *for orthotropic bodies.*

The idea to deal with this kind of problems through truss-like solutions is mainly suggested from the peculiar meaning of the optimality conditions and it also finds confirmation in the numerical solutions shown by Olhoff and Thomsen [8] and by Bendsøe [7]. Such solutions, obtained through a direct optimization procedure, without using optimality criteria, show exactly a truss like behaviour of the optimal solution, when both orientation and fiber density are adopted as design variables (see Fig. 14, borrowed from Bendsøe [7]).

As an example of application, consider a cantilever plate subject to a concentrated load P at its end. If we leave the formulation in the 2-D continuum and consider a cantilever truss, then the problem becomes a topological problem, with locations of the joint as unknowns, and the solution will result as a consequence of the choices adopted.

For instance, if we approximate the plate with a five-bars truss, then the optimal solutions (for minimum elastic compliance and for minimum weight at the limit state) are shown in Fig. 15, where it must be observed that both the solutions are characterized by the same topology.

In Fig. 15, and in the following examples, the lower bound for the structural volume in the elastic case has been assumed to be $V_{min} = kAL$.

The influence of the topology can be shown better by increasing the number of bars in the truss; in Fig. 16 the only two possible solutions characterized by constant strain (or strain rates) are illustrated for a ten-bars truss.

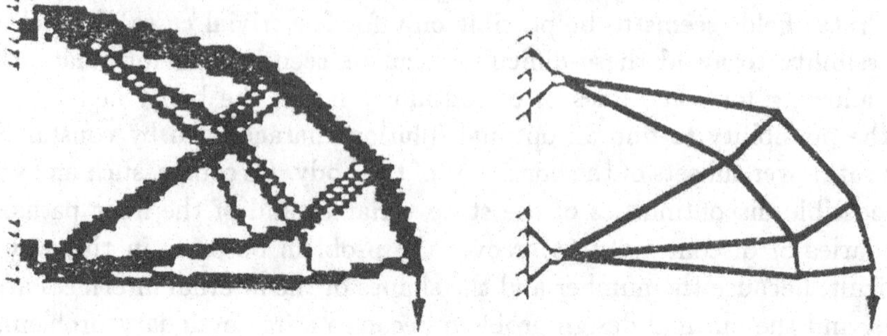

Figure 14

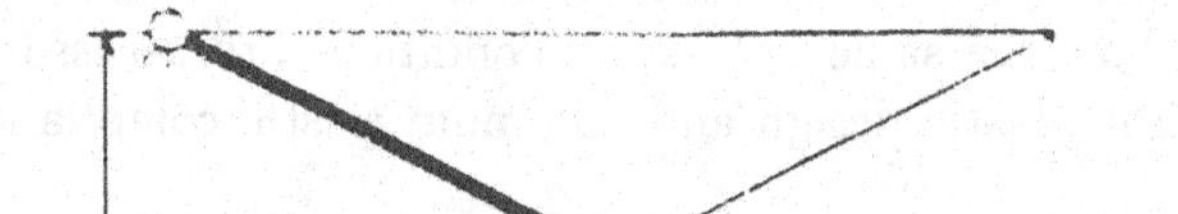

$$V = kAL$$

$$v_P = \frac{31PL}{kEA}$$

$$V_{pi} = 9AL$$

Figure 15

In this case it is well clear how the layout of the structure influence the solution. These two examples, on the other hand, constrain the solution to locate the joints in prescribed positions of the design domain. If some of such constraints are relaxed then, as shown in Fig. 17, the optimal solution can be improved.

In this way, a first approximation of the optimal structural layout has been found, and further improvements can be obtained, for instance, by increasing the number of bars. In other words, when an approximately optimal topology has been found, then starting from this tentative design it is possible to find better solutions

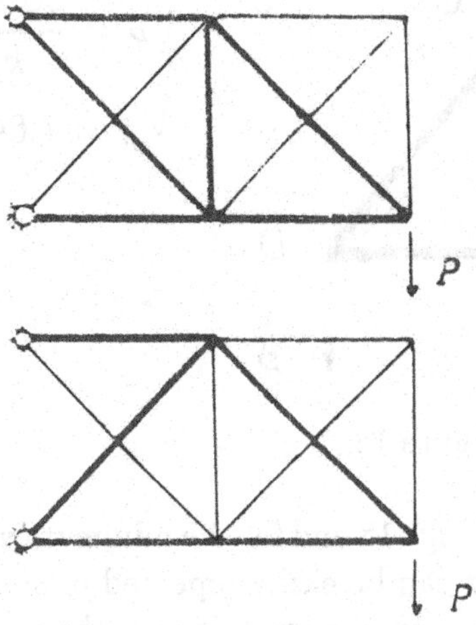

$V = kAL$

$v_P = \dfrac{81\,PL}{kEA}$

$V_{pl} = 9\,AL$

$V = kAL$

$v_P = \dfrac{64\,PL}{kEA}$

$V_{pl} = 8\,AL$

Figure 16

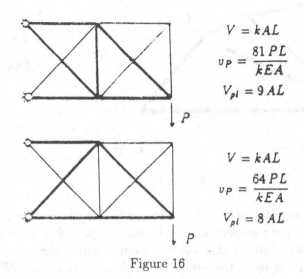

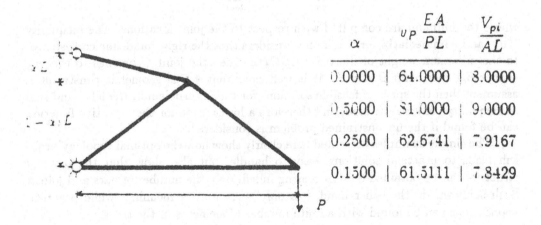

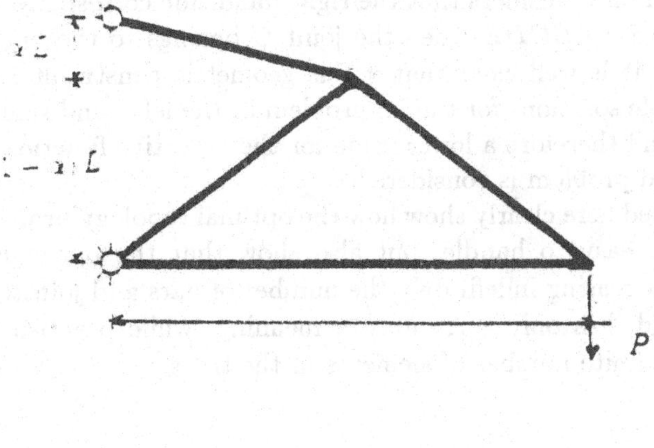

α	$v_P \dfrac{EA}{PL}$	$\dfrac{V_{pl}}{AL}$
0.0000	64.0000	8.0000
0.5000	81.0000	9.0000
0.2500	62.6741	7.9167
0.1500	61.5111	7.8429

Figure 17

transforming the initial structure in a more complex one by suitable increase of bars and joints.

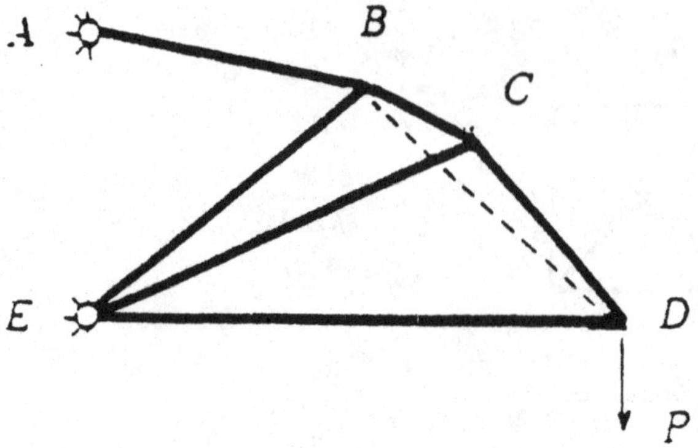

$$v_P = \frac{58.43\, PL}{kEA}$$

$$V_{pl} = 7.644\, AL$$

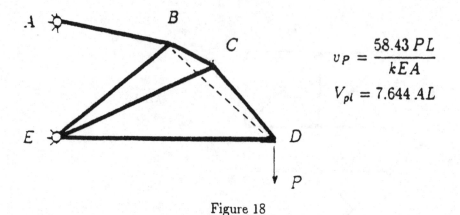

Figure 18

In fact, referring to the structure of Fig. 18 and for the minimum volume problem at the limit state (but the consideration can be easily repeated in analogous way for the elastic case), it can be shown that the minimum weight of the truss *ABCDE* (5 joints, 6 bars) is less or equal to the minimum weigh of the truss *ABDE* (4 joints, 4 bars), i.e.,

$$\inf \sum_{k=1}^{6} A_k L_k \leq \inf \sum_{k=1}^{4} A_k L_k \tag{58}$$

where the *infimum* are computed with respect to the joint locations. The inequality (58) can be immediatly verified if one considers that the right hand side corresponds to the minimum weight of the truss *ABCDE* when the joint *C* belongs to the line connecting the joints *B* and *E*. It is well clear that if this geometric constraint is assumed, then the space of feasible solutions for the *inf* problem at the left hand side of inequality (58) is restricted, and therefore a lower value for the objective function can be found if the unconstrained problem is considered.

The simple examples presented here clearly show how the optimal topology problem leads to practical solutions, easy to handle, but also show that the *optimal structure* can be obtained only increasing indefinitely the number of bars and joints. Such solution, on the other hand, has only a theoretical meaning, while practical applications can be found with a finite number of elements in the truss.

7. CONCLUDING REMARKS

The result presented seems to suggest the possibility to deal with optimization problems making use of the isotropic solution. In other words, to determine, as first step, the principal directions of stress (or strain) for the structural member seen as isotropic, then to orient the orthotropy axes along these directions; if displacement boundary conditions are absent, then the orthotropic body is able to assume a compatible configuration which shows collinearity between stress, strain and principal directions of orthotropy. If Q^0_{11} is not equal to Q^0_{22}, this practical tool furnishes a local minimum corresponding to solution (26) or (27). For instance, if we consider thin tubes made of orthotropic material subject to torsion, it is well known that the solution is given by Bredt's formula and the principal directions of stress are at $\pi/4$ with respect to the axis of the cylinder. In this case, a solution with orthotropy axes aligned along these directions can be regarded as an optimal solution.

In these lecture notes, some results on optimal design of fiber reinforced structural elements under plane states of stress were presented. The problem was dealt with both in the elastic and plastic cases, and the common features of optimal solutions are pointed out. In particular, a relevamt peculiarity of the optimal solution is found in the local collinearity between principal directions of stress, strain (or strain rate) and orthotropy (or strength). Moreover, it was pointed out that, at the optimum, the solution is featured by constant strains (or strain rates) in fiber directions.

Then, the concomitance of results between the elastic and plastic cases and the mechanical meaning of the optimality conditions, suggested the possibility of dealing with approximate solutions of the problem in a unified way. In particular, the optimality conditions were shown to be in agreement with analogous classical conditions, obtained for isotropic bodies at the limit state.

Such classical results, which date from Michell through Prager (see [12]) suggest to deal with the problem by considering the body as a truss–like continuum. In such a way, the distributed parameter problem is replaced by an optimal layout problem, where the structural topology plays the role of design variable.

Some optimal solutions of cantilever trusses are presented as an example of application. It is worth noting that such solutions, obtained through the application of the optimality criterion, find confirmation in numerical results found by means of direct optimization techniques.

Further developments in this direction, which can take into account the influence of the matrix and of angle ply composites on the optimal solutions are desirable, in order to generalize the results here obtained, and to have more realistic designs.

In conclusion, in designing fiber reinforced structural elements by suitable tayloring of fiber orientation and fiber density, the approach presented allows skipping most of the difficulties concerning optimal design of plane systems and is generally adopted as an effective way for the design of composites.

ACKNOWLEDGEMENTS

The present work has been made possible by a financial support from M.U.R.S.T. (Italian Ministry for Research), which is here gratefully acknowledged.

REFERENCES

1. Banichuck, N.V.: Optimization Problems for Elastic Anisotropic Bodies, Arch. Mech., 33 (1981), 347-363.
2. Pedersen, P.: On Optimal Orientation of Orthotropic Materials, Structural Optimization, 1 (1989), 101–106.
3. Pedersen, P.: Bound on Elastic Energy in Solids of Orthotropic Materials, DCAMM Report N. 391 (1989), The Technical Univ. of Denmark, Lyngby.
4. Pedersen, P.: Note on a Transformation combining Material and Element Rotation, DCAMM Report N. 394 (1989), The Technical University of Denmark, Lyngby.
5. Pedersen, P.: On Material Orientation for Maximum Stiffness or Flexibility, Proc. of the CEEC Course on *Progetto e Analisi di Strutture in Materiale Composito*, Milano, 28 May – 1 June 1990.
6. Bendsøe, M.Ph.: Optimal Shape Design as a Material Distribution Problem, Proc. of the COMETT Course on *Computer Aided Optimal Design of Structures*, Pavia, 4–8 September 1989.
7. Bendsøe, M.Ph.: Composites as a Basis for Topology Optimization, Proc. of The CEEC Course on *Progetto e Analisi di Strutture in Materiale Composito*, Milano, 28 May – 1 June 1990.
8. Olhoff, N. and Thomsen, J.: Optimization of Fiber Orientation and Concentration in Composites, Proc. of the CEEC Course on *Progetto e Analisi di Strutture in Materiale Composito*, Milano, 28 May – 1 June 1990.
9. Sacchi–Landriani, G. and Rovati, M.: Optimal Design for 2–D Structures Made of Composite Materials, J. Engng. Mater. and Technology (to appear).
10. Sacchi–Landriani, G. and Rovati, M.: Optimal Limit Design of Fiber–Reinforced Orthotropic Plane Systems, Proc. of the COMETT Course on *Computer Aided Optimal Design of Structures*, Pavia, 4–8 September 1989.
11. Rozvany, G.I.N.: Structural Design via Optimality Criteria, Kluwer Acad. Pub., Dordrecht (NL), 1989.
12. Prager, W.: Unespected Results in Structural Optimization, J. of Structural Mechanics, 9 (1981), 71–90 .
13. Jones, R.M.: Mechanics of Composite Materials, McGraw Hill, New York, 1975.
14. Broutman, L.J. and Krock, R.H.: Composite Materials – Structural Design and Analysis, Academic Press, New York, 1975.
15. Lekhnitskii, S.G.: Theory of Elasticity of an Anisotropic Body, MIR Publ., Moscow, 1981.
16. Lekhnitskii, S.G.: Anisotropic Plates, Gordon & Breach, New York, 1968.
17. Save, M. and Prager, W.: Structural Optimization, Plenum Press, New York, 1985.
18. Haftka, R.T. and Gürdal, Z.: Elements of Structural Optimization, Kluwer Acad. Pub., Dordrecht (NL), 1992.

OPTIMIZATION OF FIBER ORIENTATION AND CONCENTRATION IN COMPOSITES

J. Thomsen and N. Olhoff
The University of Aalborg, Aalborg, Denmark

Abstract - Linearly elastic fiber reinforced composite discs and laminates in plane stress with variable local orientation and concentration of one or two fiber fields embedded in the matrix material, are considered. The thickness and the domain of the discs or laminates are assumed to be given, together with prescribed boundary conditions and in-plane loading along the edge.

The problem under study consists in determining throughout the structural domain the optimum orientations and concentrations of the fiber fields in such a way as to maximize the integral stiffness of the composite disc or laminate under the given loading. Minimization of the integral stiffness can also be carried out. The optimization is performed subject to a prescribed bound on the total cost or weight of the composite that for given unit cost factors or specific weights determines the amounts of fiber and matrix materials in the structure. Examples are presented by the end of the paper.

1. INTRODUCTION

This paper gives a brief account of recent research reported by the first author in [1] on optimization of fiber orientation and concentration in composite discs and laminates. The research is inspired by the initial work in the field by Rasmussen [2] (reported in Danish, account in English available in Niordson and Olhoff [3]) and by important recent developments of Pedersen [4-7]. Problems concerning optimization of fiber orientation have earlier been considered by Banichuk [8], and we refer to Sacchi Landriani and Rovati [9] for other current research activities in the area. Since the present research also comprises optimization of fiber concentration, and allows for more than one field of fibers, our development can be easily augmented with appropriate constitutive material models applicable for topology optimization, cf. Bendsøe [10-11], and results for problems of this type have already been obtained.

The motivation for the work described in this paper is that fiber reinforced composite materials are ideal for structural applications, where high stiffness and strength are required at low weight. Aircraft and spacecraft are typical weight sensitive structures, in which composite materials are cost effective. To obtain the full advantage of the fiber reinforcement, fibers must be distributed and oriented optimally with respect to the actual strain field. Hence, transfer of fiber material from initially lowly stressed parts of the body in order to strengthen the parts and directions that are subjected to large internal forces is the general idea of optimization of composite structures.

Thus, relative to refs. [4-9], we in this paper both use fiber orientations and -concentrations as design variables. Based on the strain field determined by finite element analysis we construct an iterative two-level optimization procedure that consists of an optimality criterion approach as described by Pedersen [4,5,7], and a mathematical programming technique. Here,

- in the first level, the local fiber orientations corresponding to a global optimum are determined using an optimality criterion for these design variables, and

- in the second level, the local distribution of the amounts of fiber and matrix materials available within a bound on total cost or weight, are determined on the basis of analytically derived design sensitivities. In this level, the optimization is carried out by means of a dual mathematical programming technique as implemented in the optimizer CONLIN by Fleury and Braibant [12].

2. OBJECTIVE FUNCTION

The integral stiffness of the composite structure will be selected as the objective function for optimization, and we will be primarily interested in maximization. The structure of maximum integral stiffness will be defined as the structure that has minimum total elastic strain energy subject to a given loading.

We shall assume that our composite disc or laminate can be locally considered as a macroscopically homogeneous, orthotropic material. The strain energy density u will then be given by the following formula for an orthotropic laminate, see e.g. Jones [13],

$$u = \frac{1}{2}\{\epsilon\}^T[A]\{\epsilon\} = \frac{1}{2}A_{11}\epsilon_{11}^2 + \frac{1}{2}A_{22}\epsilon_{22}^2 + A_{12}\epsilon_{11}\epsilon_{22} + 2A_{66}\epsilon_{12}^2 \tag{1}$$

where $\{\epsilon\} = \{\epsilon_{11};\epsilon_{22};2\epsilon_{12}\}$ is the strain vector, and [A] the stiffness matrix.

We now use well-known formulas to express the strain component in (1) by the principal strains, ϵ_I and ϵ_{II}, and the angle ψ from the direction corresponding to the numerically largest principal strain ϵ_I ($|\epsilon_I|\geq|\epsilon_{II}|$) to the direction associated with the largest stiffness A_{11} ($A_{11}\geq A_{22}$), see Fig. 1.

Since in the finite element analysis the structure is discretized into n elements with individual constant laminate stiffness matrices $[A]_i$, the total elastic strain energy U for the structure is then given by

$$U = \sum_{i=1}^{n}\Big(\Big(\frac{1}{8}A_{11}\Big((\epsilon_I+\epsilon_{II})+(\epsilon_I-\epsilon_{II})\cos 2\psi\Big)^2 + \frac{1}{8}A_{22}\Big((\epsilon_I+\epsilon_{II})-(\epsilon_I-\epsilon_{II})\cos 2\psi\Big)^2$$
$$+ \frac{1}{4}A_{12}\Big((\epsilon_I+\epsilon_{II})^2-(\epsilon_I-\epsilon_{II})^2\cos^2 2\psi\Big) + \frac{1}{2}A_{66}(\epsilon_I-\epsilon_{II})^2\sin^2 2\psi\Big)S\Big)_i \tag{2}$$

where S_i is the area of the i-th finite element.

3. DESIGN MODEL AND COST FUNCTION

The fiber orientation and concentration within each element of the discretized structure are adopted as design variables.

Our design model is made up of elements that consist of 3 fiber plies with the fiber orientations θ, $\theta+90^\circ$ and θ, and the volumetric fiber concentrations V_{fI}, V_{fII} and V_{fI}, see Fig. 2. Introducing the variable ratio β between the thickness of fiber ply 2 (in the middle) and the total thickness h of the element, we get the symmetric and orthotropic laminate shown in Fig. 2, which can have both unidirectional ($\beta=0 \vee \beta=1$) and cross ply ($0<\beta<1$) character.

We have now defined 4 design variables for each element: V_{fI}, V_{fII}, θ and β. For these design variables, we prescribe lower and upper constraint values as follows:

$$0 \le (V_{fI})_i \le \bar{V}_f \,, \qquad 0 \le (V_{fII})_i \le \bar{V}_f \,, \qquad 0 \le \theta_i \le 180^\circ \,, \qquad 0 \le \beta_i \le 1 \,, \qquad i=1,\dots,n \tag{3}$$

Here the given upper constraint value $\bar{V}_f$ for the fiber concentrations depends on how densely the fibers can be packed in the matrix material in view of their cross-sectional shape.

We finally formulate a constraint that enforces the total cost or weight C of the structure to be less than or equal to a given upper bound $\bar{R}$ if stiffness maximization is considered,

$$\begin{aligned} C = \sum_{i=1}^{n} \Big(& c_f((V_{fI})_i h\, \beta_i + (V_{fII})_i h(1-\beta_i)) \\ & + c_m((1-(V_{fI})_i) h\, \beta_i + (1-(V_{fII})_i) h(1-\beta_i))\Big) S_i \le \bar{R} \end{aligned} \tag{4}$$

Here c_f and c_m are given so-called "unit cost factors". They denote the cost per unit volume of the fiber and matrix materials, respectively, for a cost constrained problem, whereas c_f and c_m denote the specific weights of the fiber and matrix materials, respectively, if the total weight is constrained.

4. STIFFNESS MATRIX IN TERMS OF DESIGN VARIABLES

The fiber and matrix materials will be assumed to be linearly elastic with given Young's moduli E_f and E_m and Poisson's ratios ν_f and ν_m. We now adopt the "rule of mixtures", see e.g. Jones [13], for determining the components of the tensor of elasticity for a lamina in our design model

$$\begin{aligned} & E_{Lj} = (1-V_{fj})E_m + V_{fj}E_f \,, \qquad E_{Tj} = \frac{E_f E_m}{(1-V_{fj})E_f + V_{fj}E_m} \,, \qquad j=I,II \\ & \nu_{LTj} = (1-V_{fj})\nu_m + V_{fj}\nu_f \,, \quad G_{LTj} = \frac{E_m E_f}{2(V_{fj}(E_m\nu_f - E_f\nu_m + E_m - E_f) + E_f\nu_m + E_f)} \,, \end{aligned} \tag{5}$$

Here indexes L and T refer to the longitudinal and transverse directions of the fibers, respectively, and the index j will here and in the following take on the "values" I and II that refer to the fiber layers 1 and 2, respectively.

For a composite element as shown in Fig. 2 that consists of 3 lamina with the thicknesses $0.5h\beta$, $h(1-\beta)$ and $0.5h\beta$ and the fiber orientations θ, $\theta+90^\circ$ and θ, we can easily obtain the laminate stiffness matrix [A] by means of a formula given in Tsai & Pagano [14]. We get

$$
\begin{aligned}
[A] = \frac{E_{LI} h\beta}{8\alpha_{0I}} \Bigg(& \begin{bmatrix} 8 & 0 & 0 \\ & 8 & 0 \\ s & & 4 \end{bmatrix} + \alpha_{2I} \begin{bmatrix} \cos 2\theta - 1; & 0 & ; \sin 2\theta/2 \\ & -\cos 2\theta - 1; & \sin 2\theta/2 \\ s & & -1/2 \end{bmatrix} \\
& + \alpha_{3I} \begin{bmatrix} \cos 4\theta - 1; & -\cos 4\theta & ; \quad \sin 4\theta \\ & \cos 4\theta - 1 & ; \ -\sin 4\theta \\ s & & -\cos 4\theta - 1/2 \end{bmatrix} + \alpha_{4I} \begin{bmatrix} 0 & 1 & 0 \\ & 0 & 0 \\ s & & -1/2 \end{bmatrix} \Bigg) \\
+ \frac{E_{LII} h(1-\beta)}{8\alpha_{0II}} \Bigg(& \begin{bmatrix} 8 & 0 & 0 \\ & 8 & 0 \\ s & & 4 \end{bmatrix} + \alpha_{2II} \begin{bmatrix} \cos 2\phi - 1; & 0 & ; \sin 2\phi/2 \\ & -\cos 2\phi - 1; & \sin 2\phi/2 \\ s & & -1/2 \end{bmatrix} \\
& + \alpha_{3II} \begin{bmatrix} \cos 4\phi - 1; & -\cos 4\phi & ; \quad \sin 4\phi \\ & \cos 4\phi - 1 & ; \ -\sin 4\phi \\ s & & -\cos 4\phi - 1/2 \end{bmatrix} + \alpha_{4II} \begin{bmatrix} 0 & 1 & 0 \\ & 0 & 0 \\ s & & -1/2 \end{bmatrix} \Bigg)
\end{aligned}
\qquad (6)
$$

where θ denotes the angle defining the fiber orientation, see Fig. 1, and the angle $\phi=\theta+90^\circ$ defines the orthogonal direction.

The parameters $\alpha_{0j}, \alpha_{2j}, \ldots, \alpha_{4j}$ in (6) can all be expressed explicitly in terms of the fiber concentrations V_{fj}, $j = I, II$, and the given elastic constants of the fiber and matrix materials, i.e.,

$$
\alpha_{mj} = \alpha_{mj}(V_{fI}, V_{fII}, E_f, E_m, \nu_f, \nu_m) \qquad m = 0,2,3,4 \qquad j = I, II \qquad (7)
$$

For reasons of brevity, the reader is referred to [1] for the specific expressions.

5. OPTIMIZATION TECHNIQUE

The optimization problem is solved iteratively via a two-level procedure of redesign. The stress/strain field is initially determined by finite element analysis using MODULEF [15] in each loop of redesign, and improved orientations of the fibers are subsequently determined by means of an optimality criterion in the first level of redesign. In the second level of redesign the distributions of fibers are improved via a method of sensitivity analysis and mathematical programming.

A notable feature of the present problem is that a usual gradient method may fail in determining the optimal orientation of the fibers, because local optima normally exist, see e.g. Fig. 4.4 in [7]. To circumvent this inherent difficulty in the first level of redesign we follow Pedersen [4,7] and perform an analytical investigation of the first and second derivative in order to determine the global optimum of the total strain energy with respect to fiber orientation. From (2) and (6) we get the following expression for first order sensitivities, cf. Pedersen [4,7],

$$\frac{dU}{d\theta_i} = \frac{dU}{d\psi_i} = (4A\alpha_3(\epsilon_I-\epsilon_{II})^2 \sin 2\psi(\gamma+\cos 2\psi)S)_i \quad , \quad i = 1,\dots,n \tag{8}$$

where A is a constant, and the parameter γ_i is defined by

$$\gamma_i = \left(\frac{\alpha_2}{4\alpha_3} \frac{1+\epsilon_{II}/\epsilon_I}{1-\epsilon_{II}/\epsilon_I} \right)_i \quad , \quad i = 1,\dots,n \tag{9}$$

The material parameters α_2 and α_3 are those appearing in (6) and they are dubbed as C_2 and C_3 in [7]. The results of a complete investigation of the extrema of U with respect to the key parameters ψ, α_3 and γ are summarized in a table in refs. [1],[4] and [7] (Tabel 3.1 in [7]).

As described in [1],[4] and [7], the fiber orientation θ_i for each element can be determined by means of this table and the formula

$$\theta_i = \psi_i + \eta_i \quad , \quad i = 1,\dots,n \tag{10}$$

where η_i is the angle of rotation of the principal strain or stress direction of the i-th element relative to the X_1 axis of the finite element coordinate system, see Fig. 1.

In the above solution procedure for the first level of redesign it was found that the fibers should be oriented along the principal stress directions if stiffness maximization is performed, and along the principal strain directions in problems of stiffness minimization.

The second stage in the loop of redesign consists in determining an improved distribution of the amount of fiber material, i.e., to obtain improved values of the design variables β_i, $(V_{fI})_i$ and $(V_{fII})_i$, (i=1,...,n). This is done by a dual method of mathematical programming using mixed variables as developed by Fleury and Braibant [12] and implemented in the computer code CONLIN. To this end we need the sensitivities of the objective function and constraints with respect to the aforementioned design variables.

Now, it is shown by Pedersen in [4,7] that by means of Clayperon's theorem and the principle of virtual displacements for structures with design independent loads, the gradient of the total strain energy can be determined from the gradient of the strain energy density u_i for a given element, whose strain field is considered to be fixed,

$$\frac{dU}{da_i} = -\frac{\partial u_i}{\partial a_i} S_i \ , \qquad i=1,\ldots,n \tag{11}$$

Here a_i denotes any of the design variables β_i, $(V_{fI})_i$, or $(V_{fII})_i$, i=1,...,n.

The sensitivities of the total strain energy U with respect to β_i, $(V_{fI})_i$ and $(V_{fII})_i$ can thus be determined by (2) and (11), assuming the strain field to be fixed, and restricting variation to the laminate stiffness matrix [A]. For the i-th element of the discretized geometry we then obtain the following expression for sensitivities w.r.t. the design variables a_i

$$\begin{aligned} U_{,a_i} = -\Big(\Big(\tfrac{1}{8}A'_{11}\Big[(\epsilon_I+\epsilon_{II})+(\epsilon_I-\epsilon_{II})\cos 2\psi\Big]^2 + \tfrac{1}{8}A'_{22}\Big[(\epsilon_I+\epsilon_{II})-(\epsilon_I-\epsilon_{II})\cos 2\psi\Big]^2 \\ + \tfrac{1}{4}A'_{12}\Big[(\epsilon_I+\epsilon_{II})^2-(\epsilon_I-\epsilon_{II})^2\cos^2 2\psi\Big] + \tfrac{1}{2}A'_{66}(\epsilon_I-\epsilon_{II})^2\sin^2 2\psi\Big)S\Big)_i \end{aligned} \tag{12}$$

where a_i denotes any of the design variables β_i, $(V_{fI})_i$ or $(V_{fII})_i$, and A'_{kl} is a shorthand notation for the derivatives $\partial A_{kl}/\partial a_i$ of a component A_{kl} of the stiffness matrix [A].

The derivatives of [A] with respect to β_i are easily obtained from (6), with θ taken to be equal to zero to give the material orthotropic characteristics.

Since [A] depends on α, E_L and V_f, i.e.,

$$[A_{kl}] = [A_{kl}](\alpha_{0I}(V_{fI}),\alpha_{2I}(V_{fI}),\alpha_{3I}(V_{fI}),\alpha_{4I}(V_{fI}),$$
$$E_{LI}(V_{fI}),\alpha_{0II}(V_{fII}),\alpha_{2II}(V_{fII}),\alpha_{3II}(V_{fII}), \tag{13}$$
$$\alpha_{4II}(V_{fII}),E_{LII}(V_{fII}))$$

the sensitivities of [A] with respect to V_{fj} are found by means of the chain rule, and we get

$$\frac{\partial[A_{kl}]}{\partial V_{fI}} = \sum_{m=0,2,3,4}\left(\frac{\partial[A_{kl}]}{\partial\alpha_{mI}}\frac{\partial\alpha_{mI}}{\partial V_{fI}}\right) + \frac{\partial[A_{kl}]}{\partial E_{LI}}\frac{\partial E_{LI}}{\partial V_{fI}}$$
$$\frac{\partial[A_{kl}]}{\partial V_{fII}} = \sum_{m=0,2,3,4}\left(\frac{\partial[A_{kl}]}{\partial\alpha_{mII}}\frac{\partial\alpha_{mII}}{\partial V_{fII}}\right) + \frac{\partial[A_{kl}]}{\partial E_{LII}}\frac{\partial E_{LII}}{\partial V_{fII}} \tag{14}$$

These sensitivities are derived analytically in [1], and the results are available therein. The sensitivities of the cost function (4) are readily derived analytically, and we thus have all the necessary sensitivity information that is required for the optimization in the second level of redesign.

6. EXAMPLES

We first consider three example problems of optimization of the rectangular composite disc shown in Fig. 3. The disc has one of its sides fixed against displacements in the X and Y directions, while the opposite side is subjected to a parabolically distributed shear loading. The upper constraint value $\overline{V}_f$ for fiber concentration in (3) is taken to be $\overline{V}_f$=80 pct, and we only consider cases of c_m=0 and c_f=1, which means that the fibers are dominating in the cost or weight function C in (4).

In the first example we consider maximization of the stiffness of the disc under the condition that only one fiber field is allowed in each element. This corresponds to the special case of β=0 ∨ β=1, see Chapter 3. The structure is discretized into 20×40 4-node elements (type QUAD 2Q1D, see [15]). The result of the optimization is shown in

Fig. 4, where the direction and density of the hatching within each elements illustrate the fiber orientation and concentration, respectively. We see that the lowly stressed elements do not contain any fibers. It is also noteworthy that the design contains "holes" in the fiber reinforcement in the mid part of the structure, where shear forces are dominating.

No doubt this is due to the fact that only one fiber field is allowed to exist in each element. This is not favourable in shear dominated areas with almost equal principal stresses, and the pattern obtained in the mid part may be conceived as the best possible attempt of the structure to increase its "shear force stiffness" under the given design conditions. The design shown in Fig. 4 is associated with a reduction of the total elastic energy U by 51% relative to the initial design, where all the fibers were uniformly distributed and given the orientation $\theta_i = 0^\circ$.

However, the convergence is very slow, and different designs may be obtained as a result of the optimization. In particular, the designs depend on the size of the applied FE-mesh, and it is not possible to obtain a limiting, numerically stable design by consecutively decreasing the mesh size. These features, along with the generation of "holes" in the design, indicate the necessity of a regularization of the formulation of the optimization problem (see, e.g., the survey by Olhoff and Taylor [16]).

This leads to our <u>second example</u>: Regularization of the formulation of the type of problem just considered is simply obtained by extending the design space such as to allow for formation of two orthogonal fiber fields everywhere in the disc (which is actually covered in the preceding chapters). Introducing two fiber fields, the design in Fig. 4 is replaced by the solution shown in Fig. 5, where the "shear force reinforcement" appears along the horizontal center line in agreement with the boundary and symmetry conditions. Optimizing the structure, U is reduced by 55%. Now the convergence is rapid and the design is found to be independent of the discretization, which confirms that regularization has been achieved.

In our <u>third example</u>, we consider the somewhat abnormal, but theoretically interesting problem of *minimization* of the integral stiffness, assuming two fiber fields and the total amount of fiber material to be larger than a given *lower* bound. The result of this problem is shown in Fig. 6, where we see that the design only uses one of the fiber fields, and clearly distributes and orientates this field in such a way as to avoid its properties as stiffness reinforcement. In this example U is increased by 270%, compared to an initial design where the fibers are uniformly distributed in two fiber fields with the orientations $\theta=0^\circ$ and $\theta=90^\circ$.

In a final example, we wish to demonstrate that the method and software developed here for optimization of composites has been extended to cover topology optimization as well. In the latter type of problem the structure is considered as a domain of space with a high concentration of material, see Bendsøe [10-11]. The present extension is made by introducing relationships between stiffness components and concentration of layered, second rank microstructures that replace the two fiber fields, and assigning the matrix material vanishing stiffness. The reader is referred to Bendsøe [10-11]. Fig. 7 shows the result of optimizing the topology of a structure for which the left hand side of the rectangular domain in Fig. 7 offers full fixation, and where the structure is required to carry a vertical force at the lower right hand corner. The result is clearly seen to become a truss-like structure.

Acknowledgement - The work received support from the Danish Technical Research Council (Programme of Research on Computer Aided Design).

REFERENCES

[1] Thomsen, J.: *Optimization of Composite Discs*. Rept. no. 21, Institute of Mechanical Engineering, The University of Aalborg, Aalborg, Denmark, pp. 27, June 1990. (Submitted to *Structural Optimization*).

[2] Rasmussen, S.H.: *Optimization of Fiber Reinforced Structures* (in Danish), DCAMM report no. S12, The Technical University of Denmark, Lyngby, Denmark, pp. 127, 1979.

[3] Niordson, F.; Olhoff, N.: Variational Methods in Optimization of Structures. *Trends in Solid Mechanics 1979*, Proc. W.T. Koiter Anniversary Symp. (eds.: J.F. Besseling and A.M.A. van der Heijden),

Sijthoff & Noordhoff, 177-194, 1979.

[4] Pedersen, P.: Bounds on Elastic Energy in Solids of Orthotropic Materials. *Structural Optimization*, 2, 55-63, 1990.

[5] Pedersen, P.: On Optimal Orientation of Orthotropic Materials. *Structural Optimization*. 1, 101-106, 1989.

[6] Pedersen, P.: Combining Material and Element Rotation in One Formula. *Comm. Appl. Num. Meth.* (to appear 1990).

[7] Pedersen, P.: Energy Bounds by Material Rotation. Same issue of this Journal

[8] Banichuk, N.V.: *Problems and Methods of Optimal Structural Design*. New York: Plenum Press, 181-207, 1983.

[9] Sacchi Landriani, G.; Rovati, M.: Optimal Design of 2-D Structures Made of Composite Materials. (Submitted to *J. Engrg. Materials and Technology*).

[10] Bendsøe, M.P.; Kikuchi, N.: Generating Optimal Topologies in Structural Design Using a Homogenization Method. *Computer Methods in Applied Mechanics and Engineering*. 71, 197-224, 1988.

[11] Bendsøe, M.P.: Optimal Shape Design as a Material Distribution problem. *Structural Optimization*. 4, 193- 202, 1989.

[12] Fleury, C.; Braibant, V.: Structural Weight Optimization by Dual Methods of Convex Programming. *International Journal for Numerical Methods in Engineering*. 14, 1761-1783, 1986.

[13] Jones, R.M.: *Mechanics of Composite Materials*. New York: McGraw-Hill, 1975.

[14] Tsai, S.W.; Pagano, N.J.: Invariant Properties of Composite Materials. *Composite materials workshop*. Technomic, Westport Com, 1968.

[15] Institut National de Recherche en Informatique et en Automatique (INRIA, France). Avril 1983: *Modulef, Bibliotheque D'Elasticite*. 101, 1983.

[16] Olhoff, N.; Taylor, J.E.: On Structural Optimization, *J. Appl. Mech.*, 1139-1151, 1983.

FIGURES

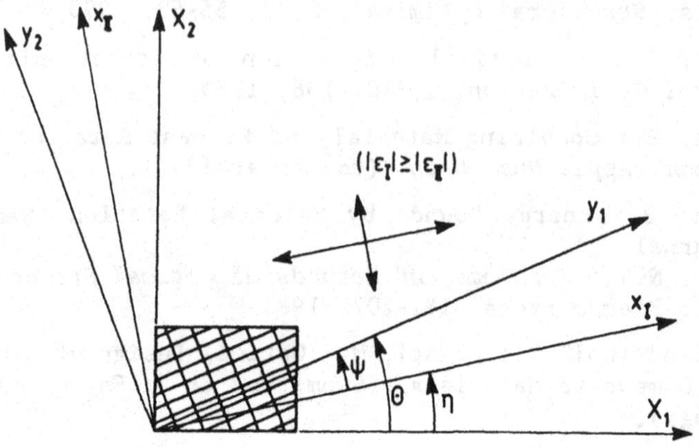

Fig. 1. Definition of the angles ψ, θ and η for mutual rotations of the finite element coordinate system X_1, X_2, the principal strain coordinate system x_I, x_{II} and the material coordinate system y_1, y_2

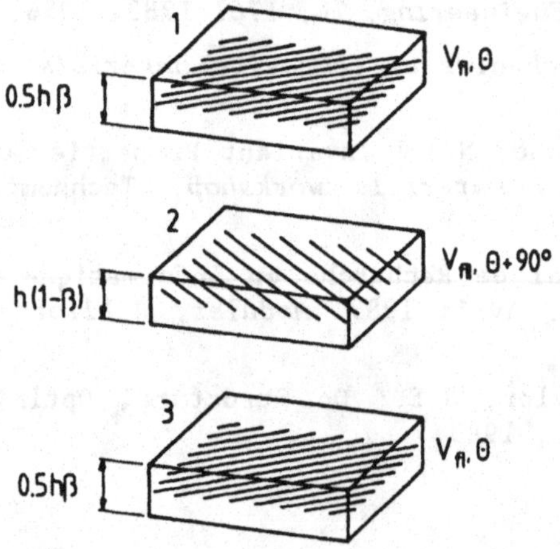

Fig. 2. Design variables of an element consisting of 3 orthogonal plies

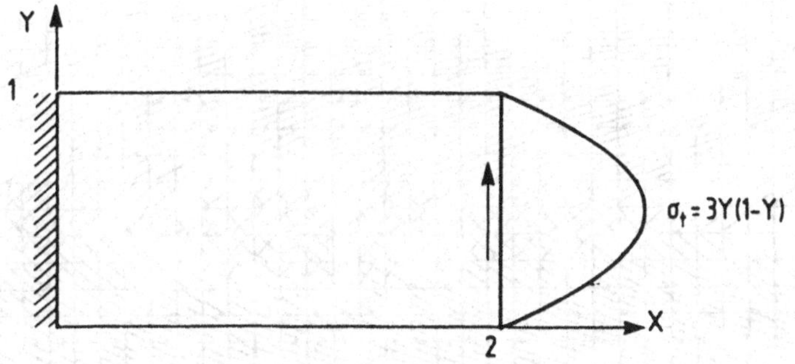

Fig. 3. Example problem for optimization

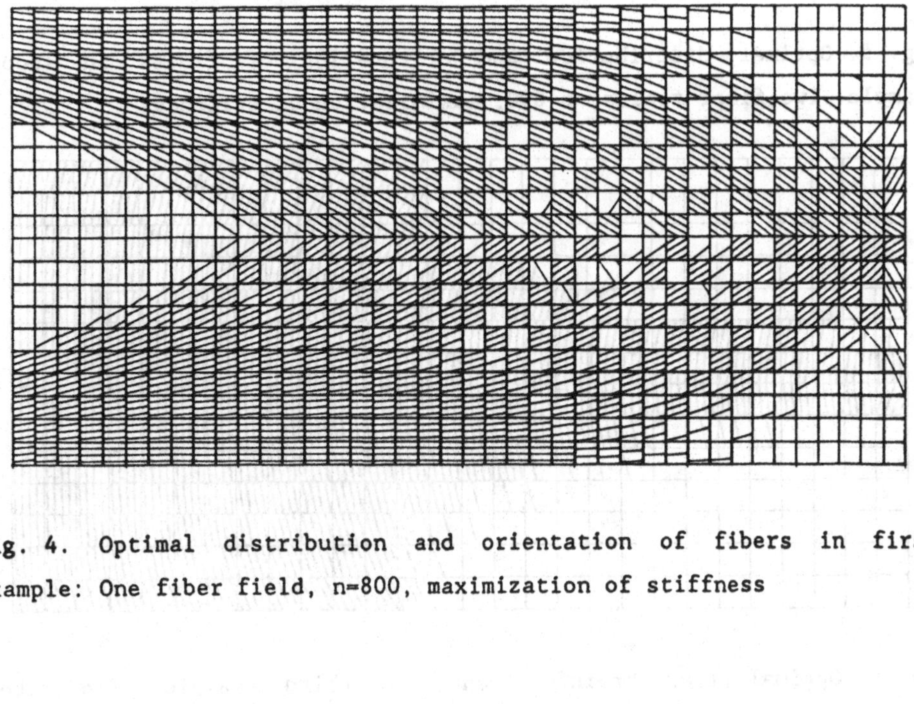

Fig. 4. Optimal distribution and orientation of fibers in first example: One fiber field, n=800, maximization of stiffness

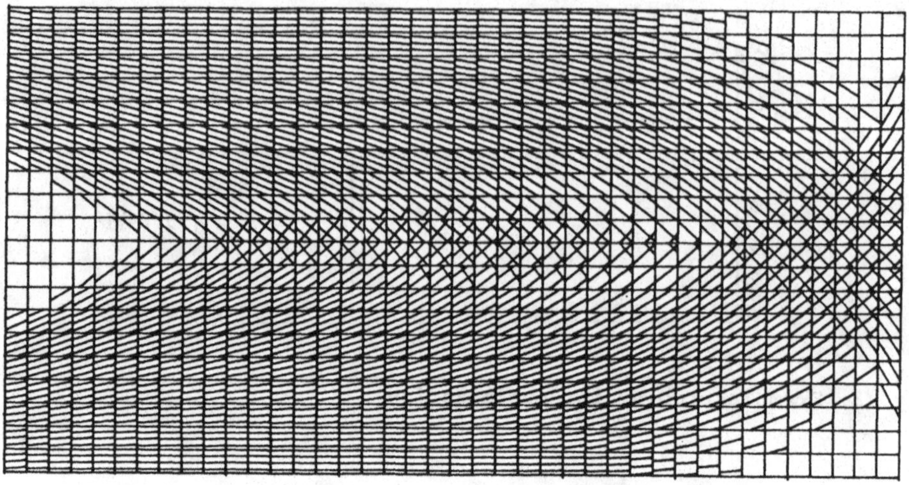

Fig. 5. Optimal distribution and orientation of fibers in second example: Two fiber fields, n=800, maximization of stiffness

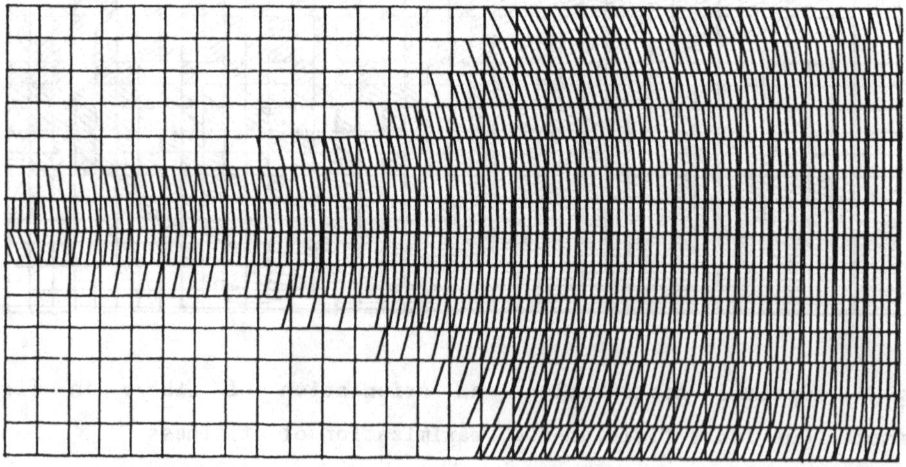

Fig. 6. Optimal fiber "reinforcement" in third example: Two fiber fields, n=392, minimization of stiffness

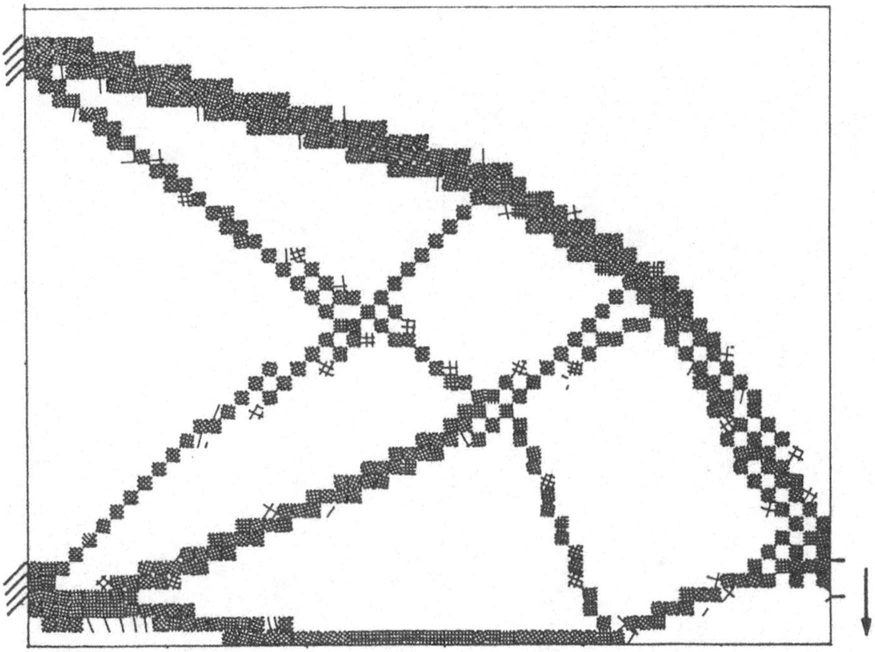

Fig. 7. Example of topology optimization (see text)

TOPOLOGY OPTIMIZATION OF STRUCTURES COMPOSED OF ONE OR TWO MATERIALS

J. Thomsen
The University of Aalborg, Aalborg, Denmark

Abstract – *Maximization of the integral stiffness of a structure composed of one or two isotropic materials of large stiffness is considered using the homogenization technique. Material is modelled by a second rank composite, and we use the concentrations and orientations of the composite as design variables. Numerical results are presented at the end of the paper.*

1 Introduction

During the last decade a number of software systems for shape optimization based on boundary variation have been developed in the field of structural optimization. The efficiency and user-friendliness of many of these systems are so satisfactory that they can be used in industries as CAE (Computer Aided Engineering) tools. The first experiments dealing with boundary variation optimization were carried out in the 'seventies. Since then, authors such as Esping (1984, 1986); Braibant and Fleury (1984); Haftka and Gandhi (1986); Rasmussen (1990) and Rodriques (1988) have published results on the subject.

The boundary variation method is however attended with some limitations. The result of the optimization process is, e.g. very dependent on the chosen initial design, as the optimized structure is topologically equivalent to this one. The user of the optimization program will have to define an initial design of the structure and define how the boundaries of the structure may change. If this is not done appropriately, the gain of the optimization process may be very limited. During boundary variation the user may besides have to manually redefine the finite element mesh in order to retain a satisfactory mesh.

Using topology optimization we avoid above mentioned problems. By this method a prescribed initial topology is not required, and topology optimization can often be applied as a most suitable preprocessor for problems of boundary variation, in which a sensible initial design is essential. Integration of topology and boundary variation optimization was successfully implemented by Bendsøe et al. (1990a); Bendsøe and Rodriques (1990b) and Olhoff et al. (1991) by manual definition of the bounds of a boundary variation model based on a topology optimized structure. An automatic interface between the two optimization methods has been developed by Papalambros and Chirehdast (1990) and implemented in the system, SAPOP.

This paper deals with topology optimization of plane, linearly elastic structures. Topology optimization is performed as a material distribution problem using a composite material. We apply a method as described in Thomsen (1991), which deals with optimization of plane composite structures consisting of fiber and matrix materials, where the concentration and the orientation of the fibers are used as design variables. Assuming the fiber concentration of such a composite material can take on values from 0 to 100%, and the stiffness of the matrix material is chosen to be very low compared to the stiffness of the fibers, we have in principle defined a topology optimization problem, where there will be "no material" and "material" in domains with 0 and 100% fibers, respectively.

In this paper, the material model is a second rank composite consisting of an isotropic material of large stiffness and a very soft material. The elastic moduli A_{ij} of the composite are obtained by homogenization. On account of the continuous nature of the composite material the optimization can be carried out with two different fiber concentrations as design variables. The concentration of the material of large stiffness determines whether there will be "material" or "no material" at a given point of a loaded structure. In the following the term "stiff" material does not imply "rigid" material. The term "stiff" denotes material of large but finite stiffness.

The purpose of introducing a composite material is not only to get a convenient, continuous material model which can be used to obtain analytical expressions for the elastic moduli. Thus, if the problem had been stated as an integer optimization problem so that either "material" or "no material" could be generated at any point of a design domain, the formulation would *not* have been *correctly posed*, and the existence of a solution (an optimal design) would not be obvious (Strang and Kohn 1986). The key would then be to reformulate the optimization problem by introducing a family of composites constructed from the basis materials of the original problem. This process is sometimes called *relaxation* and has, e.g. been studied by Murat and Tatar (1985); Lurie and Cherkaev (1986); Kohn and Strang (1986); Thomsen and Olhoff (1990) and Thomsen (1991). Relaxation implies enlarging of the design space, and tends to remove local optima (Kohn 1990). Traditionally, it was thought that one must consider the totality of all possible composites assembled from the set of originally given materials. This approach is called *full relaxation*. Recent investigations however shown that only the set of finite rank laminar composites need to be considered for many optimization problems (Avellaneda 1987; Kohn 1988a; Kohn and Lipton 1988b). This technique is called *partial relaxation* and is performed by introducing some convenient, finite-parameter micro structure. This approach also enlarges the design space but has the disadvantage that the solution may be dependent on the chosen micro structure. In addition, a partial relaxation implies the same difficulty as the discrete material/no material problem, that the problem may have no solution. Studies on bounds on the effective material properties of composite materials composed of two materials have shown that for plane elasticity the stiffest material can be obtained by a layered medium, with layering at two different micro scales (Avellaneda 1987). This means that the existence of solutions is ensured for minimum compliance shape optimization problems if such a material model is used.

This paper deals with maximization of the integral stiffness of a structure with given load and boundary conditions and with a given amount of available material. The integral stiffness of the structure for a given load case is represented by the internal elastic energy. Numerical analysis of the structure is performed by means of the finite element program MODULEF, and the concentration of the stiff material and the axes of orthotropy in each finite element are used as design variables.

Assuming the strain field to be determined at a given design stage, we apply an iterative two-level optimization procedure where:

- the distribution of the amount of available material is determined by using analytical sensitivity analysis and a convex mathematical programming technique, and

- the orientations of the axes of orthotropy are determined using a global optimality criterion method.

The above formulation of topology optimization was recently extended to cover application of two materials with different stiffnesses so that subregions with "material 1", "material 2", "no material" and "mixtures of material 1, material 2 and no material" can be generated in any part of a design domain. This model makes it is possible to generate typical sandwich structures.

2 Material model

We now formulate a material model which can describe an arbitrary plane structure consisting of one or two stiff materials. In the literature dealing with topology optimization several numerical and analytical material models have been used. In all models integer optimization is avoided by using a continuous material model, which can have intermediate values of "material" and "no material", meaning that essentially the optimization can be performed as a sizing problem. Bendsøe and Kikuchi (1988); Suzuki and Kikuchi (1989) and Diaz and Bendsøe (1990) are using a numerically determined material model based on a micro structure consisting of an isotropic material with rectangular holes, and they use the orientation and the size of the holes as design variables of the optimization problem.

In this paper we use an analytical material model, where two isotropic materials with different stiffnesses can be described by a second rank composite material. The composite is constructed in three micro levels. At the first level we model a composite using "material 1" and "material 2" by turns. The concentrations of the two materials are given in terms of the thicknesses δ_1 and $1-\delta_1$ as shown in Fig. 1.a. At the second level we construct a composite composed of the material in Fig. 1.a and "very soft material", where the ratio between these is given in terms of δ_2 and $1-\delta_2$, refer to Fig. 1.b. At the third level we construct a composite using "material 1" and the composite in Fig. 1.b. by turns, where the concentrations of these are given in terms of γ and $1-\gamma$, refer to Fig. 1.c. The three basis materials used in the material model are isotropic and defined by the stiffness matrices:

Q^{f}_{kl} : Material 1

Q^{m}_{kl} : Material 2

Q^{-}_{kl} : Very soft material

If the design variables of the material model are chosen suitable, we can define topology optimization problems for either one or two stiff materials. If $\delta_1=1$, we get the composite shown in Fig. 2, which can be used in usual problems of topology optimization, where there may be either "material" or "no material" at a given point of a design domain. This formulation has previously been used by Bendsøe (1989). If all the design variables of the material model in Fig. 1.c are allowed to vary between 0 and 1, we get a composite, which can describe "material 1", "material 2" and "no material" when suitable values of γ, δ_1 and δ_2 are chosen. We presume that the relatively simple material model in Fig. 1.c is general enough to be used in problems of topology optimization. In addition to $\gamma(x)$, $\delta_1(x)$ and $\delta_2(x)$ we apply the material orientation $\theta(x)$ in a point x as a design variable. As mentioned, structures composed of isotropic materials can be generated during the optimization. It is however possible that parts of the structure may become anisotropic, as mixtures of the materials may be formed.

3 Optimization problem

This paper deals with stiffness maximization of linearly elastic structures loaded in plane stress. Structures are analyzed using a fixed finite element model with known boundary conditions and in plane loadings. The structure of maximum integral stiffness will be defined as the structure having minimum total elastic strain energy subject to a given loading. The total internal elastic energy U is given by, see e.g. Jones (1975):

$$U = \sum_{i=1}^{n} \left\{ \left\{ \frac{1}{8} A_{11} \left[(\varepsilon_I + \varepsilon_{II}) + (\varepsilon_I - \varepsilon_{II}) \cos 2\psi \right]^2 + \frac{1}{8} A_{22} \left[(\varepsilon_I + \varepsilon_{II}) - (\varepsilon_I - \varepsilon_{II}) \cos 2\psi \right]^2 \right. \right.$$
$$\left. \left. + \frac{1}{4} A_{12} \left[(\varepsilon_I + \varepsilon_{II})^2 - (\varepsilon_I - \varepsilon_{II})^2 \cos^2 2\psi \right] + \frac{1}{2} A_{66} (\varepsilon_I - \varepsilon_{II})^2 \sin^2 2\psi \right\} S \right\}_i \qquad (1)$$

where ε_I and ε_{II} are the principal strains, $[A]$ is the matrix of extensional stiffness, ψ is the angle from the direction corresponding to the numerically largest principal strain ε_I to the direction associated with the largest stiffness A_{11} and S is the area of an element. The design variables of the optimization problem are the density and the orientation of "material" in each finite element.

By topology optimization using *one* stiff material we apply the concentrations γ and δ_2 (refer to Fig. 2) along with the orientation θ of the composite as design variables. We formulate a constraint that enforces the total amount of stiff material to be less than or equal to a given upper bound $\overline{M}$:

$$0 \le \gamma_i \le 1 \quad ; \quad \delta_{1i} = 1 \quad ; \quad 0 \le \delta_{2i} \le 1 \quad ; \quad i = 1,\dots,n$$
$$C_1 = \sum_{i=1}^{n} \left[\gamma_i + (1-\gamma_i)\delta_{2i}\right] S_i \le \overline{M} \tag{2}$$

By topology optimization using *two* stiff materials we apply γ, δ_1, δ_2 (refer to Fig. 1.c) and θ as design variables, and we enforce the total amounts of "material 1" and "material 2" to be less than or equal to $\overline{M}_1$ and $\overline{M}_2$, respectively:

$$0 \le \gamma_i \le 1 \quad ; \quad 0 \le \delta_{1i} \le 1 \quad ; \quad 0 \le \delta_{2i} \le 1 \quad ; \quad i = 1,\dots,n$$
$$C_2 = \sum_{i=1}^{n} \left[\gamma_i + (1-\gamma_i)\delta_{2i}\delta_{1i}\right] S_i \le \overline{M}_1 \quad ; \quad C_3 = \sum_{i=1}^{n} \left[(1-\gamma_i)(1-\delta_{1i})\delta_{2i}\right] S_i \le \overline{M}_2 \tag{3}$$

4 Stiffness matrix in terms of design variables

The stiffness matrix of the material shown in Fig. 1.c is determined by homogenization in three steps, where stiffnesses are found for each micro level. We follow Bendsøe (1989) in determining the constitutive matrix Q_{kl}^H of a composite:

$$Q_{11}^H = \left[M\left(\frac{1}{Q_{11}}\right)\right]^{-1} \quad , \quad Q_{22}^H = M(Q_{22}) - M\left(\frac{Q_{12}^2}{Q_{11}}\right) + \left[M\left(\frac{Q_{12}}{Q_{11}}\right)\right]^2 \left[M\left(\frac{1}{Q_{11}}\right)\right]^{-1}$$
$$Q_{12}^H = M\left(\frac{Q_{12}}{Q_{11}}\right)\left[M\left(\frac{1}{Q_{11}}\right)\right]^{-1} \quad , \quad Q_{66}^H = \left[M\left(\frac{1}{Q_{66}}\right)\right]^{-1} \tag{4}$$

where $M(f)$ is the average value of a function $f(y)$ in the interval Y:

$$M(f) = \frac{1}{|Y|}\int_Y f(y)dy \tag{5}$$

and Q_{kl} are the so-called reduced stiffnesses, which for an isotropic material with the Young's modulus E, Poisson's ratio ν and plane stress conditions are given by (6), Jones (1975):

$$Q_{11} = Q_{22} = \frac{E}{1-\nu^2} \quad ; \quad Q_{12} = \frac{\nu E}{1-\nu^2} \quad ; \quad Q_{66} = \frac{E}{2(1+\nu)} \tag{6}$$

We now consider a composite material composed of two isotropic materials with the stiffnesses Q^f_{kl} and Q^m_{kl}, refer to Fig. 1.a. To simplify the calculations both materials are presumed to have the same Poisson's ratio. The elasticity constants Q^{H1}_{kl} of the composite are found by (4)-(6)

$$Q^{H1}_{11} = J_1 \quad ; \quad Q^{H1}_{22} = J_3 \quad ; \quad Q^{H1}_{12} = \nu J_1 \quad ; \quad Q^{H1}_{66} = \frac{1-\nu}{2}J_1 \tag{7}$$

where Q^{H1}_{11} and Q^{H1}_{22} are the stiffnesses corresponding to the orientation of the 1- and 2-axes, refer to Fig. 1, and

$$\begin{aligned} J_1 &= \left(\frac{\delta_1}{Q^f_{11}} + \frac{1-\delta_1}{Q^m_{11}}\right)^{-1} \\ J_2 &= \delta_1 Q^f_{11} + (1-\delta_1)Q^m_{11} \\ J_3 &= J_2(1-\nu^2) + \nu^2 J_1 \end{aligned} \tag{8}$$

The constitutive matrix Q^{H3}_{kl} of the second rank composite shown in Fig. 1.c can be determined, by repeating twice the use of (4) and (5):

$$Q_{11}^{H3} = \left(\frac{\gamma}{Q_{11}^{f}} + \frac{1-\gamma}{Q_{22}^{H2}}\right)^{-1}$$

$$Q_{22}^{H3} = \gamma Q_{11}^{f} + (1-\gamma)Q_{11}^{H2} - \nu^2\left[\gamma Q_{11}^{f} + \left(Q_{11}^{H2}\right)^2 \frac{1-\gamma}{Q_{22}^{H2}} - Q_{11}^{H3}\left(\gamma + Q_{11}^{H2}\frac{1-\gamma}{Q_{22}^{H2}}\right)^2\right] \qquad (9)$$

$$Q_{12}^{H3} = \nu\left(\gamma + Q_{11}^{H2}\frac{1-\gamma}{Q_{22}^{H2}}\right)Q_{11}^{H3}$$

$$Q_{66}^{H3} = \frac{1-\nu}{2}\left(\frac{\gamma}{Q_{11}^{f}} + \frac{1-\gamma}{Q_{11}^{H2}}\right)^{-1}$$

where J_1, J_3, Q_{11}^{H2} and Q_{22}^{H2} are given by (8) and (10).

$$Q_{11}^{H2} = \left(\frac{\delta_2}{J_1} + \frac{1-\delta_2}{Q_{11}^{-}}\right)^{-1} \qquad (10)$$

$$Q_{22}^{H2} = \delta_2 J_3 + (1-\delta_2)Q_{11}^{-} - \nu^2\left[\delta_2 J_1 + (1-\delta_2)Q_{11}^{-}\right] + \nu^2 Q_{11}^{H2}$$

Finally, we determine the matrix of extensional stiffness A_{kl} of a disc with the thickness h (Vinson and Sierakowski 1987):

$$A_{kl} = \int_{-h/2}^{h/2} Q_{kl}^{H3}\,dz = hQ_{kl}^{H3} \qquad (11)$$

5 Optimization technique

The optimization problem is solved iteratively by a two-level procedure of redesign. The stress/strain field is initially determined by finite element analysis using MODULEF in each loop of redesign, and improved orientations θ_i (i=1,...,n) of the composite are subsequently determined by means of an optimality criterion at the first level of redesign. At the second level of redesign the material densities δ_{1i}, δ_{2i} and γ_i are improved by a method of analytical sensitivity analysis and mathematical programming.

A notable feature of the present problem is that a usual gradient method may fail in determining the optimal orientation of the composite because local optima normally exist. To circumvent this inherent difficulty, we use the results obtained by Pedersen (1989, 1990), who has performed an analytical investigation of the first and second derivative of the total strain energy with respect to the orientation of the composite. The results of the investigation are summarized in a table in Pedersen (1990) and Thomsen (1991). In an optimization problem where the stiffness of a structure is maximized using the material orthotropy directions as design variables, we may either orient the composite material relative to the principal stress or strain directions (Pedersen et al. 1991b). Numerical examples however show that the best convergence of the optimization problem is obtained, if the composite is rotated relative to the principal stress directions. Coincidence between the largest principal stress and strain directions is always found to be a result of the orientation optimization, and normally these directions will coincide with the material direction associated with the largest stiffness (unless the material has a relatively high shear stiffness, see Pedersen (1990)).

The second stage in the loop of redesign consists in determining an improved distribution of the amount of material, i.e., to obtain improved values of the design variables δ_{1i}, δ_{2i} and γ_i, (i=1,...,n). We apply a dual method of mathematical programming using mixed variables as developed by Svanberg (1987) and implemented in the computer code MMA (Method of Moving Asymptotes). To this end we need the sensitivities of the objective function and the constraints with respect to the above mentioned design variables.

Results by Pedersen (1990, 1991a) show that by means of Clayperon's theorem and the principle of virtual displacements for structures with design independent loads, the gradient of the total strain energy U can be determined from the gradient of the specific strain energy u_i of a given element, whose strain field is considered to be fixed

$$\frac{dU}{da_i} = -\frac{\partial u_i}{\partial a_i} S_i \quad ; \quad i=1,...,n \tag{12}$$

Here a_i denotes any of the design variables δ_{1i}, δ_{2i} or γ_i (i=1,...,n). Thus, the sensitivities of the total strain energy U with respect to δ_{1i}, δ_{2i} and γ_i can be determined by (1) and (12), assuming the strain field to be fixed, and restricting variation to the stiffness matrix [A]. For the i–th element of the discretized geometry we obtain the following expression for sensitivities w.r.t. the design variables a_i

$$U_{,a_i} = -\left\{\left\{\frac{1}{8}A'_{11}\left[(\varepsilon_I+\varepsilon_{II})+(\varepsilon_I-\varepsilon_{II})\cos 2\psi\right]^2+\frac{1}{8}A'_{22}\left[(\varepsilon_I+\varepsilon_{II})-(\varepsilon_I-\varepsilon_{II})\cos 2\psi\right]^2 \right.\right.$$
$$\left.\left. +\frac{1}{4}A'_{12}\left[(\varepsilon_I+\varepsilon_{II})^2-(\varepsilon_I-\varepsilon_{II})^2\cos^2 2\psi\right]+\frac{1}{2}A'_{66}(\varepsilon_I-\varepsilon_{II})^2\sin^2 2\psi\right\}S\right\}_i \tag{13}$$

$$i = 1,...,n$$

Here A'_{kl} is a shorthand notation for the derivatives dA_{kl}/da_i of a component of the stiffness matrix [A]. These sensitivities are analytically derived by Thomsen (1992), and results are available therein. Sensitivities of the constraints in (2) and (3) are readily derived analytically, and thus we have all necessary sensitivity information required for the optimization at the second level of redesign.

6 Examples of optimization using one material of large stiffness

Initially, we consider examples of topology optimization where we only use *one* material of large stiffness. The optimization method has been tested for a number of structures with various design domains and boundary conditions, and we have chosen examples where the results can be compared with analytical solutions.

The optimization is performed iteratively by choosing the *orientation* and *concentration* in each finite element which *separately* maximize the stiffness of the structure. We apply an initial geometry consisting of an anisotropic material as shown in Fig. 2 with the orientations $\theta_i=0$ and concentrations $\gamma_i=\delta_{2i}$, $(i=1,...,n)$.

Let us now consider a Michell truss, which is an analytical solution to an optimization problem. A Michell truss is obtained by volume minimization of a "truss–universe", which has prescribed upper bounds on allowable tension and compression stresses, see Hemp (1973). Fig. 3 shows such an example, where the force 2P is carried by the truss structure in the design domain ABCD. By the topology optimization we presume that the optimal structure is symmetrical about the centerline in Fig. 3, and using this we reduce the design domain as shown in Fig. 4.

Fig. 5 shows results of topology optimization where 4–node elements have been used. The available amount of material is set to be 40% of the design domain volume. From Fig. 5 it appears that the structure is very similar to the Michell truss in Fig. 3. We see that only a very few elements in the optimized structure remain anisotropic (hatched elements), which means that the design domain has been separated into sub–domains consisting of very soft material (holes) and isotropic stiff material, respectively.

Fig. 6 illustrates two other examples of Michell trusses which are simply supported and fixed against displacements in two points, respectively, and where the design domains are the upper half planes. By topology optimization of the structures with the load and boundary conditions shown in Fig. 6 we utilize the symmetry about the vertical centerlines and thus only analyze the left half of the structures assuming the right edges to be simple supported, refer to Fig. 7. Available amounts of material of the structures in Fig. 7.a and b are set to be 30% and 25%, respectively, and we obtain the optimal topologies in Fig. 8.

The Michell trusses do all follow two geometrical conditions (Hemp 1973):

- If a pair of tension and compression members meet at a point, they must be orthogonal (see e.g. the intersession, where the force is acting in Fig. 3)

- If two tension (compression) members and one compression (tension) member meet at a point, then the compression (tension) member must be orthogonal to the other two (see e.g. the upper (lowest) row of members in Fig. 3 and the intersection points along the curved lines in Fig. 6.a and b).

Considering the optimal topologies, we see that they all comply with these conditions. If we had used larger amounts of available material, the optimal topologies would have had frame character and would have differed from the Michell trusses, which have joint connections in the intersection points of the truss members.

By topology optimization we do not take stability of compression loaded truss members into consideration. To get a practically useful truss structure we could apply the optimized topologies as initial designs in a traditional truss optimization program. Truss optimization has been undertaken by Pedersen (1972), who minimized the total weight of a truss structure considering cross section areas of the bars and joints coordinates as design variables. Pedersen considers truss optimization for several simultaneous load cases taking stress constraints into consideration for bars in tension and stability constraints for bars in compression in the elastic (Euler) and plastic domain. A corresponding method is applied by Olhoff et al. (1991), who minimize the weight of a truss structure with fixed compliance using a topology optimized structure as an initial design for the truss optimization program, SCOTS.

7 Examples of optimization using two materials of large stiffnesses

Finally we consider examples of topology optimization, where the material is modelled by one very soft material with elastic moduli Q_{kl}^{-} and two stiff materials with the elastic moduli Q_{kl}^{f} and Q_{kl}^{m}, respectively. We consider optimization of the cantilever beam in Fig. 9, and once more use symmetry.

Initially, we show two examples where the stiffness ratios between "material 1" and "material 2" are set to be 10 and 75, respectively:

$$\text{(a)}\quad Q_{kl}^{f}=10Q_{kl}^{m} \qquad \text{(b)}\quad Q_{kl}^{f}=75Q_{kl}^{m} \tag{14}$$

The available amounts of "material 1" and "material 2" are set to be 20% and 65%, respectively of the design domain volume. During the optimization we choose the *orientations* θ_i and *concentrations* γ_i, δ_{1i} and δ_{2i} which maximize the stiffness of the structure.

Fig. 10 shows the optimal topologies of the two examples where the structures have been discretized into 12x96 4-node elements. The hatching densities of an element in two perpendicular directions are proportional to the elastic moduli A_{11} and A_{22}, and the orientation of the hatching indicates the corresponding directions. The optimized structures mainly consist of orthotropic material. Only a very few elements along the upper left edges consist of isotropic "material 1", and isotropic "material 2" elements have not been generated. The stiffer material along the upper edge carries the large normal stresses, whereas the shear stresses are carried by a softer orthotropic material, the stiffnesses of which are almost equal in the two principal material directions. Elements at the right, upper and the left, lower corner have small strain energy densities due to the applied load and boundary conditions, and no material is distributed in these elements.

For the examples in Fig. 10.a and b, we have chosen the material orientations θ_i such as to *maximize* the stiffness of the structure. We now show two examples where we choose the material orientations that *minimize* the stiffness of the structure, while we still choose the material concentrations δ_{1i}, δ_{2i} and γ_i that *maximize* the stiffness. When this is done, anisotropic material becomes very unfavorable, because the material stiffnesses in the directions of the principal stresses become very low relative to the material consumption. This leads to generation of topologies that only consists of "holes" and isotropic "material 1" and "material 2". Fig. 11.a and 11.b show two examples, where we have used the same elastic moduli Q^f_{kl} and Q^m_{kl} and amounts of available material as in Fig. 10, but where the orientations are chosen such as to minimize the stiffness. It should be noticed that the orientation of the material has no influence on the stiffness when all material has become isotropic. In this figure the distributions of isotropic "material 1" and "material 2" are illustrated by black and hatched domains, respectively, whereas white elements illustrate void. It appears that stiff "material 1" is distributed along the upper edge of the structure in order to carry the largest normal stresses, and that shear stresses are carried by the softer "material 2" like in ordinary sandwich structures composed of two "skins" and one "core". For the structure in Fig. 11.b it has been advantageous to use an amount of the stiff "material 1" as "shear force reinforcement", because the stiffness ratio $Q^f_{kl}/Q^m_{kl}=75$ of this example is large, and since there is relatively much "material 1" available. Note that the topologies in Fig. 11.a and 11.b represent *local* optima, and they store 3% and 26% more elastic energy than the topologies in Fig. 10.a and 10.b, respectively. The purpose of the examples in Fig. 11 has been to demonstrate how to obtain a topology only consisting of isotropic materials, which is preferable from the point of view of manufacture.

Acknowledgement – The work received support from the Danish Technical Research Council (Program of Research on Computer Aided Design).

References

Avellaneda, M. 1987: Optimal Bounds and Microgeometries for Elastic Two–phase Composites. SIAM J. Appl. Math. 47, 1216–1228.

Bendsøe, M.P.; Kikuchi, N. 1988: Generating Optimal Topologies in Structural Design using a Homogenization Method. Computer Methods in Applied Mechanics and Engineering. 71, 197–224.

Bendsøe, M.P. 1989: Optimal Shape Design as a Material Distribution Problem. Structural Optimization. Vol. 1, No. 4, 193– 202.

Bendsøe, M.P.; Rasmussen, J.; Rodriques, H.C. 1990a: Topology and Boundary Shape Optimization as an Integrated Tool for Computer Aided Design. Engineering Optimization in Design Processes. 63, Proceedings, Springer– Verlag.

Bendsøe, M.P.; Rodriques, H.C. 1990b: Integrated Topology and Boundary Shape Optimization of 2–D Solids. Comput. Meth. Appl. Mech. Engrg. (to appear).

Braibant, V.; Fleury, C. 1984: Shape Optimal Design using B–splines. Comp. Meth. Appl. Mech. Engrg. Vol. 44, 247–267.

Díaz, A.R.; Bendsøe, M.P. 1990: Shape Optimization of Multipurpose Structures by a Homogenization Method. Report, Mechanical Engineering, Michigan State University, East Lansing, MI 48824, USA.

Esping, B.J.D. 1984: Minimum Weight Design of Membrane Structures using eight node Isoparametric Elements and Numerical Derivatives. Computers & Structures, Vol. 19, No. 4, 591–604.

Esping, B.J.D. 1986: The OASIS Structural Optimization System. Computers & Structures, Vol. 23, 365–377.

Haftka, R.T.; Gandhi, R.V. 1986: Structural Shape Optimization – a Survey. Comp. Meth. Appl. Mech. Engrg. 57, 91–106.

Hemp, W.S. 1973: Optimum Structures. Clarendon Press, Oxford. 70–101.

Jones, R.M. 1975: Mechanics of Composite Materials. 85–96, New York: McGraw–Hill.

Kohn, R.; Strang, G. 1986: Optimal Design and Relaxation of Variational Problems I–III. Comm. Pure Appl. Math. 39, 113–138, 139–182, 353–377.

Kohn, R. 1988a: Recent Progress in the Mathematical Modelling of Composite Materials. In: Composite Material Response: Constitutive Relations and Damage Mechanisms, G. Sih et. al. eds, Elsevier, 155–177.

Kohn, R.; Lipton, R. 1988b: Optimal Bounds for the Effective Energy of a Mixture of Two Incompressible Materials. Arch. Rat. Mech. Anal. 102, 331–350.

Kohn, R. 1990: Composite Materials and Structural Optimization. Proc. Workshop on Smart/Intelligent Materials and Systems. Honolulu, March 1990, (Thechonomic Press).

Lurie, K.; Cherkaev, A. 1986: Effective Characteristics of Composites and Problems of Optimum Structural Design (in Russian). Uspekhi Mekhaniki 9, 1–81.

MODULEF, 1983: Institut National de Recherche en Informatique et en Automatique. Bibliotheque D'Elasticite, 101.

Murat, F; Tartar, L. 1985. Calcul des Variations et Homogenization. Le Methodes de l'Homogeneisation: Theorie et Applications en Physique, Coll. de la Dir. des Etudes et Recherche d'Electricite de France, Eyrolles, 319–369.

Olhoff, N.; Bendsøe, M.P.; Rasmussen, J. 1991: On CAD-integrated Structural Topology and Design Optimization. Computer Methods in Applied Mechanics and Engineering 89, North-Holland, 259–279.

Papalambros, P.Y.; Chirehdast, M. 1990: An Integrated Environment for Structural Configuration Design. J. Engrg. Design. 1, 73–96.

Pedersen, P. 1972: On the Optimal Layout of Multi-Purpose Trusses. Computers & Structures, Vol. 2, 695–712.

Pedersen, P. 1989: On Optimal Orientation of Orthotropic Materials. Structural Optimization. 1, 101–106, Lyngby, Denmark, pp. 127.

Pedersen, P. 1990: Bounds on Elastic Energy in Solids of Orthotropic Materials, Structural Optimization, Vol. 2, No. 1, 55–63.

Pedersen, P. 1991a: On Thickness and Orientational Design with Orthotropic Materials. Structural Optimization. Vol. 3, No. 2, 69–78.

Pedersen, P.; Bendsøe, M.P.; Nagendra, S. 1991b: Note on the 2D–Match of Coaligned Principal Stresses and Strains. Dept. of Solid Mechanics, Technical University of Denmark, Lyngby, Denmark.

Rasmussen, J. 1990: The Structural Optimization System CAOS. Structural Optimization, Vol. 2, No. 2, 109–115.

Rodriques, H.C. 1988: Shape Optimal Design of Elastic Bodies using a Mixed Variational Formulation. Comp. Meth. Appl. Mech. Engrn., Vol. 69, 29–44, 1988.

Rozvany, G.I.N.; Zhou M. 1990: Applications of the COC Algorithm in Layout Optimization. Engineering Optimization in Design Processes, Proceedings of the International Conference, Karlsruhe Nuclear Research Center, Germany, September 3–4, 1990. Springer–Verlag.

Strang, G.; Kohn, R.V. 1986: Optimal Design in Elasticity and Plasticity. International Journal for Numerical Methods in Engineering. Vol. 22, 22, 183–188.

Suzuki, K.; Kikuchi, N. 1989: A Homogenization Method for Shape and Topology Optimization. Department of Mechanical Engineering and Applied Mechanics, The University of Michigan, Ann Arbor, Michigan 48109, USA.

Svanberg, K. 1987: The Method of Moving Asymptotes – a New Method for Structural Optimization. Int. J. Num. Meth. Eng., Vol. 24, 359–373.

Thomsen, J.; Olhoff, N. 1990: Optimization of Fiber Orientation and Concentration in Composites. Control and Cybernetics, Vol. 19, No. 3–4.

Thomsen, J. 1991: Optimization of Composite Discs. Structural Optimization . Vol. 3, No. 2, 89–98.

Thomsen, J. 1992: Optimization of the Properties of Anisotropic Materials and the Topologies of Structures. (in English). Ph.D. thesis. Institute of Mechanical Engineering, Aalborg University, Denmark.

Vinson, J.R.; Sierakowski R.L. 1987. The Behavior of Structures Composed of Composite Materials. Martinus Nijhoff Publishers, Dordrecht.

Figures

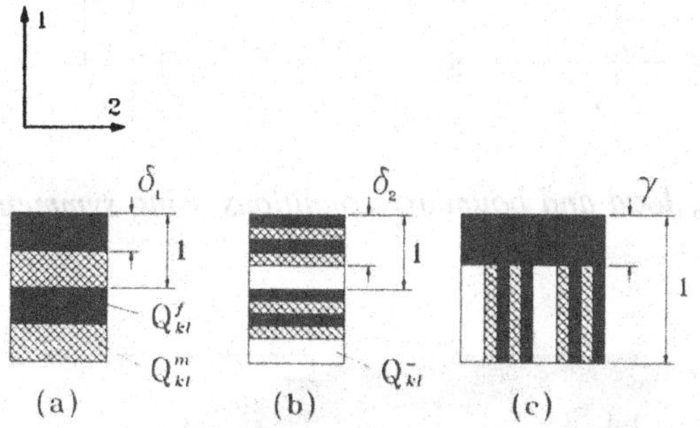

Fig. 1. Construction of composite materials. (a) *First level.* (b) *Second level.* (c) *Third level*

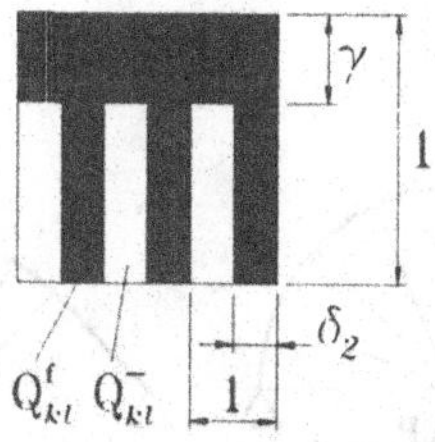

Fig. 2. Material model for optimization with one stiff material

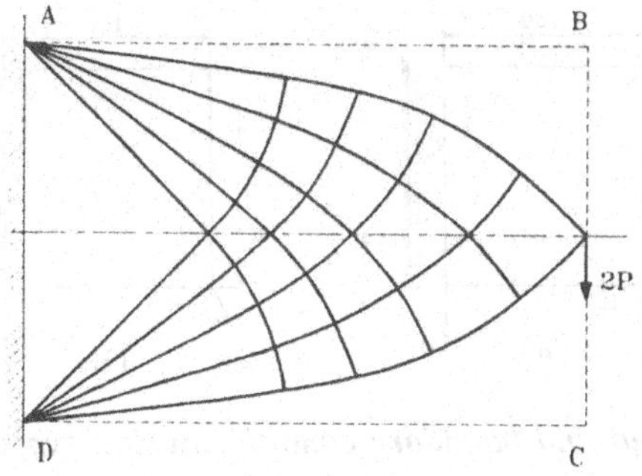

Fig. 3. Michell truss. Analytical solution for truss optimization problem in the design domain ABCD (Rozvany and Zhou 1990)

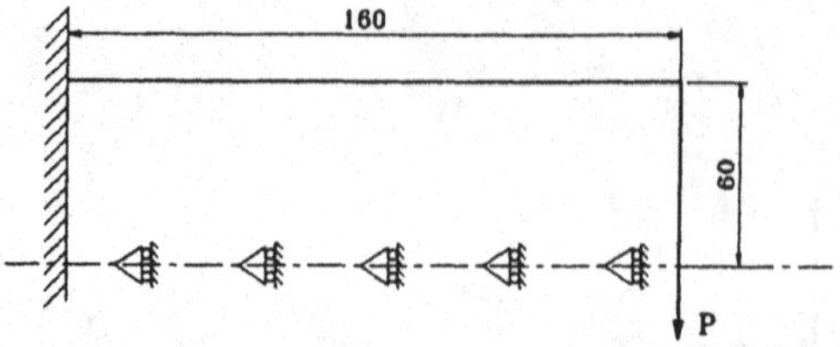

Fig. 4. Design domain, load and boundary conditions using symmetry

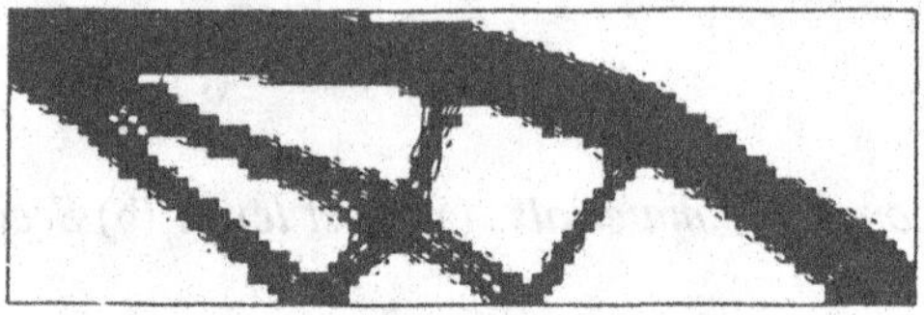

Fig. 5. Optimal topology. Volume=40% (n=28X90)

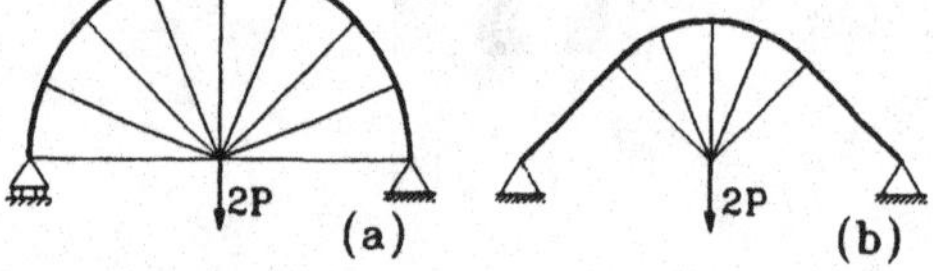

Fig. 6. Michell trusses (Hemp 1973)

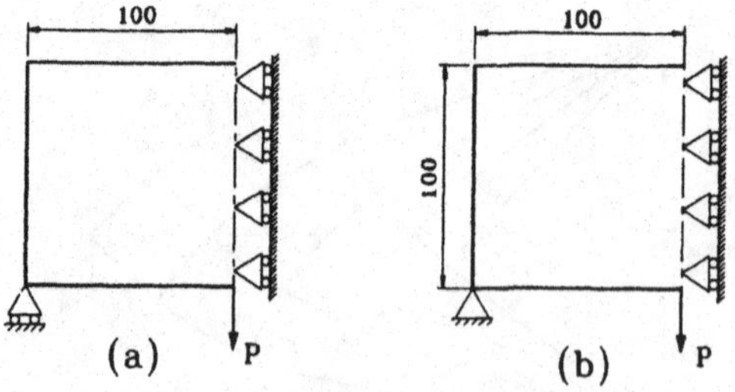

Fig. 7. Design domain, load and boundary conditions using symmetry

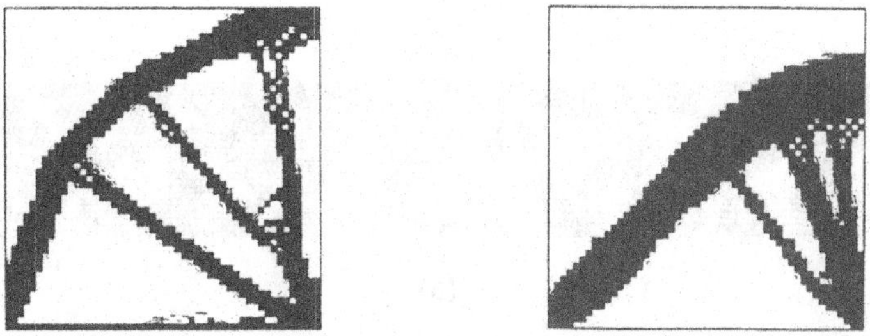

(a) Volume=30% (b) Volume=25%

Fig. 8. Optimal topologies (n=50x50)

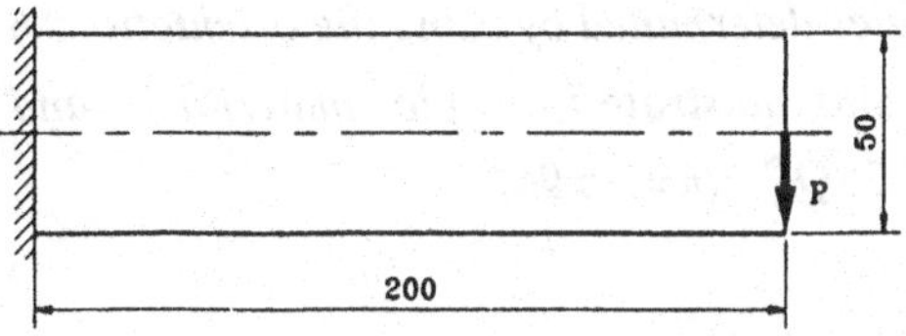

Fig. 9. Cantilever beam

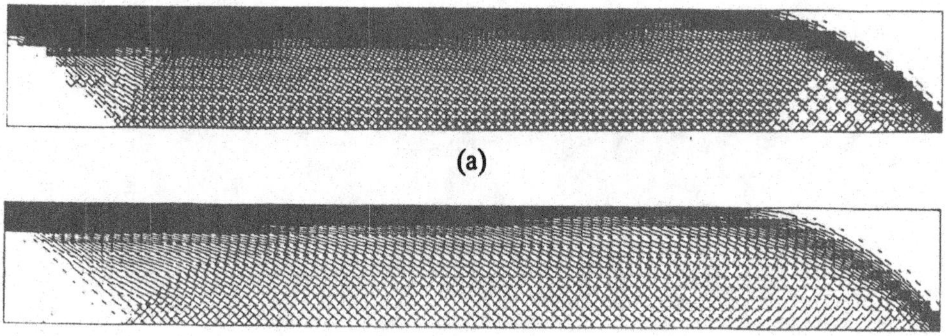

(a)

(b)

Fig. 10. Optimal topologies determined by using the orientations θ_i that maximize the stiffness. Hatching density is proportional to stifness. (a) $Q^f_{kl}=10Q^m_{kl}$. (b) $Q^f_{kl}=75Q^m_{kl}$ *(n=12x96)*

18

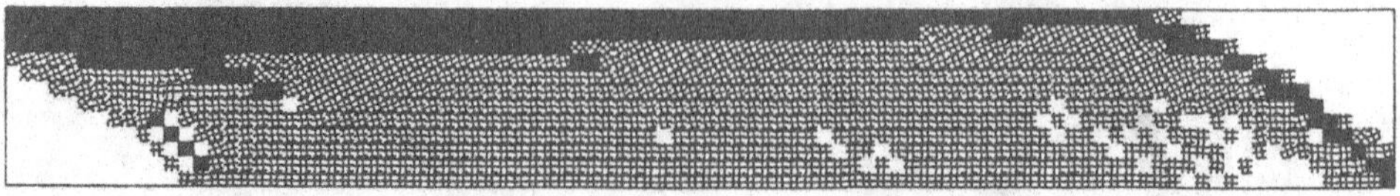

(a)

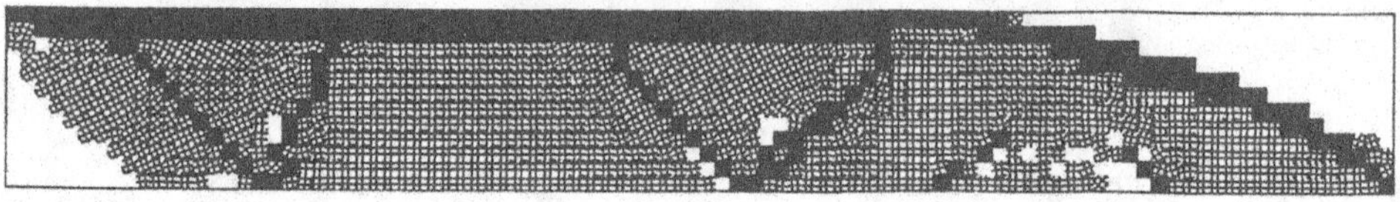

(b)

Fig. 11. Optimal topologies determined by using the orientations θ_i that minimize the stiffness. Black and hatched domains illustrate isotropic "material 1" and "material 2", respectively. (a) $Q^f_{kl}=10Q^m_{kl}$. (b) $Q^f_{kl}=75Q^m_{kl}$ *(n=12x96)*

ON CAD-INTEGRATED STRUCTURAL TOPOLOGY AND DESIGN OPTIMIZATION

N. Olhoff

The University of Aalborg, Aalborg, Denmark

M.P. Bendsøe

The Technical University of Denmark, Lyngby, Denmark

J. Rasmussen

The University of Aalborg, Aalborg, Denmark

Abstract - Concepts underlying an interactive CAD-based engineering design optimization system are developed, and methods of optimizing the topology, shape, and sizing of mechanical components are presented. These methods are integrated in the system, and the method for determining the optimal topology is used as a preprocessor for subsequent shape or sizing optimization. Some illustrative examples of application of the engineering design systems are presented.

1. INTRODUCTION

Structural optimization [1,2] can be essentially conceived as a rational search for the optimal spatial distribution of material within a prescribed admissible structural domain, assuming the loading and boundary conditions to be given. In the general case, this problem consists in determining both the optimal topology and the optimal design of the structure. Here the label "optimal design" covers the optimal shape or sizing of the design.

Recent years have witnessed a definite tendency towards development and augmentation of large, practise–oriented optimization systems, se e.g. [3–15]. However, despite many attractive features, these systems are as of yet only capable of conducting shape optimization within the framework of a given structural topology. Thus, until recently, methods for topology optimization were only available for truss–like structures composed of slender members, see e.g. [16], so only systems that include sizing optimization of such structures may, up to now, have been endowed with some sort of tool for topology optimization. This picture is being rapidly changed, however, as substantial current research efforts are devoted to integration of methods of topology optimization [17,18] and optimal design, see [19–27].

Along these lines, the present paper deals with the development of complementary methods for optimization of structural topologies and designs, and as a special feature we describe their actual integration and implementation in an interactive CAD–based structural optimization system [13–14]. The presentation is restricted to optimization problems involving linearly elastic, two–dimensional structures and components. In Chapter 2 we present a method of optimization of structural topology, and then discuss in Chapter 3 the advantages of integrating the method with methods of interactive CAD based optimal design. In the two subsequent chapters we give an account of optimal design methods, namely shape optimization in Chapter 4 and sizing optimization in Chapter 5. Finally a number of examples are presented in Chapter 6.

2. OPTIMIZATION OF TOPOLOGY

Solution of the integral structural optimization problem requires a two–level approach with the generation of an optimal topology as a kind of preprocessing for the problem of design optimization.

Here, we determine the optimal topology by means of a software system called HOMOPT [18] that is based on a newly developed method in which the structure is considered as a domain of space with a high density of material [17,18]. This problem is basically one of discrete optimization, but this difficulty is avoided by introducing relationships between stiffness components and density, based on physical modelling of porous, periodic microstructures whose orientation and density are described by continous variables over the admissible design domain.

The solution of this problem is based on a finite element discretization of the admissible domain, and the optimum values of the design variables (density and orientation of the microstructures) are determined iteratively via an optimality criterion approach.

More precisely, for the topology optimization we minimize compliance for a fixed, given volume of material, and use a density of material as the design variable. The density of material and the effective material properties related to the density is controlled via geometric variables which govern the material with microstructure that is constructed in order to relate correctly material density with effective material property.

The problem is thus formulated as:

$$\begin{aligned} &\textit{Minimize} && L(w) \\ &\textit{Subject to} && a_D(w,v) = L(v) \textit{ for all } \quad v \in H \qquad (1) \\ & && \textit{Volume} \le V \end{aligned}$$

where

$$L(v) = \int_\Omega B_i v_i d\Omega + \int_\Gamma T_i v_i d\Gamma \qquad (2)$$

$$a_D(w,v) = \int_\Omega E_{ijkl}(D)\, \varepsilon_{ij}(w)\, \varepsilon_{kl}(v) d\Omega \qquad (3)$$

Here, B_i and T_i are the body forces and surface tractions, respectively, and ε_{ij} denote linearized strains. H is the set of kinematically admissible displacements. The problem is defined on a fixed reference domain Ω and the components of the tensor of elasticity E_{ijkl} depend on the design variables used. For a so-called second rank layering constructed as in fig. 1, we have a relation

$$E_{ijkl} = E_{ijkl}(\mu,\gamma,\theta) \tag{4}$$

where μ and γ denote the densities of the layering and θ is the rotation angle of the layering. The relation (4) can be computed analytically [18] and for the volume we have

$$Volume = \int_{\Omega}(\mu + \gamma - \mu\gamma)\,d\Omega \tag{5}$$

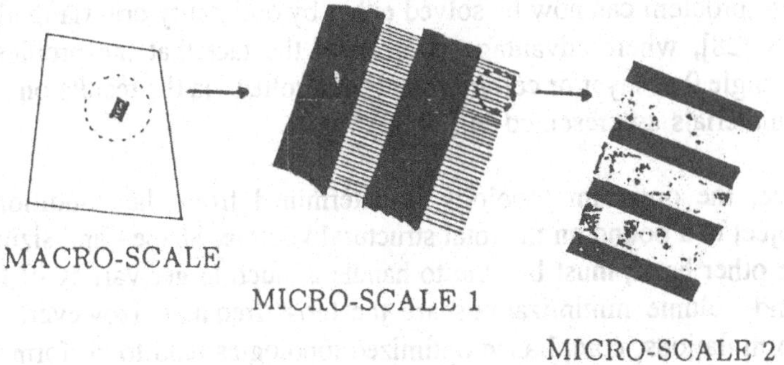

Fig. 1. Construction of a layering of second rank.

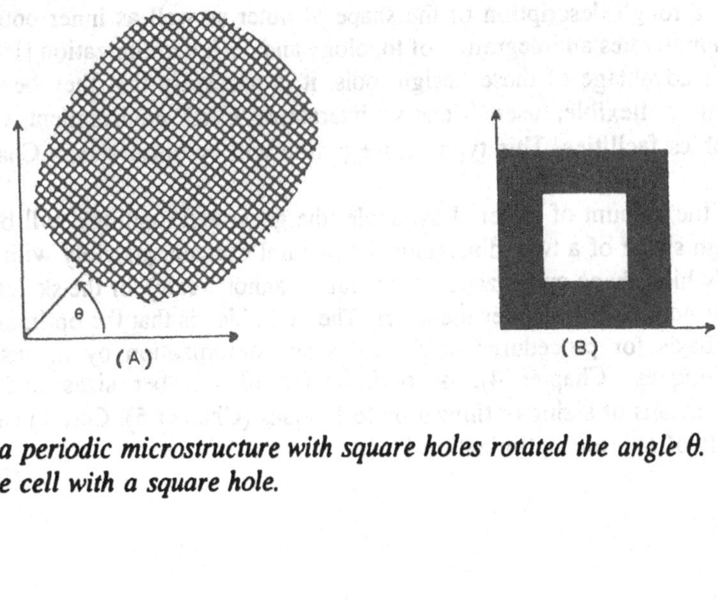

Fig. 2. (A): a periodic microstructure with square holes rotated the angle θ.
(B): a square cell with a square hole.

Layered materials (used for topology generation in the example in Section 6.1) is just one possible choice of microstructure that can be applied. The important feature is to choose a microstructure that allows density of material to cover the complete range of values from zero (void) to one (solid), and that this microstructure is periodic so that effective properties can be computed (numerically) through homogenization (theory of cells). This excludes circular holes in square cells, while square or rectangular holes in square cells, see fig. 2, are suitable choices of simple microstructures. For the case of a rectangular hole in a square cell, the volume is also given by eq.(5), with μ and γ denoting the amount of material used in the directions of axes of the cell and hole; for this microstructure the angle of rotation of the unit cell becomes a design variable. (This type of microstructure is used for topology generation in the examples of Section 6.2 og 6.3).

The optimization problem can now be solved either by optimality criteria methods [18] or by duality methods [28], where advantage is taken of the fact that the problem has just one constraint. The angle θ of layer or cell rotation is controlled via the results on optimal rotation of orthotropic materials as presented in [29,30].

As stated above, the optimum topology is determined from the condition of minimum compliance subject to a bound on the total structural volume. Shape– and sizing optimization systems, on the other hand, must be able to handle a much larger variety of formulations of which stress and volume minimizations are the most frequent. However, in spite of the incompatible formulations, compliance optimized topologies tend to perform well also from a stress minimization point of view. This is due to the fact that a relatively high amount of energy is stored in possible areas of stress concentration which then become undesirable also in the compliance minimization. The initial topology optimization can therefore in many cases lead to substantial improvements of the final result even though the actual aim is to perform an optimization of a different type.

The topology optimization thus results in a prediction of the structural type and overall lay-out, and gives a rough description of the shape of outer as well as inner boundaries of the structure. This motivates an integration of topology and design optimization [19–27]. In order to gain the full advantage of these design tools, it is necessary that they be integrated and implemented in a flexible, user–friendly, interactive CAD environment with extensive computer graphics facilities. This type of integration will be discussed in Chapter 3.

Depending on the amount of material available, the generated topology will basically either define the rough shape of a two–dimensional structural domain, possibly with macroscropic interior holes (which shape optimization procedures cannot create), or the skeleton of a truss– or beam–like structure with slender members. The main idea is that the optimal topology can be used as a basis for procedures of refined shape optimization by means of boundary variations techniques (Chapter 4), or optimization of member sizes and positions of connections by means of sizing optimization techniques (Chapter 5). Certain cases may even require methods of lay–out optimization [31].

3. INTEGRATION OF TOPOLOGY AND DESIGN OPTIMIZATION IN A CAD ENVIRONMENT

The actual topology preprocessor HOMOPT [18] is integrated with the structural shape optimization system CAOS (Computer Aided Optimization of Shapes) [13,14]. CAOS is based on the concept [32] of integration of software modules of finite element analysis, sensitivity analysis, and optimization by mathematical programming, and the system is developed in the spirit of refs. [6,7,33,34]. CAOS was originally based solely on the boundary variation method, i.e., the optimized geometry is a result of changing the shapes of the original boundaries of the structure [35,36]. The initial purpose of CAOS was to act as a framework for experiments with various structural optimization techniques, and, above all, to examine the problem of integrating shape optimization into a traditional Computer Aided Design (CAD) environment. The widely used commercial CAD system AutoCAD is used as the basis for CAOS, but the system concept is independent of the AutoCAD data structure and the techniques used in CAOS can therefore be applied in connection with most other CAD systems as well.

The geometric versatility of CAOS is achieved by an adaptation of a so-called "design element technique", see, e.g. [6], which allows the user to describe the continuous shape of the structure by a small number of variables, and solves the problem of automatic updating of the finite element model as the shape is changed. Fig. 3 shows a structure divided into design elements (for more examples, see [14]). The shape of each boundary of the design elements is determined by movements of control points along the directions of the arrows. The magnitude of these movements are treated as the design variables of the problem. This technique in connection with the user interface of the CAD system allows the designer to construct a shape optimization model of an initial structure very quickly. Any two-dimensional initial shape fit for analysis by membrane finite elements can be handled by the system.

The CAD integration of CAOS implies that all definitions of the optimization problem take place in the CAD system using the interactive facilities otherwise available for drawing and modification of geometric entities. Move directions, for instance, are inserted directly into the drawing with the pointing device of the workstation and they are subsequently treated by the CAD system exactly like any other geometric entity. The division of the structure into design elements, specification of objective- and constraint functions and all other definitions necessary to define the problem take place in the same way. Thus, modifications are very easy to perform by simply moving, erasing or otherwise modifying the specifications with the editing facilities already available in the CAD system.

As mentioned above, CAOS was originally based on boundary variation. However, via integration with the HOMOPT system the models created in CAOS can now be subjected either to topology or shape optimization.

The CAOS-HOMOPT connection was easily established because HOMOPT works directly on a finite element model of the design space in question. A system for monitoring the present result of the topology optimization has been developed taking advantage of the graphics capabilities of the CAD system. This enables the user to follow the process from within the CAD system and to transfer the optimized topology directly into the CAD model.

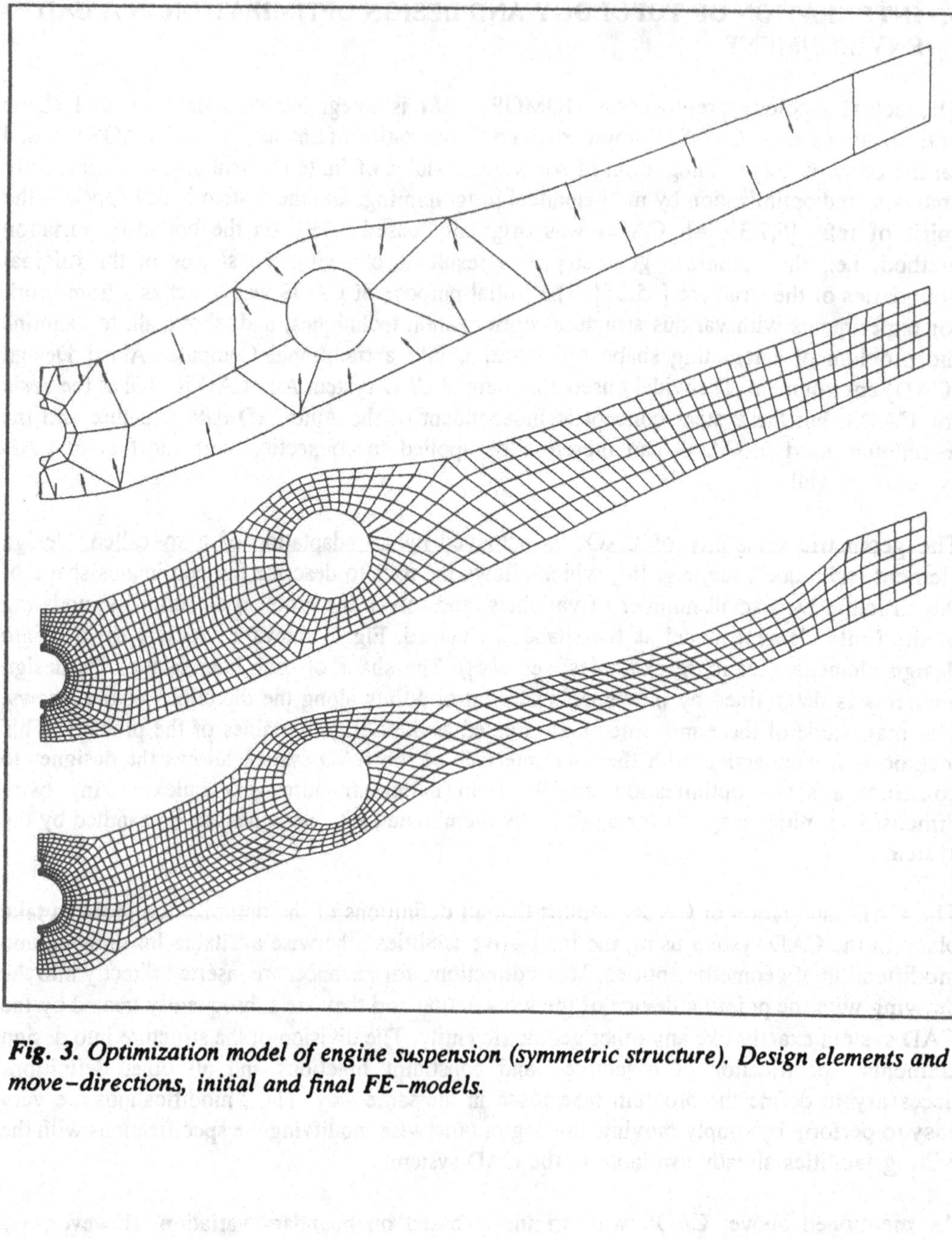

Fig. 3. ***Optimization model of engine suspension (symmetric structure). Design elements and move-directions, initial and final FE-models.***

The actual shape optimization model is generated very conveniently by simply drawing the initial structure on top of the generated topology. It is very easy for the designer to approximate the initial position and shape of holes and boundaries while taking the necessary practical constraints, stemming from function, fabrication etc., into consideration. This way, complicated image processing is unnecessary (and probably undesirable) because the designer's choice is often guided by non-geometric information like, eg., the manufacturing capabilities of the company or the relation of the component in question with the rest of the construction of which it is a part.

4. SHAPE OPTIMIZATION

There is a large number of possible formulations of shape optimization problems. One may choose to minimize weight, stress, compliance, displacement or any other property that can be derived from the geometrical model or the output from an analysis program which is usually a finite element module. The same set of possibilities should be available for specification of constraints. This requires a mathematical versatility which is achieved in CAOS with the use of the so-called "bound formulation" [37]. The present version of CAOS handles the following criteria which could act either as the objective or constraint functions:

1. Weight
2. Elastic displacement of a given material point
3. Maximum elastic displacement of any point in the structure
4. Stress (several types) at a given material point
5. Maximum stress (several types) at any point in the structure
6. Compliance

Mathematically, different entries in this list lead to very different optimization problems. Entries 2 and 4 are ordinary scalar functions which can be derived directly from the output from the finite element analysis. Entries 1 and 5 are of integral type and require some post-processing of the results to be evaluated. Entries 3 and 5 lead to min/max-problems with non-differentiable objective functions.

The usual mathematical program of structural optimization is the following:

$$\text{Minimize} \qquad f(a_i), \qquad i=1..n-1 \tag{6}$$

$$\text{Subject to} \qquad g_j(a_i) \leq G_j, \qquad i=1..n-1,\ j=1..m \tag{7}$$

$$\underline{a}_i \leq a_i \leq \bar{a}_i, \qquad i=1..n-1 \tag{8}$$

where n−1 is the number of design variables, a_i, and m is the number of constraints. Eqs. (8) are side constraints, i.e., upper and lower limits for the design variables. The functions f and g_j are specified by the user as a part of the optimization specification by interactively picking them from the list above.

The bound formulation enables the CAOS system to handle the optimization problem in a uniform way regardless of the blend of scalar–, integral– and min/max–criteria defined by the user. Given the min/max objective function $f = \max(f_j)$, $j=1..p_o$, and a number of constraints, $g_k = \max(g_{kj}) \leq G_k$, $k=1..m$, $j=1..p_k$, we get the following bound formulation of the problem:

$$\underset{a_i,\, \beta}{\text{Minimize}} \qquad \beta \tag{9}$$

$$\text{Subject to} \qquad f_j(a_i) - \beta \leq 0, \qquad j=1..p_o,\ i=1..n-1 \tag{10}$$

$$w_k g_{kj}(a_i) - \beta \leq 0, \qquad k=1..m,\ j=1..p_k,\ i=1..n-1 \tag{11}$$

$$\underline{a}_i \leq a_i \leq \bar{a}_i, \qquad i=1..n-1 \tag{12}$$

An extra design variable, β, has been introduced, rendering the total number of variables to n. The variable m is still the original number of constraints regardless of whether these are scalar–, integral– or min/max–functions. Each min/max–criterion gives rise to several actual constraints represented by nodal values, the number of which is designated by p_k for min/max–condition k. If condition k is scalar or integral, p_k is obviously 1. The weight factors w_k are imposed on the constraints to allow them to be limited by the same β–value as f. We evaluate w_k prior to the call of the optimizer from the relation:

$$G_k w_k = \beta \rightarrow w_k = \beta / G_k \tag{13}$$

and it is subsequently treated as a constant in the problem.

The tableau (9) through (12) is valid regardless of the blend of functions f and g_k, and the mathematical operations performed are therefore identical for any problem that the user could possibly define. The problem (9)–(12) is solved by sequential programming using either a SIMPLEX algorithm, the CONLIN optimizer [28] or the Method of Moving Asymptotes [38], all of which require the derivatives of the objective and constraint functions to be calculated. For this purpose, a sensitivity analysis [39,40] is performed for each design variable at each iteration. CAOS uses a semi–analytical method based on a differentiation of the finite element equation of equilibrium:

$$[K]\{u\} = \{f\} \tag{14}$$

where [K] is the global stiffness matrix of the structure, {u} is the vector of unknown nodal displacements, and {f} is the vector of external loads, see, e.g.[41]. Differentiation with respect to a design variable, say a_i, gives:

$$\frac{\partial [K]}{\partial a_i}\{u\} + [K]\frac{\partial \{u\}}{\partial a_i} = \frac{\partial \{f\}}{\partial a_i} \tag{15}$$

$$\Rightarrow [K]\frac{\partial\{u\}}{\partial a_i} = \frac{\partial\{f\}}{\partial a_i} - \frac{\partial[K]}{\partial a_i}\{u\}$$

$$= \{P_{ps}\} \tag{16}$$

The right hand side, $\{P_{ps}\}$, of (16) is often termed "Pseudo Load" because it plays the role of an extra load case in the sensitivity analysis. With $\{P_{ps}\}$ known, (16) can be solved using the factorization performed in connection with the initial analysis (14) thus returning the sensitivities of the nodal displacement vector {u} with respect to a_i.

We normally assume $\partial\{f\}/\partial a_i = \{0\}$. Furthermore, we shall approximate $\partial[K]/\partial a_i$ by forward finite differences:

$$\frac{\partial[K]}{\partial a_i} \approx \frac{[K(\mathbf{a}+\Delta a_i)] - [K(\mathbf{a})]}{\Delta a_i} \tag{17}$$

where Δa_i is a small perturbation of the i'th component of **a**.

$\{P_{ps}\}$ is thus calculated by:

$$\{P_{ps}\} = -\frac{[K(\mathbf{a}+\Delta a_i)] - [K(\mathbf{a})]}{\Delta a_i}\{u\} \tag{18}$$

The sensitivities of the nodal stresses are easily found from $\partial\{u\}/\partial a_i$. This calculation is also based on finite differences. For each element, the vector of nodal stresses is given by

$$\{\sigma\} = [C][B(x,y)]\{u^e\} \tag{19}$$

where [C] is the constitutive matrix connecting nodal strains with nodal stresses. The matrix [C] depends only on the material characteristics and remains unchanged by a perturbation of the design variables. The matrix [B] is the geometrical condition connecting nodal displacements with strains. This matrix is a function of the element node coordinates (x,y). The vector $\{u^e\}$ is the part of {u} concerning the element in question.

Knowing the derivatives of the element nodal displacements, we can estimate nodal displacements of the perturbed geometry:

$$\{u^{e*}\} \approx \{u^e\} + \frac{\partial\{u^e\}}{\partial a_i} \Delta a_i \tag{20}$$

Using these displacements, we can calculate the stresses of the perturbed geometry directly:

$$\{\sigma^*\} = [C][B(x^*,y^*)]\{u^{e*}\} \tag{21}$$

where (x^*,y^*) are the perturbed node coordinates. The stress derivative is now approximated by finite differences:

$$\frac{\partial\{\sigma\}}{\partial a_i} \approx \frac{\{\sigma^*\} - \{\sigma\}}{\Delta a_i} \tag{22}$$

For most problems, the calculation of stress derivatives is surprisingly stable considering the approximations involved. However, for geometries involving large rigid body rotations, the evaluation of displacement sensitivities becomes inaccurate. This, in turn, means that the stress sensitivities loose their reliability and absolute convergence of the problem is unattainable.

5. SIZING OPTIMIZATION

Our structural optimization system possesses capabilities for optimization of 2-D and 3-D truss structures under multiple loading conditions, using cross-sectional areas of bars and positions of joints as design variables. The system is called SCOTS (Sizing and Configuration Optimization of Truss Structures). The development is inspired by [42-44]. In the current setting, weight minimization is the design objective, and constraints include stresses, displacements, and elastic as well as plastic buckling of bars in compression.

The mathematical programming formulation is:

$$\underset{A_i, x_k}{\textit{Minimize}} \quad \sum_{i=1}^{n} \rho_i A_i \ell_i \tag{23}$$

$$\text{Subject to} \quad \frac{\sigma_i}{\sigma_i^{yt}} \leq 1, \quad \frac{\sigma_i}{\sigma_i^{c}} \leq 1, \qquad i = 1..n \tag{24a,b}$$

$$\frac{|d_j|}{\overline{d}_j} \leq 1 \qquad j = 1..J \tag{25}$$

$$\underline{A}_i \leq A_i \leq \overline{A}_i \ , \ \ i = 1..n \ \ ; \underline{x}_k \leq x_k \leq \overline{x}_k \ , \ \ 1..k \tag{26a,b}$$

where

$$\sigma_i^c = -\frac{\pi^2 \alpha_i^2 E_i A_i}{\ell_i^2 S_i} \quad \text{for} \quad \ell_i \geq \pi\alpha_i \sqrt{\frac{2E_i A_i}{\sigma_i^{yc} S_i}} \tag{27a}$$

$$\sigma_i^c = -\sigma_i^{yc} + \frac{(\sigma_i^{yc})^2 S_i \ell_i^2}{4\pi^2 E_i \alpha_i^2 A_i} \quad \text{for} \quad \ell_1 \leq \pi\alpha_i \sqrt{\frac{2E_i A_i}{\sigma_i^{yc} S_i}} \tag{27b}$$

Here, A_i, ℓ_i, ρ_i and E_i denote cross-sectional area, length, specific weight and Young's modulus for the i-th bar of the truss. The design variables of the problem, see Eq. (23), are A_i and x_k, where the latter symbol represents an element of the total set of variable components of position vectors of joints in the structure. Eqs. (26a,b) are side constraints for the design variables. For simplicity in notation, the remaining constraints are written for a single loading case. Eqs. (25) are constraints that may be specified for any displacement component for the joints of the truss.

Bar stresses are denoted by σ_i and eqs. (24a,b) express constraints for tensile and compressive stresses, respectively. In (24a), σ_i^{yt} represents the yield stress or some other specified upper stress limit. When buckling constraints are considered, the lower limit $-\sigma_i^c$ for compressive stresses is design dependent and given by eq. (27a) or (27b), respectively, where the former represents dimensioning against Euler Buckling, and the latter plastic buckling on the basis of Ostenfeld's formula. In (27a,b), we have $\alpha_i^2 = r_i^2/A_i$, where r_i is the radius of inertia of the bar cross-section, S_i the factor of safety against buckling, and σ_i^{yc} (>0) the compressive yield stress of the i-th bar. In the literature on optimization of trusses, it is often the case that yielding rather than buckling constraints are considered for compressive bars. This simplification is obtained, if we set $\sigma_i^c = -\sigma_i^{yc}$.

The analyses associated with the mathematical program (23–27) are based on a finite element formulation of the type (14), and the determination of displacement sensitivities follows eqs. (15) and (16) with the exceptions that the design derivatives of the stiffness matrix [K] are derived analytically, and that selfweight loading can be taken into account, whereby $\partial\{f\}/\partial a_i \neq 0$.

6. EXAMPLES

In this chapter we present examples where topology optimization as described in Chapter 2 has been used as a preprocessor for subsequent refined sizing or shape optimization depending on the type of structure predicted. In Section 6.1 an example of a truss-like structure will be discussed, and in Sections 6.2 and 6.3 we shall present some detailed examples of postprocessing topology results by refined shape optimization, taking advantage of the CAD-integrated boundary variations techniques of CAOS [13,14].

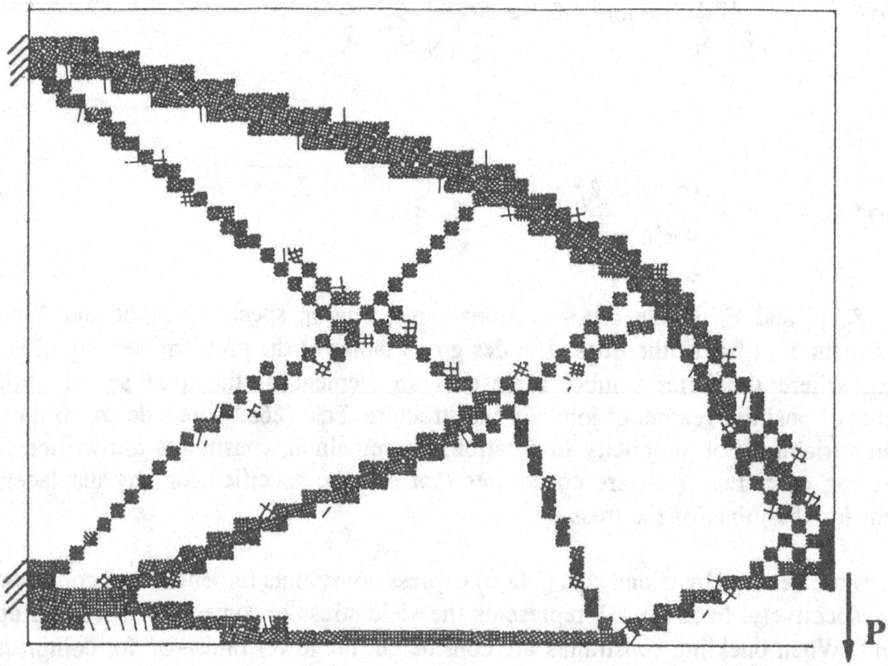

Fig. 4. Solution of topology optimization problem.

6.1. Truss-like structure

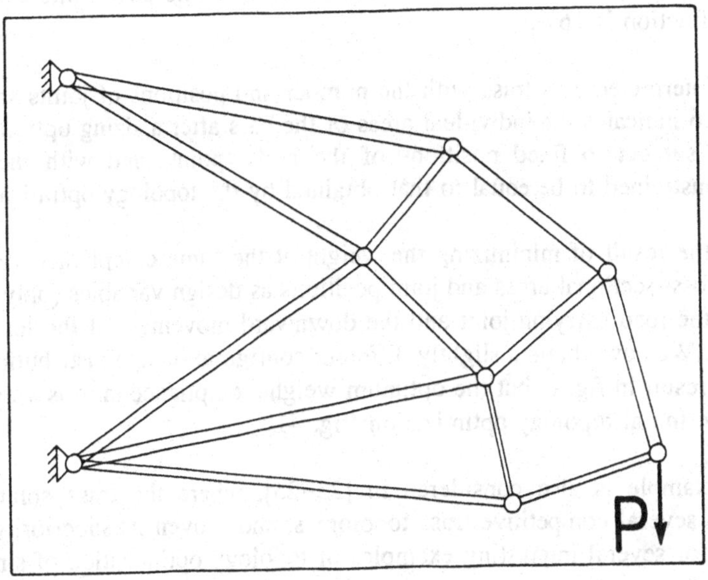

Fig. 5. Truss interpretation of result in Fig. 4 with bar areas determined by sizing for fixed positions of joints.

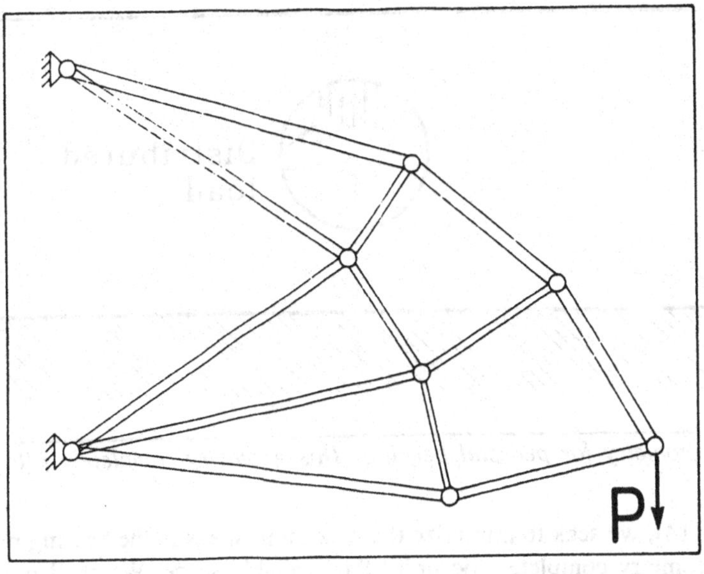

Fig. 6. The result when both bar cross-sectional areas and positions of unrestrained joints are used as design variables.

Fig. 4 presents the result of optimizing the topology of a structure within the rectangular design domain shown in the figure. The structure is required to carry a vertical load P, see fig. 4, and is offered support (displacement constraint) along the two hatched parts of the left hand side of the design domain. The domain is subdivided into 2610 finite elements, and the solid volume fraction is 16%.

The result is interpreted as a truss with the number and positions of joints shown in fig. 5. This figure also indicates the individual areas of the bars after a sizing optimization (weight minimization) subject to fixed positions of the nodal points and with the value of the compliance constrained to be equal to that obtained by the topology optimization.

Fig. 6 shows the result of minimizing the weight at the same compliance value, but using both the bar cross-sectional areas and joint positions as design variables (only the horisontal movement of the load carrying joint and the downward movement of the lower-most joint are restrained). We now obtain a slightly different configuration and distribution of bar areas relative to the result in fig. 6, but the optimum weight/compliance ratio is very close to that obtained by the initial topology optimization (fig. 4).

The present example is also considered in [23–25], where the truss solution has been compared with several competitive truss topologies, and proven its superiority. We refer to Refs. [19–25] for several interesting examples of topology optimization of similar type.

6.2 Example: Bearing Pedestal

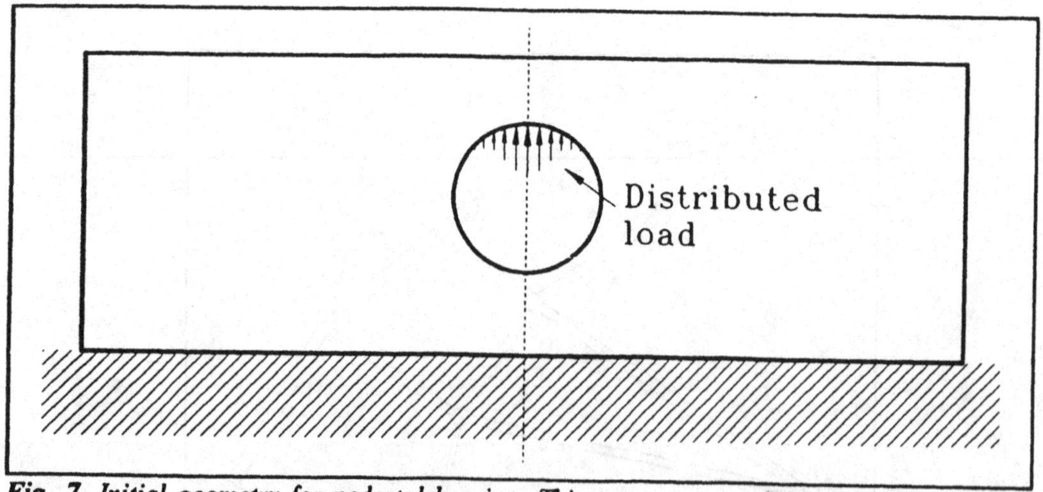

Fig. 7. Initial geometry for pedestal bearing. This geometry completely fills the available space.

In this example [14], we seek to minimize the maximum stress in the bearing pedestal of fig. 7. The initial geometry completely occupies the available space. We shall use 20% of this area as the upper limit for the volume, i.e., the topology optimization starts with an evenly distributed density of 20% in the entire structure with the exception of the rim of the hole

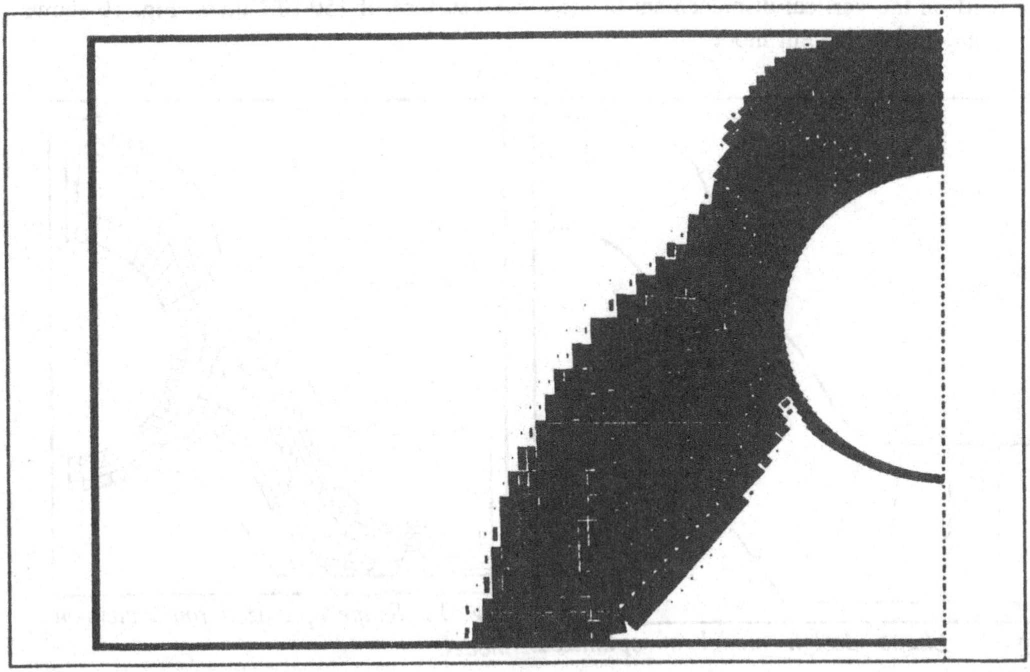

Fig. 8. Optimized topology illustrated by filling the elements by lumps of material corresponding to their final density.

which is required to have 100% density in order to properly position the bearing. Due to the symmetry of the geometry and loading, only the left half of the structure is considered.

Fig. 8 is an illustration of the optimized topology. Graphically, each element is filled with a lump of material corresponding to its final density. This creates the impression that, in some regions, isolated lumps of material remain outside the solid part of the structure. This is not necessarily the case. The lumps are merely a convenient way of illustrating the porous material. When using CAOS for the actual shape optimization based on an optimized topology, the user has to decide upon the actual position of interfaces between material and void based on the filling of elements by lumps of material. In the present case, the position of interfaces is relatively clear and leads to the shape optimization design element configuration shown in fig. 9. The initial shape has been slightly modified in comparison with the optimized geometry of fig. 8. This is out of practical considerations. The additional material provides a basis for the possible attachment of the pedastal to the underlying surface by, e.g., a bolt joint. The shape optimization model is defined to ensure a minimum thickness of this region. Furthermore, as in the topology model, a minimum thickness of the material surrounding the hole is required.

Due to limited analysis facilities, CAOS is incapable of handling contact problems. Thus, the analysis model presumes that the joint, regardless of its type, provides full contact with the underlying surface in all cases.

a bound on the vertical displacement of the loaded surface of $150 \cdot 10^{-4}$ units. Fig. 10 shows The final finite element model.

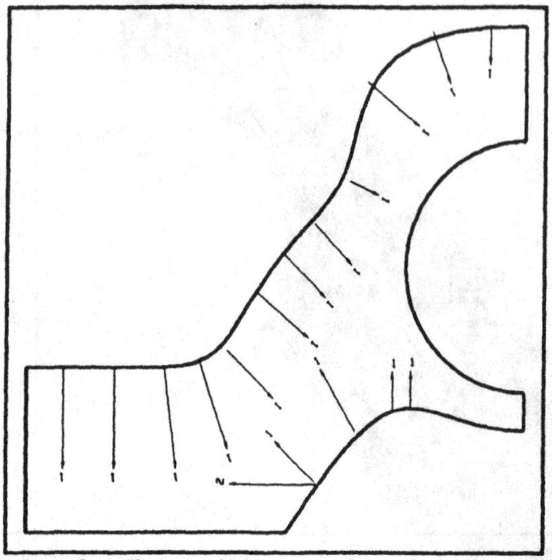

Fig. 9. Shape design model of optimized topology.

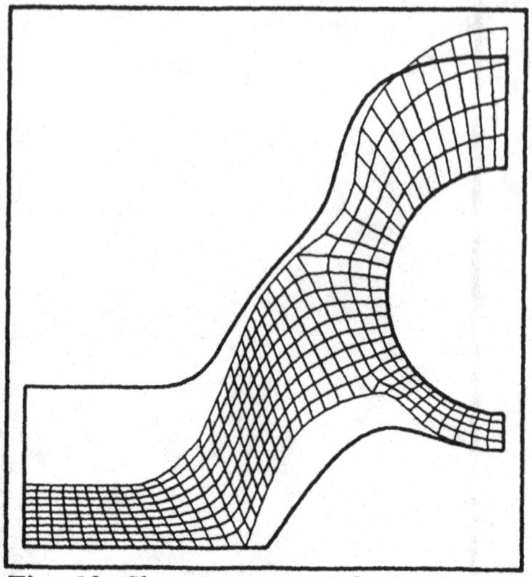

Fig. 10. Shape optimized finite element model.

The optimized geometry can be transferred back into the CAD system where the final geometrical adjustments are easily performed by the designer, yielding for instance the geometry of fig. 11. Compared to real-life structures, the geometry of Fig. 11 seems somewhat fragile.

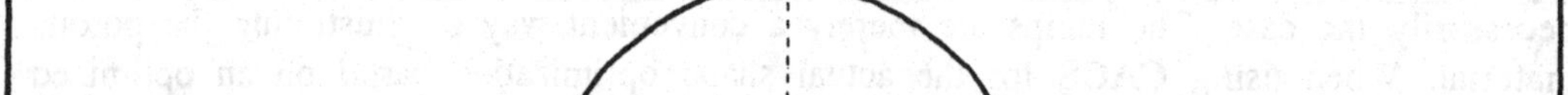

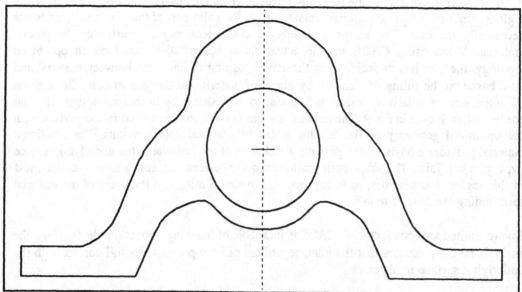

Fig. 11. The final geometry could look something like this.

This frequently happens when the optimization is performed with respect to a single well-defined load case. Real-life structures are usually designed to handle multiple load cases, and for practical use, the shape optimization system must meet this demand. CAOS presently handles only single load case problems but the design of the system allows for an expansion to multiple load case problems.

The problems of fabrication are often mentioned as a serious drawback of shape optimized structures. The CAD integration of CAOS to a large extent solves this problem. Several excellent CAM interfaces are available that will provide automatic numerical machining based on the CAD model of fig. 11.

6.3 Example: MBB-Beam

This section describes the optimization of a support beam from a civil aircraft produced by Messerschmitt-Bölkow-Blohm GmbH, München, BRD. The structure (see fig. 12) has the function of carrying the floor in the fuselage of an Airbus passenger carrier and must meet the following requirements:

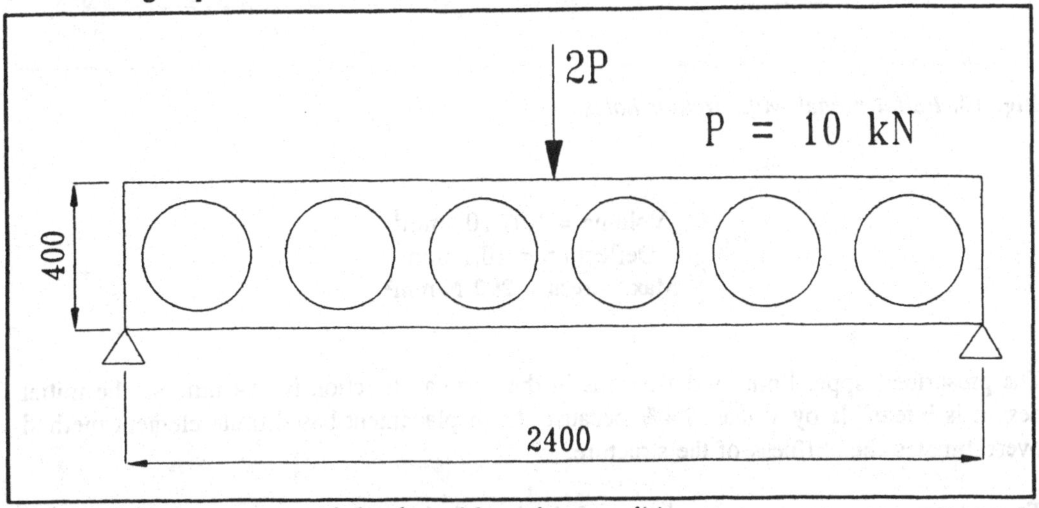

Fig. 12. Initial geometry with loads and boundary conditions.

1. The upper and lower surfaces must be planar and the distance between them cannot be changed.

2. The maximum deflection of the beam must not exceed 9.4 mm under the given load.

3. The maximum von Mises stress should not exceed 385 N/mm².

4. There must be a number of holes in the structure to allow for wires, pipes etc. to pass through.

The purpose of the optimization is to find the shapes of the holes that minimize the weight of the beam while not violating any of the requirements mentioned above. Because of the symmetry of the structure, we shall analyze only the right hand half of the beam with boundary conditions as indicated in fig. 13 that also shows the finite element mesh used for analysis of the initial structure. The data for the initial structure are found to be:

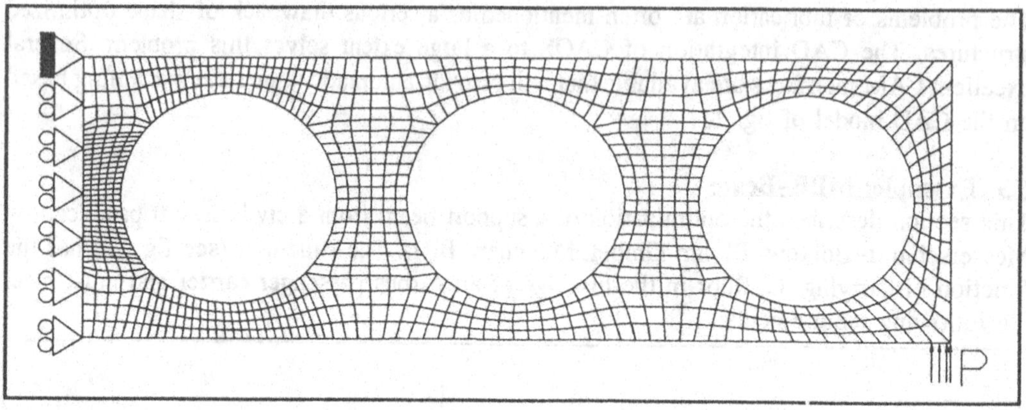

Fig. 13. Initial model with circular holes.

$$\text{Volume} = 1.07 \cdot 10^6 \text{ mm}^3$$
$$\text{Deflection} = 10.1 \text{ mm}$$
$$\text{Max. Stress} = 292 \text{ N/mm}^2$$

The prescribed upper limit on deflections in the vertical direction is 9.4 mm, so the initial design is infeasible by at least 7.4% because the displacement based finite element method overestimates the stiffness of the structure.

To perform optimization via variation of the boundaries of the holes, we represent these boundaries by b–splines, and introduce a number of master nodes in order to give the system sufficiently many design parameters for the optimization. We shall require symmetry about the horizontal mid–axis of the geometry and utilize link facilities implemented in CAOS to link the movements of master nodes above this line to the corresponding master nodes below. The design model is illustrated in fig. 14.

During the optimization of this problem, the stress constraint never becomes active. It is therefore interesting to notice that the final design (fig. 15) is one of very smooth shapes. This is unexpected because there is no smoothness requirement imposed on the transitions between the individual b–splines that make up the holes. This, in combination with the absence of active stress constraints, would in most cases lead to the generation of sharp vertices.

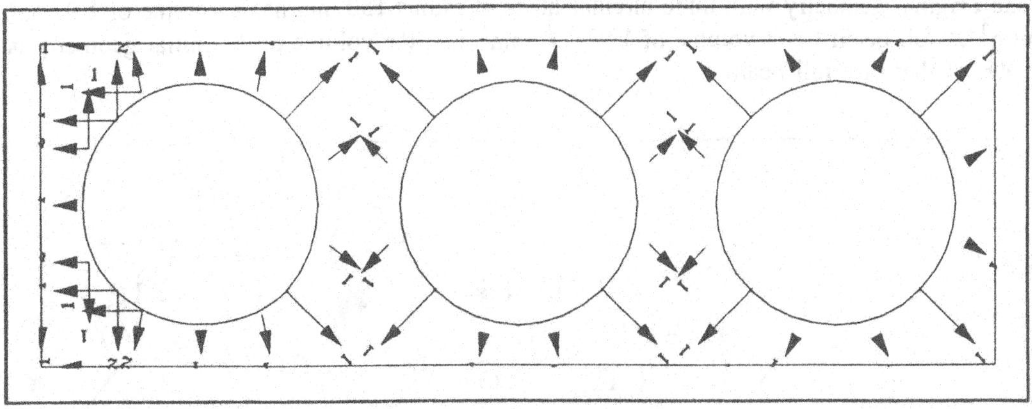

Fig. 14. Design model with b-splines as hole boundaries.

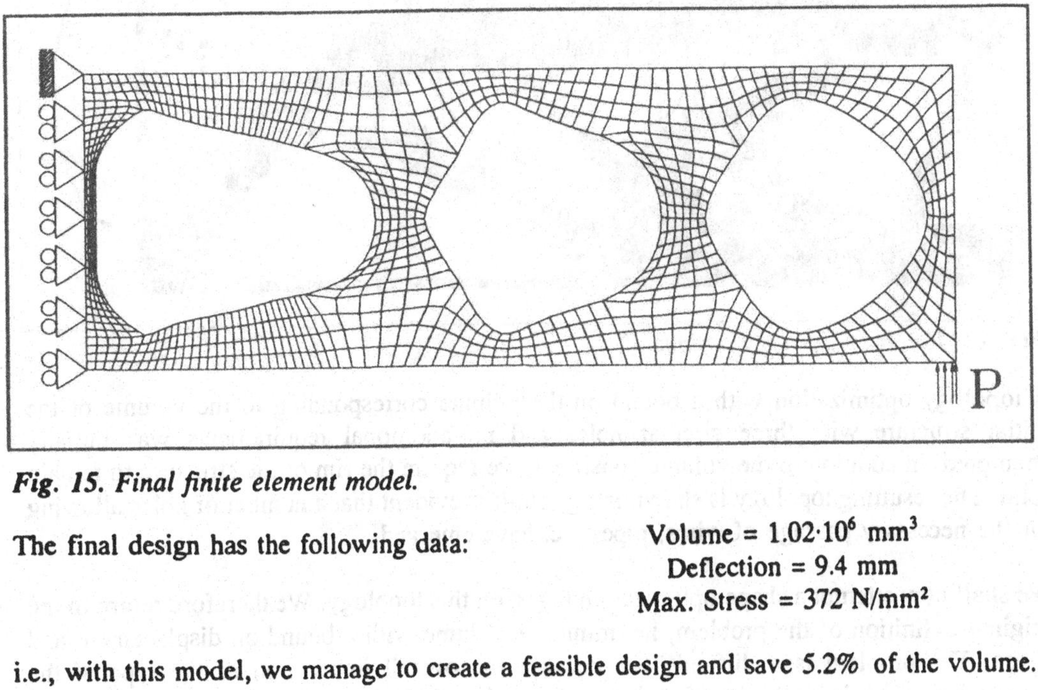

Fig. 15. Final finite element model.

The final design has the following data:

Volume = $1.02 \cdot 10^6$ mm^3
Deflection = 9.4 mm
Max. Stress = 372 N/mm^2

i.e., with this model, we manage to create a feasible design and save 5.2% of the volume.

The fact that the possible volume reduction even with a b-spline model is rather modest leads to the suspicion that the three-hole topology is not well suited for a structure of this type. It is therefore tempting to start the redesign procedure by a topology optimization.

As discussed in Chapter 2, the topology optimization requires a volume constraint to be defined. The topology optimization system will then distribute the available volume in the available domain such that the stiffness is maximized. The system enables the user to specify regions or boundaries which are required to be solid, that is, of density 1. We shall use this facility in the present example because the function of the structure requires that the outer contour, except for the left vertical symmetry boundary, remains unchanged.

The original geometry with three circular holes of radius 150 mm has a volume of $1.07 \cdot 10^6$ mm^3. A full beam has a volume of $1.92 \cdot 10^6$ mm^3, i.e., the volume of the initial geometry is 56% of that of a full beam.

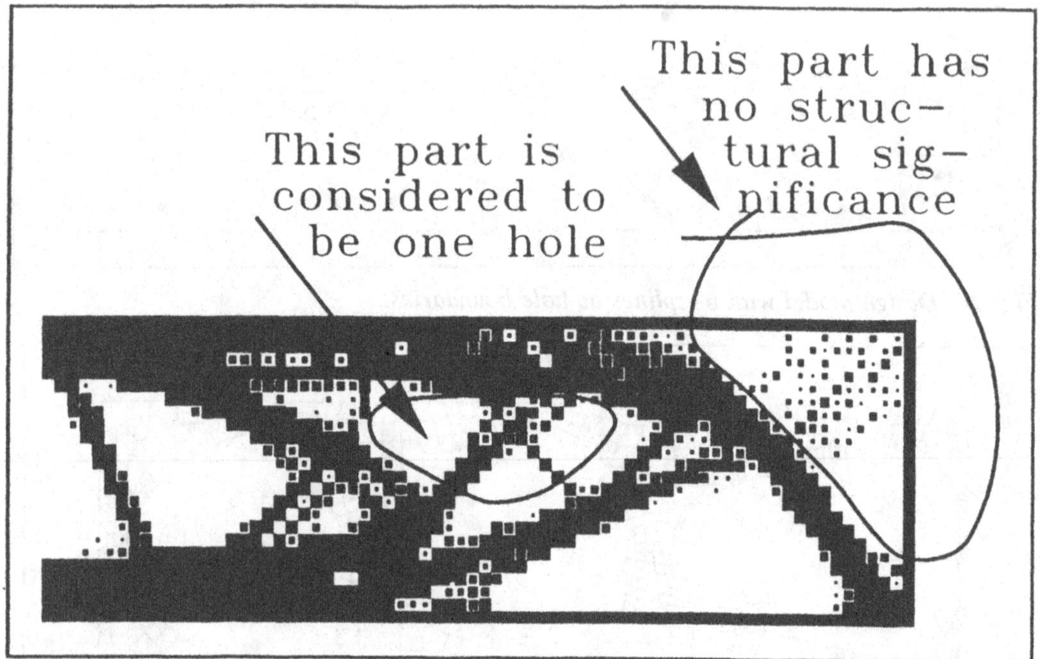

Fig. 16. Result of topology optimization.

A topology optimization with a bound on the volume corresponding to the volume of the initial structure with three circular holes and no additional requirements was initially attempted. In addition to the volume constraint, we require the rim of the structure to remain solid. The resulting topology is shown in fig. 16. It is evident that a number of holes allowing for the necessary passage of wires, pipes etc. have emerged.

We shall now attempt a shape optimization based on this topology. We therefore return to the original definition of the problem, i.e. minimize volume with a bound on displacement and stress. The problem is a difficult one because the overall impression of the type of the structure is that it is on the interface between a disk and a frame or truss structure. Thus, the division of the geometry into design elements is relatively complicated. Furthermore, while creating the shape optimization model, we shall have to take a number of practical considerations into account:

1. Due to the cost of manufacture, the complexity of the geometry should be kept at a minimum, i.e., there is a limit to the number of holes that are practical for a structure like this.

2. The sizes of the individual holes should be comparable to the holes of the initial structure in order to allow for the passage of the same components.

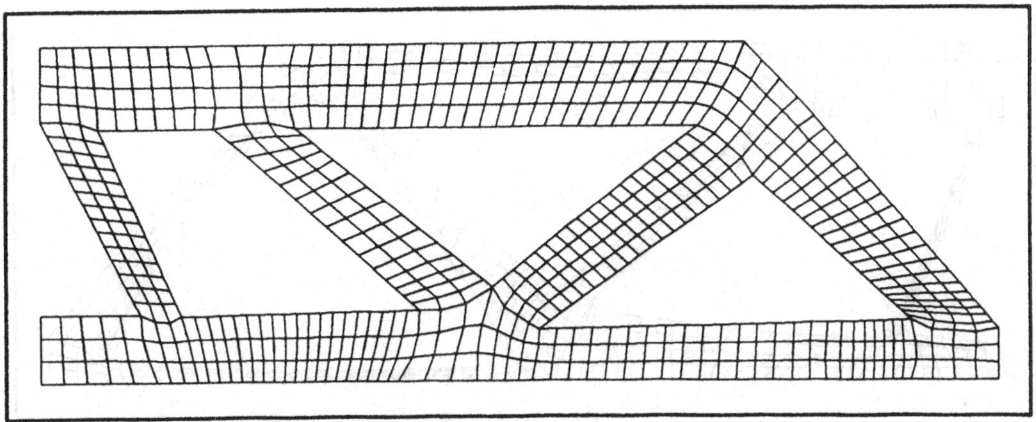

Fig. 17. Initial finite element mesh of optimized topology.

The upper right corner of the frame has been removed. This part of the geometry has a function, but it is structurally insignificant, and can therefore be excluded from the shape optimization and added to the modified structure afterwards. This simplification greatly facilitates the generation of the design model. Fig. 16 illustrates the modifications that have been imposed on the optimized topology and the resulting initial finite element model is shown in fig. 17. This structure has the data:

$$\text{Volume} = 1.10 \cdot 10^6 \text{ mm}^3$$
$$\text{Deflection} = 6.0 \text{ mm}$$
$$\text{Max. Stress} = 227 \text{ N/mm}^2$$

The volume of this geometry is slightly larger than the volume of the initial geometry with three circular holes. However, due to the topology optimization, this geometry has significantly larger stiffness. The maximum stress has also been reduced, but this value is unreliable because of the vertices of this geometry. Mathematically, the stress is infinite at sharp concave vertices, but the stress functions of the finite element model in question are unable to model such a state correctly. From a physical point of view, neither vertices of infinite curvature nor infinite stresses exist.

CAOS proceeds by reducing the volume significantly. Because of the structure's resemblance with a truss structure, small relocations of the master nodes may lead to large distortions of the finite element mesh. It has therefore been necessary to perform a redefinition of the finite element mesh topology on the half way between the initial and the final designs. The final design is illustrated in fig 18. It has the data:

$$\text{Volume} = 0.624 \cdot 10^6 \text{ mm}^3$$
$$\text{Deflection} = 9.4 \text{ mm}$$
$$\text{Max. Stress} = 305 \text{ N/mm}^2$$

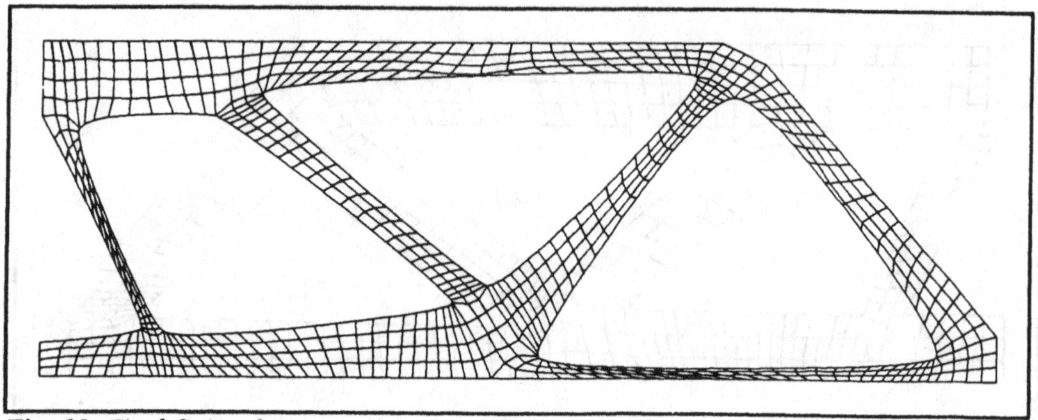

Fig. 18. Final finite element model.

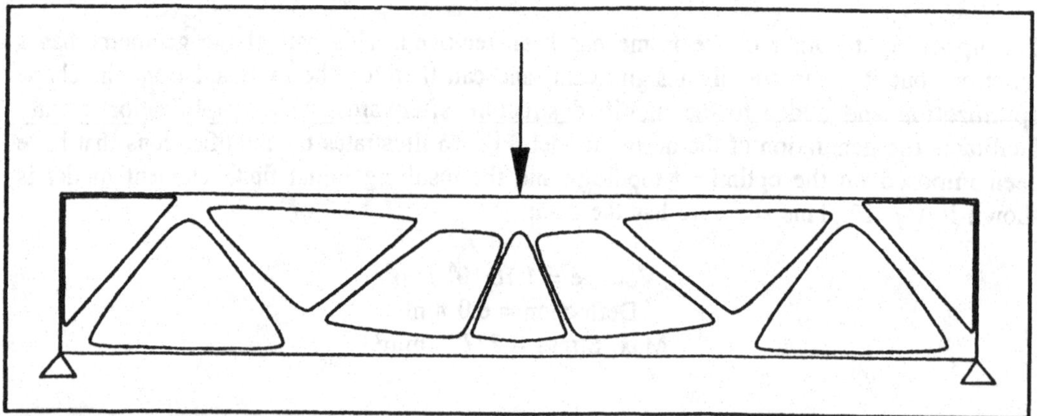

Fig 19. Example of final geometry slightly modified by the designer. The upper right corner has been added again.

The volume is reduced by 42% in comparison with the initial design with circular holes. The final geometry is a frame-like structure. The stress constraint is not active because, for practical reasons, a minimum thickness is specified for the members of the resulting geometry.

Unfortunately, the geometry of the solution introduces the problem of stability which is not covered by CAOS. It is also a problem that the generation of a suitable finite element mesh for a very thin-webbed structure like this is difficult and often requires the topology of the mesh to be redefined. However, the use of membrane elements for a problem like this creates a result that is rich in the sense that it has details that could never have been found by the use of a dedicated system for truss or frame optimization.

Based on the final design of fig. 18 the designer can update his geometrical model and perform the final adjustments, e.g. add the structurally insignificant upper right corner that was removed in order to facilitate the generation of an analysis model, and thereby yield the final design of fig. 19.

CONCLUDING REMARKS

This paper substantiates that initial topology optimization allows shape or sizing optimization to arrive at much better final results. Particularly for problems where there are large possibilities for geometrical variations, the topology optimization is an invaluable tool in the design process. It is the experience from the last example in this paper that topology optimization should be used in the early stages of the development in order to inspire the designer and lead him/her in a beneficial direction. The result of the topology optimization is merely a crude guess and can therefore safely be modified by the designer to meet practical requirements, before the more detailed shape optimization is performed.

It has been the aim of our paper to demonstrate the importance of setting a toolbox of various design facilities at the disposal of the designer. Structural optimization enforces rather than removes the creative aspect of designing, and the final result is therefore very difficult to predict. The collection of structural optimization facilities must be versatile enough to allow the designer to continue work no matter what type of structure emerges. The final design must be a product of creativity rather than availability or lack of analysis facilities.

ACKNOWLEDGEMENTS – The authors are indebted to MSc.s Jan Thomsen and Oluf Krogh, Institute of Mechanical Engineering, Aalborg University, for providing examples for Section 6.1. The work presented in this paper received support from the Danish Technical Research Council (Programme of Research on Computer Aided Design).

REFERENCES

[1] L.A. Schmit, Structural Synthesis – Its Genesis and Development (AIAA Journal 20, pp.992–1000, 1982)

[2] N. Olhoff and J.E. Taylor, On Structural Optimization (J. Appl. Mech. 50, pp.1134–1151, 1983)

[3] H.A. Eschenauer, P.U. Post and M. Bremicker, Einsatz der Optimierungsprozedur SAPOP zur Auslegung von Bauteilkomponenten (Bauingenieur 28, pp.2–12, 1988)

[4] P. Bartholomew and A.J. Morris, STARS: A Software Package for Structural Optimization (in: Proc. Int. Symp. Optimum Structural Design, Univ. of Arizona, USA, 1981)

[5] J. Sobieszanski–Sobieski and J.L. Rogers, A Programming System for Research and Applications in Structural Optimization (in: E. Atrek et. al.: New Directions in Optimum Structural Design, pp.563–585, Wiley, Chichester, 1984)

[6] C. Fleury and V. Braibant, Application of Structural Synthesis Techniques (in: C.A. Mota Soares: Preprints NATO/NASA/NSF/USAF Conf. Computer Aided Optimal Design, Troia, Portugal, 1986, Vol.2, pp.29–53, Techn. Univ. Lisbon, 1986)

[7] B. Esping and D. Holm, Structural Shape Optimization Using OASIS (in: G.I.N. Rozvany and B.L. Karihaloo: Structural Optimization, Proc. IUTAM Symp., Melbourne, Australia, 1988, pp.93–101, Kluwer, Dordrecht, 1988)

[8] R.T. Haftka and B. Prasad, Programs for Analysis and Resizing of Complex Structures (Computers and Structures 10, pp.323–330, 1979)

[9] G. Kneppe, W, Hartzheim and G. Zimmermann, Development and Application of an Optimization Procedure for Space and Aircraft Structures (in: H.A. Eschenauer and G. Thierauf: Discretization methods and structural Optimization – Procedures and Applications, Proc. of a GAMM-Seminar, Siegen, FRG, 1988, pp.194–201, Springer-Verlag, Berlin, 1989)

[10] L.X. Qian, Structural Optimization Research in China (in: Proc. Int. Conf. Finite Element Methods, Shanghai, China, pp.16–24, 1982)

[11] G. Lecina and C. Petiau, Advances in Optimal Design with Composite Materials (in: loc.cit.[6], Vol.3, pp.279–289]

[12] J.S. Arora, Interactive Design Optimization of Structural Systems (in: loc. cit. [9], pp.10–16)

[13] J. Rasmussen, The Structural Optimization System CAOS (Structural Optimization, 2, pp. 109–115, 1990)

[14] J. Rasmussen, Collection of Examples, CAOS Optimization System, 2nd Edition (Special Report No. 1c, Institute of Mechanical Engineering, Aalborg University, Denmark, 1990)

[15] S. Kibsgaard, N. Olhoff and J. Rasmussen, Concept of an Optimization System (in: C.A. Brebbia and S. Hernandez: Computer Aided Optimum Design of Structures: Applications, pp.79–88, Springer-Verlag, Berlin, 1989)

[16] U. Ringertz, A. Branch and Bound Algoritm for Topology Optimization of Truss Structures (Engineering Optimization, 10, pp.111–124, 1986)

[17] M.P. Bendsøe and N. Kikuchi, Generating Optimal Topologies in Structural Design Using a Homogenization Method (Comp. Meths. Appl. Mechs. Engrg. 71, pp.197–224, 1988)

[18] M.P. Bendsøe, Optimal Shape Design as a Material Distribution Problem (Structural Optimization, 1, pp.193–202, 1989)

[19] M.P. Bendsøe and H.C. Rodrigues, Integrated Topology and Boundary Shape Optimization of 2-D Solids, (Rept. No. 14, 31 pp. Mathematical Institute, Technical Univ. Denmark, 1989)

[20] K. Suzuki and N. Kikuchi, A Homogenization Method for Shape and Topology Optimization (Comp. Meths. Appl. Mechs. Engrg., submitted, 1989)

[21] K. Suzuki and N. Kikuchi, Generalized Layout Optimization of Shape and Topology in Three-Dimensional Shell Structures (Rept. No. 90-05, Dept. Mech. Engrg. and Appl. Mech., Comp. Mech. Lab., University of Michigan, USA, 1990)

[22] N. Kikuchi and K. Suzuki, Mathematical Theory of a Relaxed Design Problem in Structural Optimization (Paper for 3rd Air Force/NASA Symp. Recent Advances in Multidisciplinary Analysis and Optimization, San Francisco, USA, Sept. 1990)

[23] P.Y. Papalambros and M. Chirehdast, An Integrated Environment for Structural Configuration Design (J. Engrg. Design, 1, pp. 73-96, 1990)

[24] M. Bremicker, M. Chirehdast, N. Kikuchi and P.Y. Papalambros, Integrated Topology and Shape Optimization in Structural Design (Techn. Rept. UM-MEAM-DL-90-01, Design Laboratory, College of Engrg., Univ. of Michigan, USA, 1990)

[25] M. Bremicker, Ein Konzept zur Integrierten Topologie - und Gestaltoptimierung von Bauteilen (in: H.H. Müller-Slany: Beiträge zur Maschinentechnik, pp.13-39, Festschrift für Prof. H. Eschenauer, Research Laboratory for Applied Structural Optimization, University of Siegen, FRG, 1990)

[26] J.M. Guedes and N. Kikuchi, Pre and Postprocessings for Materials based on the Homogenization Method with Adaptive Finite Element Methods (Rept., Dept. Mech. Engrg. and Appl. Mech., University of Michigan, USA, 1989)

[27] J.M. Guedes and N. Kikuchi, Computational Aspects of Mechanics of Nonlinear Composite Materials (Rept., Dept. Mech. Engrg. and Appl. Mech., University of Michigan, USA, 1989)

[28] C. Fleury and V. Braibant, Structural Optimization: A New Dual Method Using Mixed Variables (Int. J. Num. Meth. Engrg. 23, pp.409-428, 1986)

[29] P. Pedersen, On Optimal Orientation of Orthotropic Materials (Structural Optimization, 1, pp. 101-106, 1989)

[30] P. Pedersen, Bounds on Elastic Energy in Solids of Orthotropic Materials (Structural Optimization, 2, pp. 55-63, 1990)

[31] G.I.N. Rozvany, Structural Layout Theory - the Present State of Knowledge (in: loc.cit. [5], Chapter 7)

[32] P.Pedersen, A Unified Approach to Optimal Design (in: H. Eschenauer and N. Olhoff: Optimization Methods in Structural Design, Proc. Euromech - Colloquium 164, Univ. Siegen, FRG, 1982, pp.182–187, Bibliographishes Institut, Mannheim, FRG, 1983)

[33] V. Braibant and C. Fleury, Shape Optimal Design Using B-splines (Comp. Meths. Appl. Mech. Engrg. 44, pp. 247–267, 1984)

[34] J.A. Bennett and M.E. Botkin, Structural Shape Optimization with Geometric Description and Adaptive Mesh Refinement (AIAA Journal 23, pp. 458–464, 1985)

[35] R.T. Haftka and R.V. Gandhi, Structural Shape Optimization – A Survey (Comp. Meths. Appl. Mech. Engrg. 57, pp.91–106, 1986)

[36] Y. Ding, Shape Optimization of Structures: A Literature Survey (Computers and Structures 24, pp.985–1004, 1986)

[37] N. Olhoff, Multicriterion Structural Optimization via Bound Formulation and Mathematical Programming (Structural Optimization 1, pp.11–17, 1989)

[38] K. Svanberg, The method of Moving Asymptotes - A new Method for Structural Optimization (Int. J. Num. Meth. Engrg. 24, pp. 359–373, 1987)

[39] E.J. Haug, K.K. Choi and V. Komkov, Design Sensitivity of Structural Systems (Academic Press, New York, 1986)

[40] R.T. Haftka and H.M. Adelmann, Recent Developments in Structural Sensitivity Analysis (Structural Optimization, 1, pp. 137–151, 1989)

[41] G. Cheng and L. Yingwei, A New Computation Scheme for Sensitivity Analysis (Eng. Opt. 12, pp. 219–234, 1987)

[42] P. Pedersen, On the Minimum Mass Layout of Trusses (Advisory Group for Aerospace Research and Development, Conf. Proc. No. 36, Symposium on Structural Optimization, Istanbul, Turkey, AGARD-CP-36-70, 1970)

[43] P. Pedersen, On the Optimal Layout of Multi-Purpose Trusses (Computers and Structures, 2, pp. 695–712, 1972)

[44] P. Pedersen, Optimal Joint Positions for Space Trusses (Journal of The Structural Division, ASCE 99, pp. 2459–2476, 1973)

Printed in the United States
By Bookmasters